HUMAN EVOLUTION

HUMAN EVOLUTION

An Illustrated Introduction

FOURTH EDITION

Roger Lewin

Associate of the Peabody Museum
Harvard University
Cambridge, Massachusetts

**Blackwell
Science**

©1999 by Blackwell Science, Inc.

Editorial Offices:
Commerce Place, 350 Main Street, Malden, Massachusetts 02148, USA
Osney Mead, Oxford OX2 0EL, England
25 John Street, London WC1N 2BL, England
23 Ainslie Place, Edinburgh EH3 6AJ, Scotland
54 University Street, Carlton, Victoria 3053, Australia

Other Editorial Offices:
Blackwell Wissenschafts-Verlag GmbH, Kurfürstendamm 57, 10707 Berlin, Germany
Blackwell Science KK, MG Kodenmacho Building, 7–10 Kodenmacho Nihombashi, Chuo-ku, Tokyo 104, Japan

Distributors:

USA
Blackwell Science, Inc.
Commerce Place
350 Main Street
Malden, Massachusetts 02148
(Telephone orders: 800–215–1000 or 781–388–8250; fax orders: 781–388–8270)

Canada
Login Brothers Book Company
324 Saulteaux Crescent
Winnipeg, Manitoba, R3J 3T2
(Telephone orders: 204–224–4068)

Australia
Blackwell Science Pty, Ltd.
54 University Street
Carlton, Victoria 3053
(Telephone orders: 03–9347–0300; fax orders: 03–9349–3016)

Outside North America and Australia
Blackwell Science, Ltd.
c/o Marston Book Services, Ltd.
P.O. Box 269
Abingdon
Oxon OX14 4YN
England
(Telephone orders: 44–01235–465500; fax orders: 44–01235–465555)

Acquisitions: Nancy Hill-Whilton
Production: Irene Herlihy
Manufacturing: Lisa Flanagan
Text and cover design by Leslie Haimes
Typeset by Northeastern Graphic Services, Inc.
Printed and bound by Edwards Brothers, Inc.

Printed in the United States of America

00 01 02 5 4 3 2

The Blackwell Science logo is a trade mark of Blackwell Science Ltd., registered at the United Kingdom Trade Marks Registry

Library of Congress Cataloging-in-Publication Data

Lewin, Roger.
    Human evolution : an illustrated introduction / Roger Lewin. — 4th ed.
        p. cm.
    Includes bibliographical references and index.
    ISBN 0-632-04309-1
    1. Human evolution.   I. Title.
GN281.L49 1999
    599.93′B—DC21
                                                    98-35384
                                                        CIP

CONTENTS

PREFACE

The five years since the third edition of *Human Evolution: An Illustrated Introduction* have been an extraordinarily productive time for paleoanthropology. Many important fossil discoveries have been made, including some that have been named new species; new dates for existing fossils have changed our perception of the history of *Homo erectus*; and molecular anthropology continues to produce important data that bear on the issue of modern humans, a continuing hot topic in paleoanthropology. This new edition reflects these developments, and includes additions and expansions of certain topics as a response to reader comments.

One of the most important developments in paleoanthropology in recent years has been a shift in perspective from assuming that the evolution of humans must have involved special circumstances, to recognizing that, special though *Homo sapiens* is in many ways, much of the species' evolutionary history was in the context of humans as animals. In this spirit, *Human Evolution* examines in some detail the various correlates of body size and shape.

The past three years have witnessed an extraordinarily rich crop of fossil discoveries, including remains of three new species of early human. Two of these are older than previously known human fossils and confirm that the early stages of human phylogeny were more like a bush than a tree, as previous editions of *Human Evolution* had predicted. One of the new species was discovered 2500 kilometers west of the Rift Valley, showing that the geographical range of early humans was much broader than had been assumed.

Important new evidence on the origin of modern humans has emerged in recent years, including the following: the discovery of yet more fossils at the site of Atapuerca in Spain; the redating of certain fossils in Java; new genetic data from modern populations and new interpretations of existing genetic data; and, most stunning of all, the extraction of DNA from the fossilized bones of the first Neanderthal specimen ever found. These developments are incorporated in a comprehensive treatment of the question of modern human origins, including a separate chapter on the Neanderthals themselves.

The treatment of modern evolutionary theory is much more extensive in this book compared with its predecessor. This is placed in the context of the physical circumstances of evolution, including human evolution, about which much has been learned in recent years. A chapter in this section looks at patterns of extinction and evolution throughout the history of complex forms of life, thus providing a context for human evolution, a recent event in this picture.

Systematics has gained even further in importance during the past few years, as anthropologists strive for methods to resolve issues of evolutionary relationship. The treatment of this topic is extensive, and includes molecular as well as morphological systematics. The former is, of course, pertinent to the question of the relationship between humans and the African great apes and the origin of modern humans.

The fact that so much has changed in just a few years is testimony to a science that is healthy, robust, and making progress, even though it retains foibles that derive from the goal of the science: an attempt to understand how we came to be the way we are. With such a goal, no science can be completely objective; subjectivity is certain to insinuate itself, and does. Armed with this knowledge, the student is better prepared to assess what is being said in one debate or another in the science.

Once again I should like to express the privilege I feel as a spectator of this exciting science, and the pleasure I enjoy through contact with its practitioners—too numerous to list—who have always helped and encouraged me. I should like to thank those who were kind enough to review parts of the last edition and suggest ways of improving it. They were Brenda Baker, Peter Brown, Iain Davidson, Dean Falk, Robert Foley, Morris Goodman, Andrew Hill, Phillip Habgood, Kenneth A. R. Kennedy, Richard Klein, Misia Landau, Phyllis Lee, Paul Leslie, Lawrence Martin, Sally McBrearty, R. D. McCall, Henry McHenry, David Meltzer, Maryellen Ruvolo, Kathy Schick, Jeffrey Schwartz, Pat Shipman, Chris Stringer, Ian Tattersall, Nick Toth, Jane Underwood, Mary Voigt, Elisabeth Vrba, Alan Walker, and Bernard Wood. Their advice was always good, and I hope I have done at least some small justice to it.

Lastly, I should like to acknowledge Birute, an inspiring partner in life and work. Thank you.

RL

PART 1

HUMAN EVOLUTION IN PERSPECTIVE

MAN'S PLACE IN NATURE

1

The title of this introductory unit derives from a landmark book published in 1863 by Charles Darwin's friend and champion, Thomas Henry Huxley: *Evidence as to Man's Place in Nature.* The book, which appeared a little more than three years after Darwin's *Origin of Species,* was based on evidence from comparative anatomy among apes and humans, embryology, and fossil evidence of early humans (of which little was available at the time). Huxley's conclusion, that humans share a close evolutionary relationship with the great apes, particularly the African apes, was a key element in a revolution in the history of Western philosophy: humans were to be seen as being *a part of* nature, no longer as *apart from* nature.

Although Huxley was committed to the idea of the evolution of *Homo sapiens* from some type of ancestral ape, he nevertheless considered humans to be a very special kind of animal. "No one is more strongly convinced than I am of the vastness of the gulf between . . . man and the brutes," wrote Huxley, "for, he alone possesses the marvellous endowment of intelligible and rational speech [and] . . . stands raised upon it as on a mountain top, far above the level of his humble fellows, and transfigured from his grosser nature by reflecting, here and there, a ray from the infinite source of truth."

The explanation of this "gap" between humans and the rest of animate nature has always exercised the minds of Western intellectuals, both in pre- and post-evolutionary eras. One difference between the two eras was that, after Darwin, naturalistic explanations had to account not only for the human physical form but also for humans' exceptional intellectual, spiritual, and moral qualities. Previously, these qualities had been regarded as God-given.

As a result, said Glynn Isaac, "Understanding the literature on human evolution calls for the recognition of special problems that confront scientists who report on this topic," a remark he made at the 1982 centenary celebration of Darwin's death. "Regardless of how scientists present them, accounts of human origins are read as replacement materials for Genesis. They . . . do more than cope with curiosity, they have allegorical content, and they convey values, ethics and attitudes." In other words, in addition to reconstructing phylogenies—or evolutionary family trees—paleoanthropological research also addresses "Man's place in nature" in more than just the physical sense. As we shall see, that "place" has long been regarded as being special in some sense.

The revolution wrought by Darwin's work was, in fact, the second of two such intellectual upheavals within the history of Western philosophy. The first revolution oc-

curred three centuries earlier, when Nicolaus Copernicus replaced the geocentric model of the universe with a heliocentric model. Although the Copernican revolution deposed humans from being the cosmic center of all of God's creation and transformed humans into the occupants of a small planet cycling in a vast universe, humans nevertheless remained the pinnacle of God's works. From the sixteenth through the mid-nineteenth centuries, those who studied humans and nature as a whole were coming close to the wonder of those works.

This pursuit—known as **natural philosophy**—positioned science and religion in close harmony, with the remarkable design so clearly manifested in creatures great and small being seen as evidence of God's hand. In addition to design, a second feature of God's created world was natural hierarchy, from the lowest to the highest, with humans being near the very top, just a little lower than the angels. This continuum, known as the **Chain of Being**—was not a statement of evolutionary relationships between organisms, reflecting historical connections and evolutionary derivations. Instead, notes Harvard biologist Stephen Jay Gould, "The chain is a static ordering of unchanging, created entities, a set of creatures placed by God in fixed positions of an ascending hierarchy."

Powerful though it was, the theory faced problems—specifically, some unexplained gaps. One such discontinuity appeared between the world of plants and the world of animals. Another separated humans and apes.

Knowing that the gap between apes and humans should be filled, eighteenth and early nineteenth century

THE ANTHROPOMORPHA OF LINNAEUS: In the mid-eighteenth century, when Linnaeus compiled his *Systema Naturae,* Western scientific knowledge about the apes of Asia and Africa was sketchy at best. Based on tales of sea captains and other transient visitors, fanciful images of these creatures were created. Here, produced from a dissertation of Linnaeus' student Hoppius, are four supposed "manlike apes," some of which became species of *Homo* in Linnaeus' *Systema Naturae.* From left to right: *Troglodyta bontii,* or *Homo troglodytes,* in Linnaeus; *Lucifer aldrovandii,* or *Homo caudatus; Satyrus tulpii,* a chimpanzee; and *Pygmaeus edwardi,* an orangutan.

PTOLEMY'S UNIVERSE: Before the Copernican revolution in the sixteenth century, scholars' views of the universe were based on ideas of Aristotle. The Earth was seen as the center of the universe, with the Sun, Moon, stars, and planets fixed in concentric crystalline spheres circling it.

scientists tended to exaggerate the humanness of the apes while overstating the simianness of some of the "lower" races. For instance, some apes were "known" to walk upright, to carry off humans for slaves, and even to produce offspring after mating with humans. By the same token, some humans were "known" to be brutal savages, equipped with neither culture nor language.

This perception of the natural world inevitably became encompassed within the formal classification system, which was developed by Carolus Linnaeus in the mid-eighteenth century. In his *Systema Naturae,* published first in 1736 until a tenth edition in 1758, Linnaeus included not only *Homo sapiens,* the species to which we all belong, but also the little-known *Homo troglodytes,* which was said to be active only at night and to speak in hisses, and the even rarer *Homo caudatus,* which was known to possess a tail. "Linnaeus worked with a theory that anticipated such creatures," notes Gould, "since they should exist anyway, imperfect evidence becomes acceptable." This concept did not represent scientific finagling, but rather proved that honest

scientists saw what they expected to see, a human weakness that has always operated in science—in all sciences—and always will.

The notion of evolution, the transmutation of species, had been in the air for a long time when, in 1859, the power of data and argument in the *Origin of Species* proved decisive. Geological ideas had been changing as well, with the notion of Cuvier's **"catastrophism"** giving way to the **"uniformitarianism"** of Hutton and Lyell. In parallel with this development was a revision of the accepted age of the Earth, from the 6000 years implied by calculations based on biblical chronologies to many millions of years implied by the idea of the slow, steady change of uniformitarianism.

Interestingly, although the advent of the evolutionary era brought an enormous shift in intellectual perceptions of the *origin* of humankind, many elements concerning the *nature* of mankind remained unassailed. For instance, humans were still regarded as being "above" other animals and endowed with special qualities, those of intelligence, spirituality, and moral judgment. And the gradation from "lower" races to "higher" races that had been part of the Chain of Being was now explained by the process of evolution.

"The progress of the different races was unequal," noted Roy Chapman Andrews, a researcher at the American Museum of Natural History in the 1920s and 1930s. "Some developed into masters of the world at an incredible speed. But the Tasmanians . . . and the existing Australian aborigines lagged far behind, not much advanced beyond the stages of Neanderthal man." Such overtly racist comments were echoed frequently in literature of the time and were reflected in the evolutionary trees published then.

In other words, inequality of races, with blacks on the bottom and whites on the top, was explained away as the natural order of things: before 1859 as the product of God's creation, and after 1859 as the product of natural selection.

In the same vein, eighteenth-century discussions of human evolution incorporated the notion of progress, and specifically the inevitability of *Homo sapiens* as the ultimate aim of evolutionary trends. "Much of evolution looks as if it had been planned to result in man, and in other animals and plants to make the world a suitable place for him to dwell in," observed Robert Broom in 1933 (Broom was responsible for some of the more important early human fossil finds in South Africa during the 1930s and 1940s).

Evolution as progress, the inexorable improvement to more complex, more intelligent life, has always been a seductive notion. "Progress, or what is the same thing, Evolution, is [Nature's] religion," wrote Britain's Sir Arthur Keith in 1927. The notion of progress as a driving ethos of nature, and society, has been a characteristic of Western philosophy, but not of all intellectual thought. "The myth of progress" is how Niles Eldredge and Ian Tattersall characterize this idea. "Once evolved, species with their own peculiar adaptations, behaviors, and ge-

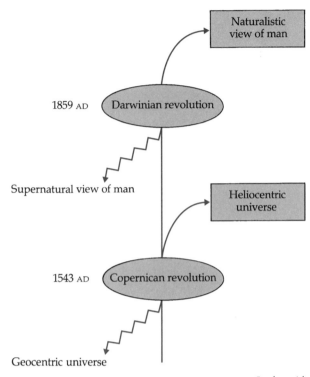

TWO GREAT INTELLECTUAL REVOLUTIONS: In the mid-sixteenth century, the Polish mathematician Nicolaus Copernicus proposed a heliocentric rather than a geocentric view of the universe. "The Earth was not the center of all things celestial," he said, "but instead was one of several planets circling a sun, which was one of many suns in the universe." Three centuries later, in 1859, Charles Darwin further changed Man's view of himself, arguing that humans were a part of nature, not apart from nature.

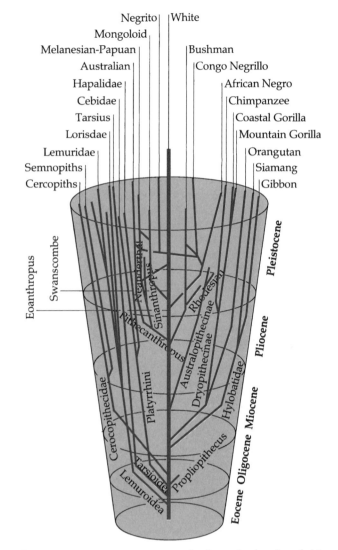

RACISM IN ANTHROPOLOGY: In the early decades of this century, racism was an implicit part of anthropology, with "white" races considered to be superior to "black" races, through greater effort and struggle in the evolutionary race. Here, the supposed ascendancy of the "white" races is shown explicitly, in Earnest Hooton's *Up from the Ape,* second edition, 1946.

netic systems are remarkably conservative, often remaining unchanged for several million years. In this light it is wrong to see evolution, or for that matter human history, as a constant progression, slow or otherwise."

Some species later in evolutionary time are clearly more complex in certain ways than many found earlier in time. This development can, however, be explained simply as the ratchet effect—the fact that evolution builds on what existed before. For the most part, the world has not become a strikingly more complex place biologically as a whole. Although most organisms remain simple, we remain blinded by the exceptions, particularly the one with which we are most familiar.

Even this brief historical sketch clearly illustrates the **anthropocentric** spectacles through which paleoanthropologists have viewed the natural world in which we evolved. Such a perception is probably inescapable in some degree, as Glynn Isaac's earlier remark implied. In 1958, for instance, Julian Huxley, grandson of Thomas Henry, suggested that mankind's special intellectual and social qualities should be recognized formally by assigning *Homo sapiens* to a new grade, the Psychozoan. "The

new grade is of very large extent, at least in magnitude to all the rest of the animal Kingdom, though I prefer to regard it as covering an entirely new sector of the evolutionary process, the psychosocial, as against the entire non-human biological sector."

The ultimate issue is "the long-held view that humans are unique, a totally new type of organism," as Cambridge University's Robert Foley points out. This type of thinking leads to the notion that human origins therefore "requires a special type of explanation, different from that used in understanding the rest of the biological world." ✻

KEY QUESTIONS

- Did the intellectual framework provided by the Great Chain of Being lead naturally to the idea of the evolution of species?
- Why did the perception of Man's place in nature not change much in some ways between pre- and post-Darwinian eras?
- Why has the notion of progress become such an integral part of evolutionary thinking within Western philosophy, particularly in relation to human evolution?
- Does the evolution of qualitatively novel characteristics require qualitatively novel explanations?

KEY REFERENCES

Cartmill M. Human uniqueness and theoretical content in paleoanthropology. *Int J Primatol* 1990;11:173–192.

Eldredge N, Tattersall I. The myths of human evolution. New York: Columbia University Press, 1982.

Gould SJ. Vision with a vengeance. *Nat Hist* Sept 1980:16–20.

———. Bound by the great chain. *Nat Hist* Nov 1983:20–24.

———. Chimp on a chain. *Nat Hist* Dec 1983:18–26.

———. Spin doctoring Darwin. *Nat Hist* July 1995:6–9, 69–71.

Lovejoy AO. The Great Chain of Being. Cambridge, MA: Harvard University Press, 1936 (new reprint).

Richards RJ. The meaning of evolution. Chicago: University of Chicago Press, 1992.

HUMAN EVOLUTION AS NARRATIVE

2

"One of the species specific characteristics of *Homo sapiens* is a love of stories," noted Glynn Isaac, "so that narrative reports of human evolution are demanded by society and even tend toward a common form." Isaac was referring to the work of Boston University anthropologist Misia Landau, who has analyzed the **narrative** component of professional—not just popular—accounts of human origins.

"Scientists are generally aware of the influence of theory on observation," concludes Landau. "Seldom do they recognize, however, that many scientific theories are essentially narratives." Although this comment applies to all sciences, Landau identifies several elements in **paleoanthropology** that make it particularly susceptible to being cast in narrative form, both by those who tell the stories and by those who listen to them.

First, in seeking to explain human origins, paleoanthropology is apparently faced with a sequence of events through time that transformed apes into humans. The description of such a sequence falls naturally into narrative form. Second, the subject of that transformation is ourselves. Being egotistical creatures, we tend to find stories about ourselves more interesting than stories about, for instance, the behavior of arthropods or the origin of flowering plants.

Traditionally, paleoanthropologists have recognized four key events in human evolution: the origin of **terrestriality** (coming to the ground from the trees), **bipedality** (upright walking), **encephalization** (brain expansion in relation to body size), and **culture** (or civilization). While these four events have usually featured in accounts of human origins, paleoanthropologists have disagreed about the order in which they were thought to have occurred.

For instance, Henry Fairfield Osborn, director of the American Museum of Natural History in the early decades of this century, considered the order to be that given above, which, incidentally, coincides closely with Darwin's view. Sir Arthur Keith, a prominent figure in British anthropology in the 1920s, considered bipedalism to have been the first event, with terrestriality following. In other words, Keith's ancestral ape began walking on two legs while it was still a tree dweller; only subsequently did it descend to the ground. For Sir Grafton Elliot Smith, a contemporary of Keith, encephalization led the way. His student, Frederic Wood Jones, agreed with Smith that encephalization and bipedalism developed while our ancestors lived in trees, but thought that bipedalism preceded rather than followed brain expansion. William King Gregory, like his colleague

Osborn, argued for terrestriality first, but suggested that the adoption of culture (tool use) preceded significant brain expansion. And so on.

Thus, we see these four common elements linked together in different ways, with each narrative scheme purporting to tell the story of human origins. And "story" is the operative word here. "If you analyze the way in which Osborn, Keith and others explained the relation of these four events, you see clearly a narrative structure," says Landau, "but they are more than just stories. They conform to the structure of the hero folk tale." In her analysis of paleoanthropological literature, Landau drew upon a system devised in 1925 by the Russian literary scholar Vladimir Propp. This system, published in Propp's *Morphology of the Folk Tale,* included a series of 31 stages that encompassed the basic elements of the hero myth. Landau reduced the number of stages to nine, but kept the same overall structure: hero enters; hero is challenged; hero triumphs.

In the case of human origins, the hero is the ape in the forest, who is "destined" to become us. The climate changes, the forests shrink, and the hero is cast out on the savannah where he faces new and terrible dangers. He struggles to overcome them, by developing intelligence, learning to use tools, and so on, and eventually emerges triumphant, recognizably you and me.

"When you read the literature you immediately notice not only the structure of the hero myth, but also the language," explains Landau. For instance, Elliot Smith writes about ". . . the wonderful story of Man's journey-

SIR GRAFTON ELLIOT SMITH: A leading anatomist and anthropologist in early twentieth century England, Elliot Smith often wrote in florid prose about human evolution. (See following figure.) (Courtesy of University College, London.)

ings towards his ultimate goal . . ." and ". . . Man's ceaseless struggle to achieve his destiny." Roy Chapman Andrews, Osborn's colleague at the American Museum, writes of the pioneer spirit of our hero: "Hurry has always been the tempo of human evolution. Hurry to get out of the primordial ape stage, to change body, brains, hands and feet faster than it had ever been done in the history of creation. Hurry on to the time when man could conquer the land and the sea and the air; when he could stand as Lord of all the Earth."

Osborn wrote in similar tone: "Why, then, has evolutionary fate treated ape and man so differently? The one has been left in the obscurity of its native jungle, while the other has been given a glorious exodus leading to the domination of earth, sea, and sky." Indeed, many of Osborn's writings explicitly embodied the notion of drama: "The great drama of the prehistory of man . . . ," he wrote, and "the prologue and opening acts of the human drama . . . ," and so on.

Of course, it is possible to tell stories with similar gusto about nonhuman animals, such as the "triumph of the reptiles in conquering the land" or "the triumph of birds in conquering the air." Such stirring tales are readily found in accounts of evolutionary history—look no further than every child's hero, the dinosaur. The fact that the hero of the paleoanthropology tale is *Homo sapiens*—ourselves—makes a significant difference, however. Although dinosaurs may be lauded as lords of the land in their time, only humans have been regarded as the inevitable product of evolution—indeed, the ultimate purpose of evolution, as we saw in the previous unit. Not everyone was as explicit about this as Broom was (unit 1), but most authorities betrayed the sentiment in the hero worship of their prose.

These stories were not just accounts of the ultimate triumph of our hero; they carried a moral tale, too—namely, triumph demands effort. "The struggle for existence was severe and evoked all the inventive and

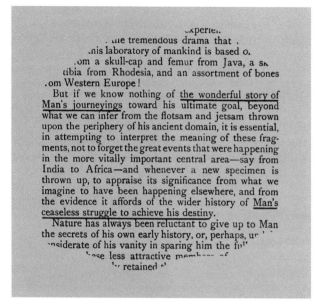

ADVENTURES IN ANTHROPOLOGY: Here, a short passage from Sir Grafton Elliot Smith's *Essays on the Evolution of Man,* published in 1924, illustrates the storytelling tone in which anthropological writing was often couched. Even modern prose is not always entirely free of this influence.

resourceful faculties and encouraged [Dawn Man] to the fashioning and first use of wooden and then stone weapons for the chase," wrote Osborn. "It compelled Dawn Man . . . to develop strength of limb to make long journeys on foot, strength of lungs for running, and quick vision and stealth for the chase."

According to Elliot Smith, our ancestors ". . . were impelled to issue forth from their forests, and seek new sources of food and new surroundings on hill and plain,

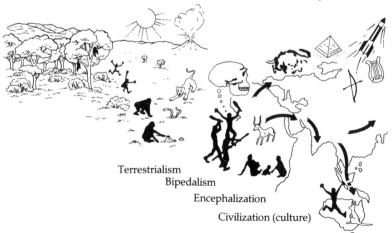

1 Initial situation   3 Change   5 Struggle/test   7 Transformation   9 Triumph !

2 Hero introduced   4 Departure   6 (Donor)   8 Tested again

Terrestrialism
Bipedalism
Encephalization
Civilization (culture)

THE HERO-MYTH FRAMEWORK: Like folk tales ancient and modern, accounts of human origins have often followed the structure of hero myth. The hero (an ancient ape) sets off on a journey, during which he faces a series of challenges and opportunities that shape his final triumph (civilization). Recounting the evolution of any species is, of course, equivalent to telling a tale of a series of historical events. The effect, in the case of *Homo sapiens,* is to see the events as if, from the beginning, the journey was inevitable. (Courtesy of Misia Landau.)

where they could obtain the sustenance they needed." The penalty for indolence and lack of effort was plain for all to see, because the apes had fallen into this trap: "While man was evolved amidst the strife with adverse conditions, the ancestors of the Gorilla and Chimpanzee gave up the struggle for mental supremacy because they were satisfied with their circumstances."

In the literature of Elliot Smith's time, the apes were usually viewed as evolutionary failures, left behind in the evolutionary race. This sentiment prevailed for several decades, but eventually became transformed. Instead of evolutionary failures, the apes came to be viewed as evolutionarily primitive, or relatively unchanged from the common ancestor they shared with humans. In contrast, humans were regarded as much more advanced. Today, anthropologists recognize that both humans and apes display advanced evolutionary features, and differ equally (but in separate ways) from their common ancestor.

Although modern accounts of human origins usually avoid purple prose and implicit moralizing, one aspect of the narrative structure lingers in current literature. Paleoanthropologists still tend to describe the events in the "transformation of ape into human" as if each event were somehow a preparation for the next. "Our ancestors became bipedal in order to make and use tools and weapons. . . . tool-use enabled brain expansion and the evolution of language . . . thus endowed, sophisticated societal interactions were finally made possible . . ." Crudely put, to be sure, but this kind of reasoning was common in Osborn's day and persists in some current narratives.

Why does it happen? "Telling a story does not consist simply in adding episodes to one another," explains Landau. "It consists in creating relations between events." Consider, for instance, our ancestor's supposed "coming to the ground"—the first and crucial advance on the long road toward becoming human. It is easy to imagine how such an event might be perceived as a courageous first step on the long journey to civilization: the defenseless ape faces the unknown predatory hazards of the savannah. "There is nothing inherently transitional about the descent to the ground, however momentous the occasion," says Landau. "It only acquires such value in relation to our overall conception of the course of human evolution."

If evolution were steadily progressive, forming a program of constant improvement, the transformation of ape to human could be viewed as a series of novel adaptations, each one naturally preparing for and leading to the next. Such a scenario would involve continual progress through time, going in a particular direction. From our vantage point, where we can view the end-product, it is tempting to view the process in that way because we can actually see that all those steps did actually take place. This slant, however, ignores the fact that evolution tends to work in a rather halting, unpredictable fashion, shifting abruptly from one "adaptive plateau" to another. These adaptive plateaux are species, of course, and each was adaptively successful and persisted for a considerable

DIFFERENT VIEWS OF THE STORY: Even though anthropologists saw the human journey as involving the same fundamental events—terrestriality, bipedalism, encephalization, and civilization—different authorities sometimes placed these steps in slightly different orders. For instance, although Charles Darwin envisaged an ancient ape first coming to the ground and then developing bipedalism, Sir Arthur Keith believed that the ape became bipedal before leaving the trees. (Courtesy of Misia Landau/*American Scientist*.)

time (several million years in some cases) before a rapid evolutionary shift, perhaps propelled by external forces, yielded a new species with a new adaptation (see unit 4).

For instance, one cannot say that the first bipedal ape would inevitably become a stone-tool maker. In fact, if the current **archeological** record serves as any guide, those two events—bipedality and the advent of stone-tool making—were separated by approximately 2.5 million years (see unit 23). The brain expanded at this time as well (see unit 21). In addition, a more humanlike body structure emerged abruptly (see unit 24). The origin of anatomically modern humans after another 2 million or so years was also probably a punctuational event (see units 27 through 30). Thus, although many writers proclaim that our ancestors were propelled inexorably along an evolutionary trajectory that ended with *Homo sapiens,* that scenario simply describes what did happen; it ignores the many other possibilities that did not transpire. As Landau remarks: "There is a tendency in theories of hominid evolution to define origins in terms of endings."

For paleoanthropology, language represents an important scientific tool that is used for the technical description of fossils and for the serious explication of evolutionary scenarios. All scientists should step back and scrutinize the language they use, because intertwined within it will be the elements of many unspoken assumptions. For human origins research, where narrative becomes a particularly seductive vehicle for assumptions, it is especially important that one carefully examines what one says and the way one says it. ❀

KEY QUESTIONS

- What is implied by the fact that, although paleoanthropologists in Osborn's time employed the same set of events to describe the transformation of ape to human, those events were linked in many different combinations?
- Is paleoanthropology particularly susceptible to the invocation of the hero myth?
- Why do evolutionary scenarios tend to lend themselves to narrative treatment?
- In what context were apes considered to be evolutionary failures?

KEY REFERENCES

Eldredge N, Tattersall I. The myths of human evolution. New York: Columbia University Press, 1982.

Isaac G. Aspects of human evolution. In: David Bendall. Evolution from molecules to men. Cambridge: Cambridge University Press, 1983:509–543.

Landau M. Human evolution as narrative. *Am Scientist* 1984;72:262–268.

———. Paradise lost: the theme of terrestriality in human evolution. In: Nelson JS, Megill A, McClosky DN. The rhetoric of the human sciences. Madison, WI: University of Wisconsin Press, 1987:111–124.

———. Narratives of human evolution. New Haven, CT: Yale University Press, 1991.

Medawar P. Pluto's Republic. Oxford: Oxford University Press, 1984.

HISTORICAL VIEWS

3

Debate over human origins has advanced substantially in recent years, particularly in broadening the scientific basis of the discussions. Nevertheless, many of the issues addressed in current research have deep historical roots. A brief sketch of the subject's progress during the past 100 years or so will put modern debates into historical context.

Two principal themes have been recurrent in this century of paleoanthropology, each of which has been seen to be more or less important at different times, depending on the ebb and flow of intellectual tides. First is the relationship between humans and apes: how close, how distant? Second is the "humanness" of our direct ancestors, the early **hominines**. (*Hominine* is the term now generally used to describe species in the human family, or clade; until recently, the term *hominid* was used, as discussed in unit 8.)

During the past century, the issue of our relatedness to the apes has gone full cycle. From the time of Darwin, Huxley, and Haeckel until soon after the turn of the century, humans' closest relatives were regarded as being the African apes, the chimpanzee and gorilla, with the Asian great ape, the orangutan, being considered to be somewhat separate. From the 1920s until the 1960s, humans were distanced from the great apes, which were said to be an evolutionarily closely knit group. Since the 1960s, however, conventional wisdom has returned to its Darwinian cast.

This shift of opinions has, incidentally, been paralleled by a related shift in ideas on the location of the "cradle of mankind." Darwin plumped for Africa, Asia became popular in the early decades of the twentieth century, and Africa has once again emerged as the focus.

While this human/African ape wheel has gone through one complete revolution, the question of the humanness of the hominine lineage has been changing as well—albeit in a single direction. Specifically, hominines—with the exception of *Homo sapiens* itself—have been gradually perceived as less humanlike in the eyes of paleoanthropologists, particularly in the last two decades. The different views on the origin of modern humans are, however, imbued with different perspectives of this issue (see unit 27).

Once Darwin's work firmly established evolution as part of mainstream nineteenth-century intellectual life, scientists had to account for human origins in naturalistic rather than supernatural terms. More importantly, as we saw in the previous unit, they had to account for the special qualities of humankind, those that appear to separate us from the world of nature. This issue posed a formidable challenge—and the response to it set the intellectual tone in paleoanthropology for a very long time.

In his *Descent of Man,* Darwin identified those characteristics that apparently make humans special—intelligence, manual dexterity, technology, and uprightness of posture—and argued that an ape endowed with minor amounts of each of these qualities would surely possess an advantage over other apes. Once the earliest human forebear became established upon this evolutionary trajectory, the eventual emergence of *Homo sapiens* appeared almost inevitable because of the continued power of natural selection. In other words, hominine origins became explicable in terms of human qualities, and *hominine* origin therefore equated with *human* origin. It was a seductive formula, and one that persisted until quite recently.

In the early decades of the twentieth century two opposing views of human origins were current:

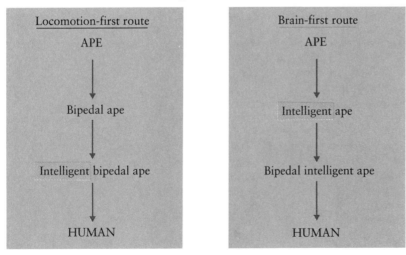

CONFLICTING VIEWS: One of the key differences of opinion regarding the history of human evolution was the role of the expanded brain: was it an early or a late development? The "brain-first" notion, promoted by Elliot Smith, was important in paving the way for the acceptance of the Piltdown man fraud.

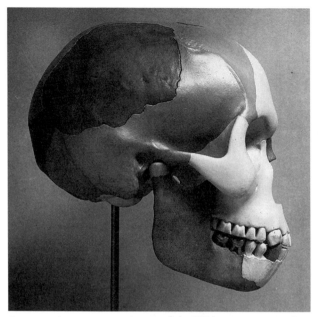

A FOSSIL CHIMERA: A cast of the Piltdown reconstruction, based on lower jaw, canine tooth, and skull fragments (shaded dark). The ready acceptance of the Piltdown forgery—a chimera of a modern human cranium and the jaw of an orangutan—derived from the British establishment's adherence to the brain-first route. (Courtesy of the American Museum of Natural History.)

At the turn of the century several interrelated intellectual debates were brewing, one of which focused on the order in which the major anatomical changes occurred in the human lineage. One notion was that the first step on the road to humanity was the adoption of upright locomotion. A second held that the brain led the way, produc-

ing an intelligent but still arboreal creature. It was into this intellectual climate that the perpetrator of the famous Piltdown hoax—a chimera of fragments from a modern human cranium and an orangutan's jaw, both doctored to make them look like ancient fossils—made his play from 1908 to 1913. (In mid-1996 the first material clues as to the identity of the Piltdown forger came to light, pointing to Martin Hinton, Arthur Smith Woodward's colleague at the Natural History Museum, London.)

The Piltdown "fossils" appeared to confirm not only that the brain did indeed lead the way, but also that something close to the modern sapiens form was extremely ancient in human history. The apparent confirmation of this latter fact—extreme human antiquity—was important to both the prominent British anthropologist Sir Arthur Keith and Henry Fairfield Osborn, because their theories demanded it. One consequence of Piltdown was that Neanderthal—one of the few genuine fossils of the time—was disqualified from direct ancestry to *Homo sapiens,* because it apparently came later in time than Piltdown and yet was more primitive. British anthropologists were of course happy to believe that Britain was now firmly on the anthropological map, apparently overshadowing German and French claims.

For Osborn, Piltdown represented strong support for his Dawn Man theory, which stated that mankind originated on the high plateaus of Central Asia, not in the jungles of Africa. During the 1920s and 1930s, Osborn was locked in constant but gentlemanly debate with his colleague, William King Gregory, who carried the increasingly unpopular Darwin/Huxley/Haeckel torch for a close relationship between humans and African apes—the Ape Man theory.

Although Osborn was never very clear about what the earliest human **progenitors** might have looked like, his ally Frederic Wood Jones espoused firmer ideas. Wood

A DISCUSSION OF THE PILTDOWN SKULL: Back row, left to right: F. G. Barlow, Grafton Elliot Smith, Charles Dawson, and Arthur Smith Woodward. Front row, left to right: A. S. Underwood, Arthur Keith (examining the skull), W. P. Pycraft, and Ray Lankester. The Piltdown man fossil, discovered in 1912 and exposed as a fraud in 1953, fitted so closely with British anthropologists' views of human origins that it was accepted uncritically as being genuine. (Courtesy of the American Museum of Natural History.)

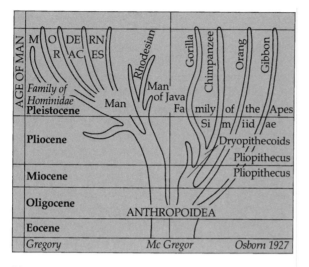

(a)

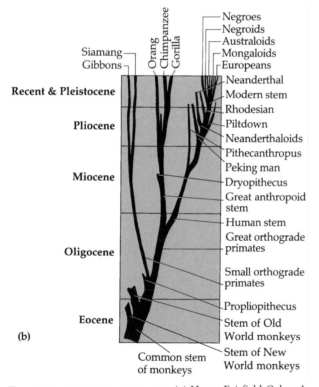

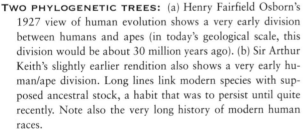

(b)

TWO PHYLOGENETIC TREES: (a) Henry Fairfield Osborn's 1927 view of human evolution shows a very early division between humans and apes (in today's geological scale, this division would be about 30 million years ago). (b) Sir Arthur Keith's slightly earlier rendition also shows a very early human/ape division. Long lines link modern species with supposed ancestral stock, a habit that was to persist until quite recently. Note also the very long history of modern human races.

Jones, a British anatomist, interpreted key features of ape and monkey anatomy as specializations that were completely absent in human anatomy. In 1919, he proposed his "tarsioid hypothesis," which sought human antecedents very low down in the primate tree, with a creature like the modern tarsier. In today's terms, this proposal would place human origins in the region of 50 to 60 million years ago, close to the origin of the primate radiation, while Keith's notion of some kind of early ape would date this development to approximately 30 million years ago.

During the 1930s and 1940s, the anti-ape arguments of Osborn and Wood Jones were lost, but Gregory's position did not immediately prevail. Gregory had argued for a close link between humans and the African apes on the basis of shared anatomical features. Others, including Adolph Schultz and D. J. Morton, claimed that, although humans probably derived from apelike stock, the similarities between humans and modern African apes were the result of parallel evolution. This position remained dominant through the 1960s, firmly supported by Sir Wilfrid Le Gros Clark, Britain's most prominent primate anatomist of the time. Humans, it was argued, came from the base of the ape stock, not later in evolution with the specializations developed by the African apes.

During the 1950s and 1960s, the growing body of fossil evidence related to early apes appeared to show that these creatures were not simply early versions of modern apes, as had been tacitly assumed. This idea meant that those authorities who accepted an evolutionary link between humans and apes, but rejected a close human/African ape link, did not have to retreat back in the history of the group to "avoid" the specialization of the modern species. At the same time, those who insisted that the similarities between African apes and humans reflected a common heritage, not parallel evolution, were forced to argue for a very recent origin of the human line. Prominent among proponents of this latter argument was Sherwood Washburn, of the University of California, Berkeley.

One of the fossil discoveries of the 1960s—in fact, a rediscovery of a specimen unearthed three decades earlier—that appeared to confirm the notion of parallel evolution to explain human/African ape similarities was made by Elwyn Simons, then of Yale University. The fossil specimen was *Ramapithecus*, an apelike creature that lived in Eurasia approximately 15 million years ago and appeared to share many anatomical features (in the teeth and jaws) with hominines. Simons, later supported closely by David Pilbeam, proposed *Ramapithecus* as the beginning of the hominine line, thus excluding a human/African ape connection.

Arguments about the relatedness between humans and African apes were mirrored by a reconsideration of the relatedness among the apes themselves. In 1927, G. E. Pilgrim had suggested that the great apes be treated as a natural group (that is, evolutionarily closely related), with humans viewed as more distant. This idea eventually became popular and remained the accepted wisdom until molecular biological evidence undermined it in 1963, via

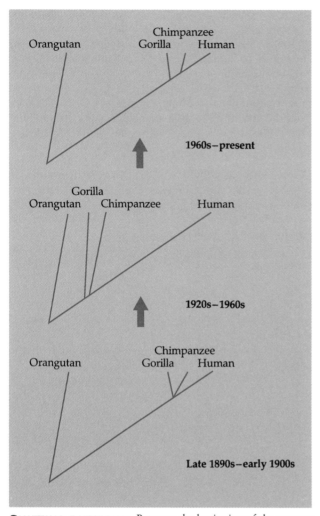

SHIFTING PATTERNS: Between the beginning of the century and today, ideas about the relationships among apes and humans have moved full circle.

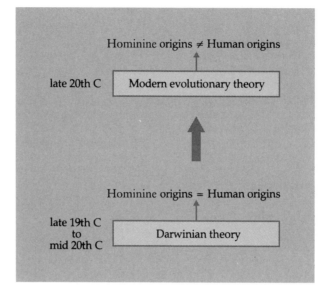

HOMININES AS HUMANS: Until quite recently anthropologists frequently thought about humanlike characteristics while considering hominine origins, a habit that can be traced back to Darwin. The humanity of hominines is now seen as a rather recent evolutionary development.

the work of Morris Goodman at Wayne State University. Goodman's molecular biology data on blood proteins indicated that humans and the African apes formed a natural group, with the orangutan more distant (see unit 15).

As a result, the Darwin/Huxley/Haeckel position returned to prominence, with first Gregory and then Washburn emerging as its champion. Subsequent molecular biological—and fossil—evidence appeared to confirm Washburn's original suggestion that the origin of the human line is quite recent, close to 5 million years ago. *Ramapithecus* was no longer regarded as the first hominine, but simply one of many early apes. (The nomenclature and evolutionary assignment of *Ramapithecus* subsequently was modified, too, as described in unit 16.)

Meanwhile, discoveries of fossil hominines, and the stone tools they apparently made, had been accumulating at a rapid pace from the 1940s through 1970s, first in

South Africa and then in East Africa. Culture—specifically, stone-tool making and tool use in butchering animals—became a dominant theme, so much so that hominine was considered to imply a hunter-gatherer lifeway. The most extreme expression of culture's importance as *the* hominine characteristic consisted of the single-species hypothesis, promulgated during the 1960s principally by C. Loring Brace and Milford Wolpoff, both of the University of Michigan.

According to this hypothesis, only one species of hominine existed at any one time; human history was viewed as progressing by steady improvement up a single evolutionary ladder. The rationale relied upon a supposed rule of ecology: the principle of competitive exclusion, which states that two species with very similar adaptations cannot coexist. In this case, culture was viewed as such a novel and powerful behavioral adaptation that two cultural species simply could not thrive side by side. Thus, because all hominines are cultural by definition, only one hominine species could exist at any one time.

The single-species hypothesis collapsed in the mid-1970s, after fossil discoveries from Kenya undisputedly demonstrated the coexistence of two very different species of hominine: *Homo erectus* (or *Homo ergaster*), a large-brained species that apparently was ancestral to *Homo sapiens*, and *Australopithecus boisei*, a small-brained species that eventually became extinct. Subsequent discoveries and analyses implied that several species of hominine coexisted in Africa some 2 million or so years ago (see unit 22), suggesting that several different ecological niches were being successfully exploited. These findings implied that to be hominine did not necessarily mean being cultural. Thus, no longer could hominine origins be equated with human origins.

During the past decade, not only has an appreciation of a spectrum of hominine adaptations—including the simple notion of a bipedal ape—emerged, but the lineage that eventually led to *Homo sapiens* has also come to be perceived as much less human. Gone is the notion of a scaled-down version of a modern hunter-gatherer way of life. In its place has appeared a rather unusual African ape adopting some novel, un-apelike modes of subsistence (see unit 26).

Today, hominine origins are completely divorced from any notion of human origins. Questions about the beginning of the hominine lineage are now firmly within the territory of behavioral ecology and do not draw upon those qualities that we might perceive as separating us from the rest of animate nature. Questions of hominine origins must now be posed within the context of primate biology. ✳

KEY QUESTIONS

- Why were post-evolutionary theory explanations of human origins considered "self-explanatory"?
- What is the effect of sparse fossil evidence on theories of human evolution?
- Was the notion of parallel evolution of similar anatomical features among humans and African apes a reasonable explanation?
- Why was "culture" so dominant a theme in explanations of human origins?

KEY REFERENCES

Bowler PJ. Theories of human evolution. Baltimore: Johns Hopkins Press, 1986.

Cartmill M. Human uniqueness and theoretical content in paleoanthropology. *Int J Primatol* 1990;11: 173–192.

Cartmill M, Pilbeam DR, Isaac GL. One hundred years of paleoanthropology. *Am Scientist* 1986;74: 410–420.

Fleagle JG, Jungers WL. Fifty years of higher primate phylogeny. In: Spencer F, ed. A history of American physical anthropology. New York: Academic Press, 1982.

Gee H. Box of bones "clinches" identity of Piltdown palaeontology hoaxer. *Nature* 1996;381:261–262.

Lewin R. Bones of contention. New York: Touchstone, 1988.

Tobias PV. An appraisal of the case against Sir Arthur Keith. *Curr Anthropol* 1992;33:243–294.

MODERN EVOLUTIONARY THEORY

4

One of the most important phenomena that a successful theory of evolution must explain is **adaptation**—that is, the way that species' anatomy, physiology, and behavior appear to be well suited to the demands of their environments. Adaptation is pervasive in nature and in pre-Darwinian times was viewed as the product of divine creation. Moreover, once so created, species were believed to change little, if at all, through time. In his *Origin of Species*, published in November 1859, Darwin explained the purpose of the book as follows: "I had two distinct objects in view; firstly to show that species had not been separately created, and secondly, that natural selection had been the chief agent of change." **Natural selection**, Darwin believed, explained how species became adapted to their environments.

The notion that species do, in fact, change through time was already in the air in 1859. Consequently, Darwin readily succeeded with his first goal, given the volume of evidence he presented in *Origin* in support of the reality of evolution. The second goal, showing that natural selection was an important engine of evolutionary change, remained elusive until the 1930s, when it became the central pillar of newly established evolutionary thinking, known as **NeoDarwinism**.

In addition to adaptation, evolutionary theory must explain the origin of new species and major trends within groups of related species, such as the increase in body size and the reduction of the number of toes among horses in that group's 50 million years of evolution and the increase in the size of the brain in human evolution. The origin of species and the pattern of trends among groups of species are collectively known as **macroevolution**. Despite the title of his most famous book, Darwin did not address the origin of species in detail in *Origin*. As stated above, his principal focus was directed toward change within species, through natural selection, which was viewed as a slow, steady process built on minute modifications through time. This process is known as **microevolution**. Macroevolution was assumed to represent the outcome of microevolutionary processes accumulating over very long periods of time within populations, an assumption that was central to NeoDarwinism as well.

During the past several decades, the validity of this assumption has been challenged. Although adaptation through natural selection remains an important part of modern evolutionary theory, the patterns of change at levels higher than the individual organism (that is, at the level of species and groups of species) are now viewed as being more complex. This unit will address the mecha-

nisms of microevolution and macroevolution and their roles in the overall pattern of life as seen in the fossil record. Unit 6 will discuss the role of extinctions—particularly mass extinctions—in creating this pattern.

THE POWER OF NATURAL SELECTION

Natural selection, as enunciated by Darwin, is a simple and powerful process that depends on three conditions. First, members of a species differ from one another, and this variation is heritable. Second, all organisms produce more offspring than can survive (although some organisms, most notably large-bodied species and those that bestow a lot of parental care, produce few offspring while others may produce thousands or even millions, the same rule applies). Third, given that not all offspring survive, those that do are, on average, likely to have an anatomy, physiology, or behavior that best prepares them for the demands of the prevailing environment. The principle of natural selection came to be known (inaccurately) as **survival of the fittest**, even though Darwin did not use that term.

Natural selection, then, is **differential reproductive success**, with heritable favorable **traits** bestowing a survival advantage on those individuals that possess them. Generation by generation, favorable traits will become ever more common in the population, causing a microevolutionary shift in the species. Such traits will remain favored, however, only if prevailing conditions remain the same. A species' environment usually does not remain constant in nature. A change in a species' physical or biological environment (see unit 5) may alter a population's **adaptive landscape**, perhaps rendering a previously advantageous trait less beneficial or making a less advantageous trait more favorable. Natural selection, or an individual's "struggle for existence" as Darwin put it, is a local process, consisting of a generation-by-generation adjustment to local conditions.

The power of natural selection can be seen in the phenomenon of **convergent** (or **parallel**) **evolution**, in which distantly related species come to resemble one another very closely by adapting to similar ecological **niches**. The anatomical similarity of the North American wolf and the Tasmanian wolf is a good example. The former is a placental mammal and the latter is a marsupial, making the two species extremely distant genetically, having been evolutionarily separate for at least 100 million years. The anatomical similarities between the two distant species of wolf reflect convergent evolution, or **analogy**—not shared ancestry. Anatomical similarities that result from shared ancestry are examples of **homology**. Homologous structures are especially important in the reconstruction of evolutionary history based on morphological characters (see unit 8).

ESTABLISHMENT OF POPULATION GENETICS

Darwin was well aware that members of a species vary, and that these variations are heritable: his observations of natural populations and experiments with domestic breeding were proof of that ability. He was not familiar with the basis of inheritance, however. Although the rules

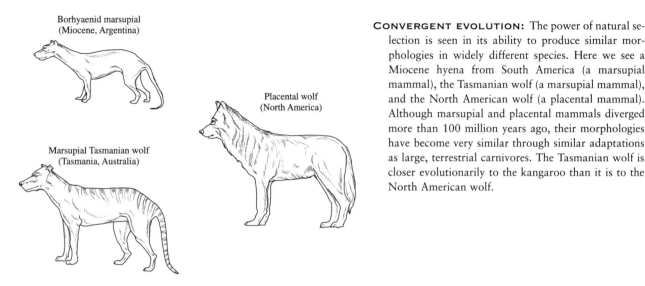

Borhyaenid marsupial
(Miocene, Argentina)

Placental wolf
(North America)

Marsupial Tasmanian wolf
(Tasmania, Australia)

CONVERGENT EVOLUTION: The power of natural selection is seen in its ability to produce similar morphologies in widely different species. Here we see a Miocene hyena from South America (a marsupial mammal), the Tasmanian wolf (a marsupial mammal), and the North American wolf (a placental mammal). Although marsupial and placental mammals diverged more than 100 million years ago, their morphologies have become very similar through similar adaptations as large, terrestrial carnivores. The Tasmanian wolf is closer evolutionarily to the kangaroo than it is to the North American wolf.

of inheritance were discovered by the Austrian monk Gregor Mendel in the early 1860s, the results of his work remained generally unknown until the turn of the century, two decades after Darwin's death.

From observations on the progeny from experimental crossing of pea plants, Mendel discovered that physical traits are determined by stable inheritance factors (what we now call genes). He also found that each plant has two genes for each trait, one from the female parent and one from the male. The variants of each gene, or **alleles**, may be identical (in which case the individual is **homozygous**) or different (the individual is **heterozygous**). When the two alleles differ, one form may be **dominant** and the other **recessive** (in humans, for instance, the allele for brown eyes is dominant relative to the blue allele). Gametes, or sex cells, receive one or the other of the two alleles with equal probability.

Mendel's experiments were very simple from a genetic standpoint, with just one or two genes affecting one trait. Before long it became apparent that most traits are influenced by many genes, not just one or two. Nevertheless, the system was amenable to mathematical analysis, and the selection of favored physical, physiological, or behavioral traits (the **phenotype**) could be studied in terms of the selection of genes that underlay them (the **genotype**).

THE EMERGENCE OF THE MODERN SYNTHESIS

The change in frequency of particular alleles within a population as a result of natural selection on them provides the basis of microevolution. From time to time, however, the DNA sequence that represents the information encoded in a gene becomes changed, often when a "mistake" occurs as the gene is copied within the **germline**. Such a **mutation** introduces the potential for further genetic variation within the population.

No simple relationship exists between a mutation and the degree of phenotypic change it might produce. For instance, a single base mutation in the gene of a serum albumin might marginally modify the physical chemistry of the blood, perhaps with some impact on adaptation or perhaps not. On the other hand, a similar mutation in a gene that affects the timing of the program of embryological development might have dramatic consequences for the mature organism. The slowing of embryological development and subsequent prolongation of the growth period, a phenomenon known as **neoteny**, was apparently important in the evolution of humans from apes.

The fate of mutations, and therefore their importance in future evolution, was the topic of intense debate in the early years of population genetics. (In this discipline, it is important to distinguish between the mutation rate of a gene, which may be quite common, and the retention, or fixation, of those mutations in the species' populations, which is much less common.) In Darwinian evolution, natural selection was viewed as retaining beneficial traits (alleles) and was therefore a creative process, not just a cleaning-up process that eliminated disadvantageous traits.

Until the mid-1940s, evolutionary theory remained distinctly at odds with strict Darwinism, and many different views were put forth to explain how the pattern of life was shaped. Then, following the creative melding of natural history, population genetics, and paleontology, a consensus of sorts appeared, known as the **modern synthesis**. This theory encompassed three principal tenets. First, evolution proceeds in a gradual manner, with the accumulation of small changes over long periods of time. Second, this change results from natural selection, with the differential reproductive success founded on favorable traits, as described earlier. Third, these processes explain not only changes within species but also higher-level processes, such as the origin of new species, producing the great diversity of life, extant and extinct. Darwinism had triumphed.

MECHANISMS OF MACROEVOLUTION

Our discussion so far has focused on microevolution, or changes within species. We will now turn to macroevolution—that is, the origin of new species and trends among groups of related species.

New species may arise in two ways. First, an existing species may be transformed by gradual change through time, so that the descendant individuals are sufficiently differentiated from their ancestors as to be recognized as a separate species. This mode is known as **anagenesis**, and it results in one species evolving into another over time. Second, a population of an existing species may become reproductively isolated from the parent species, producing a second, distinct species. This mode is known as **cladogenesis**, and comprises a splitting event that yields two species where previously only one existed. This process has obviously been important in the history of life because the fossil record shows that biodiversity has increased steadily (with fluctuations and occasional mass extinctions, as discussed in unit 6) since complex forms of life evolved, a little more than half a billion years ago. (Cladogenesis is also called **speciation**.)

On a shorter time scale, cladogenesis plays an important role in **adaptive radiation**. Adaptive radiation is a characteristic pattern of evolution following the origin of an evolutionary novelty, such as feathered flight (for birds), placental gestation (for eutherian mammals), or bipedal locomotion (in hominines). The original species bearing the evolutionary novelty very quickly yields descendant species, each representing a variant on the new adaptation. The result, drawn graphically, is an evolutionary bush, with an increasing number of coexisting species through time that have all descended from the same ancestor. The sum total of descendants of that common ancestor is known as a **clade** (see unit 8)—hence the term "cladogenesis."

Cladogenesis is most likely to occur when a small, peripheral population of a species is separated from the parental population. Such small populations, which contain less genetic variation and are less stable genetically than large populations, may become established in one of several ways, such as through the origin of new physical barriers, the colonization of islands, or the rapid crash of a subpopulation to small numbers. When a small population becomes established in one of these ways and then expands, it exhibits what is termed as a **founder effect**. A founder population that gives rise to a new species in separation from other populations of the same species produces **allopatric speciation** ("allopatric" means "in another place"). Allopatric speciation is the most common means by which new vertebrate species arise. When a new species arises from a subpopulation that is not separated from the main population, the process is termed **sympatric speciation** ("sympatric" means "in the same place").

So much for the mode of the origin of new species; what of the tempo and its mechanism? The modern synthesis decreed that macroevolution was simply an ex-

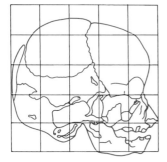

Chimp fetus

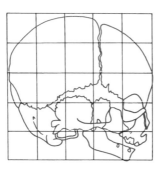

Human fetus

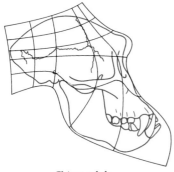

Chimp adult

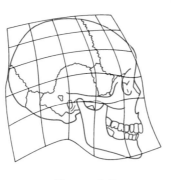

Human adult

NEOTENY IN HUMAN EVOLUTION: Although the shape of the cranium in human and chimpanzee fetuses is very similar, a slowdown in development through human evolution has produced adult crania of very different forms, varying principally in the shape of the face and the size of the brain case. The changes in grid shapes indicate the orientation of growth.

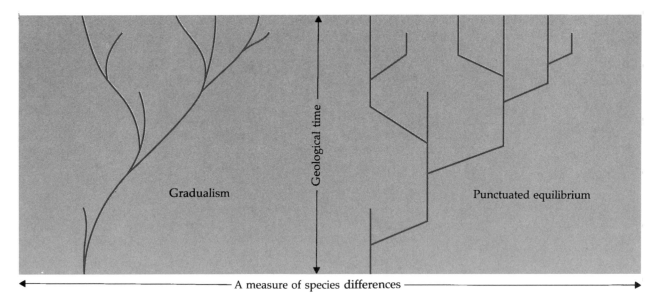

TWO MODES OF EVOLUTION: Gradualism and punctuated equilibrium. Gradualism views evolution as proceeding by the steady accumulation of small changes over long periods of time. In contrast, punctuated equilibrium sees morphological change as being concentrated in "brief" bursts of change, usually associated with the origin of a new species. Evolutionary history reflects the outcome of a combination of these two modes of change, although considerable debate has arisen as to which mode is the more important.

trapolation of microevolutionary processes: an accumulation of small changes over a long period of time, leading to large resulting changes. This process is known as **phyletic gradualism**, which, given a large enough resultant change, may yield a new species.

Because phyletic gradualism is driven by the gradual process of natural selection, it creates new adaptations that, when sufficiently different from those in the ancestral species, may lead to a new species that is characterized by those adaptations. In principle, this gradual change should be evident in the fossil record, whether anagenesis or cladogenesis is the end-result. Typically, gradual change is *not* seen in the record, however. Instead, the new species usually appears abruptly, either replacing the parental species (anagenesis) or appearing concurrently with it (cladogenesis), with no transitional forms present.

Proponents of the modern synthesis adopted Darwin's explanation for the absence of transitional forms, which was that the fossil record is incomplete. In the early 1970s, Niles Eldredge, of the American Museum of Natural History, and Stephen Jay Gould, of Harvard University, challenged this interpretation. They argued that, incomplete though the fossil record may be, it presents an accurate view of the tempo of evolutionary change. Instead of undergoing continual, gradual change, species remain relatively static for long periods of time; when change comes, it occurs rapidly ("rapidly" means a few thousand years). Apart from rare occasions in unusual geological circumstances, the bursts of change go unrecorded in the fossil record. Eldredge and Gould gave this tempo of evolution—that is, long periods of stasis interspersed with brief intervals of rapid change—the name of **punctuated equilibrium**.

An important difference between punctuated equilibrium and the traditional explanation of species formation relates to the nature of change that occurs at that time. The modern synthesis saw adaptation as the *cause* of speciation, through the accumulation of such changes through time; whereas punctuated equilibrium sees it as a potential *consequence*, as change accumulates after populations are separated geographically and genetically.

THE ORIGIN OF EVOLUTIONARY TRENDS

Punctuated equilibrium leads to another insight of macroevolution, that of trends within groups of species. Mentioned earlier was the evolutionary history of the horse clade, in which body size increased and the number of toes decreased. A second example involves the increase in brain size during human evolution, at least once the genus *Homo* had evolved, some 2.5 million years ago.

With horses, the evolutionary trend was long interpreted as a progressive improvement, as if increased body size and a reduced number of toes represented a more efficient way of being a horse. Similarly, the brain size increase evident with the first species of *Homo* is often described as the beginning of brain enlargement, as if it were a progressive process that was nurtured steadily by natural selection. Through the lens of the modern synthesis, the trends could be explained as progressions that resulted from directional natural selection. Punctuated equilibrium, however, provides a different explanation.

If, as noted earlier, species persist unchanged for most

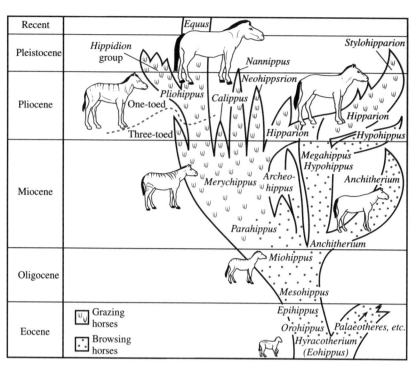

Recent		
Pleistocene		
Pliocene		
Miocene		
Oligocene		
Eocene		

EVOLUTIONARY TRENDS: The evolutionary history of horses was once considered as a series of evolutionary trends (to larger body size, more complex teeth, and fewer toes) that marked steady, directional progression. In fact, the evolution of horses is more like a bush than a directional ladder. The differential survival rates of certain species with certain characters merely gives the impression of steady progression, but does not represent reality.

of their duration, then evolution is *not* directional in this sense. Trends within groups of member species with a certain characteristic are less likely to go extinct. Many factors can influence species' tendencies for extinction (and speciation) because the two trends are linked (see units 5 and 6).

One such factor is the nature of a species' adaptation. The fossil record shows that species with highly specialized environmental and subsistence requirements are more likely to speciate and become extinct than those with much broader adaptations. The reason is that any change in prevailing environment is likely to push specialists beyond the limits of their tolerances, promoting both speciation and extinction. Clearly, generalists can accommodate much broader shifts in conditions, making speciation and extinction rarer for them.

Suppose, for example, that horse species with large body size survive longer, for some reason. The differential survival rates of species along these lines would produce a trend toward larger horses, not because it made better horses in the sense of adaptation but as a consequence of

the properties of species. Similarly for hominine species and large brain size, no persuasive evidence indicates increasing encephalization within species, merely a trend toward larger brain size within the clade. If large brain size endowed species with greater longevity, a history of increased brain size within the group would result.

In thinking about the shape of human evolution, an interesting question is how many hominine species might have existed at any one time. Adaptive radiation leads to a bushy family tree, with multiple species existing at any point, rather than linear, with just one species existing at any one time. Hominines and horses are unusual in nature in that each group is represented in today's world by a single genus. The fossil record of horses has shown, however, that this group was once a luxuriant evolutionary bush. How bushy human history was remains to be established, but calculations based on the estimated number of fossil primate species imply that in the 5 million years that the hominine group has existed, at least 16 species would have arisen. So far some 12 have been identified, most of which existed early in the group's history. ❋

KEY QUESTIONS

- Why are mutations important in evolution, and how do they become fixed in a population?
- Why is macroevolution not considered to be merely an extrapolation of microevolutionary processes operating over long periods of time?
- Why is adaptive radiation so common a pattern in evolution?
- What evolutionary factors are most important in shaping the history of human evolution?

KEY REFERENCES

Gould SJ. Darwinism and the expansion of evolutionary theory. *Science* 1982;216:380–387.

Gould SJ, Eldredge N. Punctuated equilibrium comes of age. *Nature* 1993;366:223–227.

Fitch W, Ayala FJ, eds. Tempo and mode in evolution. Washington DC: National Academy Press, 1995.

Somit A, Peterson SA, eds. The dynamics of evolution. Ithaca, NY: Cornell University Press, 1992.

Stebbins GL, Ayala FJ. The evolution of Darwinism. *Sci Am* July 1985:72–80.

Tattersall I. How does evolution work? *Evol Anthropol* 1994;3:2–3.

Weiner J. The beak of the finch. New York: Alfred A. Knopf, 1994.

THE PHYSICAL CONTEXT OF EVOLUTION

5

Two factors are recognized as influencing the evolution of new species and the extinction of existing species. First is the biotic context—that is, the interactions between members of a species and between different species, principally in the form of competition and resulting natural selection. Second is the physical context, such as geography and climate, which determines the types of species that can thrive in particular regions of the world, according to their climatic adaptations.

Biologists have long debated the relative contributions of these two factors in driving evolutionary change. Not surprisingly, Darwin emphasized the power of biotic interaction, because it lies at the core of natural selection. He did not ignore the effects of the physical environment, but saw them as merely tightening the screws of competition.

This viewpoint was central to the modern synthesis (introduced in unit 4), with physical context being granted a very secondary role. Even in the absence of change in the physical environment, it was assumed, evolution would continue, driven by the constant struggle for existence. When one individual (or species) gained a slight adaptive advantage over others, the Darwinian imperative to catch up would fuel the evolutionary engine. Predators and prey, for instance, were viewed as being engaged in a constant battle, or evolutionary arms race. In the early 1970s, the Chicago University biologist Leigh van Valen termed this idea the Red Queen hypothesis; the name is derived from the character in *Alice Through the Looking Glass*, who tells Alice that it is necessary to run faster and faster in order to stay in the same place. The same evolutionary dynamic would apply to the effect of competition among species for resources.

In recent years, however, interest has grown in the physical context of life and its possible role in evolution at all levels, from promoting change within species to being a forcing agent in speciation, and even shaping the entire flow of life. This shift in perspective comes from two sources. The first, which flows from the broad acceptance of allopatric speciation as the principal mechanism of the evolution of new species, will be the topic of this unit. The second source is the growing understanding that mass extinction is more than simply an interruption in the flow of life, and instead is a creative influence in that flow; this idea is discussed in unit 6.

THE INFLUENCE OF PLATE TECTONICS

If new species preferentially arise in small, isolated populations (allopatric speciation) rather than in large, continuous populations (sympatric speciation), as modern evolutionary theory holds, then processes that promote the establishment of such populations can be regarded as a potential engine of evolution. The physical environment provides two means by which this process might occur. First, topography on local and global scales may change, principally through the mechanism of **plate tectonics**. Second, global climate change may be driven by many factors, including some of the effects of plate tectonics.

The Earth's crust is a patchwork of a dozen or so major plates whose constant state of creation and destruction keeps them in continual motion relative to one another. Continental landmasses, which are less dense than crustal rock, ride passively atop these plates; as a result, they are also in a constant state of (extremely slow) motion, shuffling around the globe like a mobile jigsaw puzzle. Continents occasionally come together or separate, sometimes producing smaller fragments. As a result, biotas that were once united have been divided, and previously independent biotas have been brought together.

For instance, Old World and New World monkeys derive from a common stock, but followed independent paths of evolution as South America and Africa drifted apart some 50 million years ago. Australia's menagerie of marsupial mammals evolved in isolation from placental mammals, as the island continent lost contact with Old World landmasses more than 60 million years ago, before placental mammals were introduced by humans. By contrast, when the Americas joined some 3 million years ago via the Panamanian Isthmus, an exchange mingled biotas that had evolved separately for tens of millions of years. Indian and Asian species migrated into one another's lands when the continents united approximately 45 million years ago. India's continued northward movement eventually caused the uplift of the massive Himalayan range, producing further geographic and climatic modification on a grand scale. Africa and Eurasia exchanged species when the landmasses made contact approximately 18 million years ago; in the process, apes joined species making the journey from south to north and many species of antelope moved in the opposite direction.

Whenever landmasses become isolated as a result of plate tectonics, the environment—and therefore the evolutionary fate—of the indigenous species is influenced by the simple fact of isolation. The isolation of ancestral mammalian species some 100 million years ago, when landmasses were particularly fragmented, has recently been suggested to have prompted the development of the modern mammal orders. Based as it is on a comparison of gene sequences in a handful of modern mammals, this conclusion is at odds with currently accepted views of mammalian evolution. This theory posited the origination of modern orders of mammals as a result of ecological niches having been opened up following the extinction of the dinosaurs 65 million years ago.

When previously isolated landmasses unite, a complex evolutionary dynamic ensues, with some species becoming extinct. This fate befell many South American mammals during the Great American Interchange. Other species may enjoy a burst of speciation during this process, as did many of the North American mammals when they populated South America, the apes as they spread into Eurasia, and the antelopes as they thrived in Africa.

In addition to influencing evolution by shuffling land-masses, plate tectonics can modify the environment within individual continents. A prime example of this phenomenon occurred in Africa, where it may have affected the evolution of the hominine clade. Broadly speaking, 20 million years ago, the African continent was topographically level and carpeted west to east with tropical forest; tectonic activity greatly modified this pattern.

A minor tectonic plate margin runs south-to-north under East Africa. Beginning 15 million years ago, it produced localized uplift that yielded tremendous lava-driven highlands that reached 2000 meters and were centered near Nairobi in Kenya and Addis Ababa in Ethiopia. These highlands were the Kenyan and Ethiopian domes. Weakened by the separating plates, the continental rock then collapsed in a long, vertical fault, snaking several thousand kilometers from Mozambique in the south to Ethiopia in the north, and on to the Red Sea. The immediate effect of the newly elevated highlands was to throw the eastern part of the continent into a rain shadow, dramatically altering the vegetation there. Continuous forest was replaced by a patchwork of open woodlands and, eventually, grassland savannah. Such a habitat fragmentation and transformation would have fragmented the range of forest-adapted animal species living there, encouraging allopatric speciation. More important, the once topographically even terrain became extremely diverse, ranging from hot, arid lowland desert to cool, moist highlands, and a range of different types of habitat in between.

All species can tolerate only a limited range of environmental conditions, as defined by temperature, availability of water, and type of terrain. For animal species, the kinds of plant species that are available influence their ability to occupy any particular **biome**. Some species' range of tolerance is greater than that of other species; such species will, therefore, be able to live across several biomes. Over-all, however, a topographically diverse terrain will also be biologically diverse.

In addition, topographical diversity creates barriers to population movement. For instance, a species that is adapted to the conditions of high elevation may be prevented from migrating from one highland to another because the intervening terrain is inhospitable to it. As a result, a region that is topographically diverse harbors small, isolated populations and therefore represents a potential factory of the evolution of new species. The tectonic uplift and vertical faulting that formed the Great Rift Valley in East Africa produced such a topography, and may well have created conditions conducive to the evolution of hominines from an apelike ancestor.

CLIMATE CHANGE AND HABITAT THEORY

A considerable body of data has been amassed during the past decade relating to the Earth's climate during the Cenozoic, from 65 million years ago to the present, and particularly for the time period most relevant to human evolution, the last 5 million years. The climatic picture is one of continual and sometimes dramatic change within a net cooling trend. Superimposed on this pattern are global cooling and warming cycles, the so-called Milankovitch cycles, with periodicities of approximately 100,000, 41,000, and 23,000 years. Each of these cycles dominates climate fluctuation at different times in Earth history. For example, prior to 2.8 million years ago, the shortest cycle was dominant; between 2.8 and 1 million years ago, the 41,000-year cycle prevailed; from 1 million years onward, the dominant cycle has been 100,000 years.

During the 5 million years since the first appearance of the hominine clade, several major global cooling episodes occurred within this overall trend and against the background of the frequent Milankovitch cycles; the existence of these episodes has been inferred from oxygen isotope data and more recently from measures of wind-blown dust in the oceans around Africa. The first event, appearing at 5 million years, involved significant cooling. The

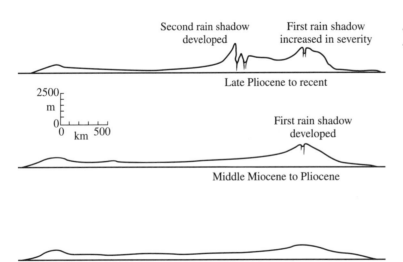

Second rain shadow developed

First rain shadow increased in severity

Late Pliocene to recent

First rain shadow developed

Middle Miocene to Pliocene

Late Oligocene to mid-Miocene

TOPOGRAPHIC SECTION OF AFRICA ALONG THE EQUATOR: During the past 20 million years, tectonic activity beneath East Africa caused uplift and subsequent faulting, forming the modern Great Rift Valley. The effect was twofold. First, it threw the continent east of the uplifted highlands into a rain shadow, causing once-continuous forest cover to shrink and fragment. Second, it produced great topographic diversity, which generated a mosaic of fragmented habitats. These effects are thought to have been influential in the evolution of the hominines, among other evolutionary changes. (Courtesy of T. Partridge *et al.*)

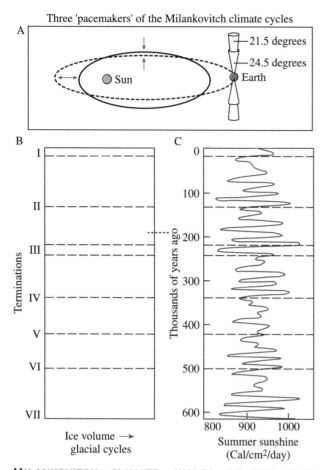

Three 'pacemakers' of the Milankovitch climate cycles

A

21.5 degrees
24.5 degrees
Sun   Earth

B   C

I   0

II   100

III   200

IV   300

V   400

VI   500

VII   600

Terminations

Thousands of years ago

Ice volume →
glacial cycles

800   900   1000
Summer sunshine
(Cal/cm²/day)

MILANKOVITCH CLIMATE CYCLES OF THE PAST 600,000 YEARS: Superimposed on long-term global climate change are regular cycles driven by three pacemakers: (A) changes in orbital eccentricity, and tilt and orientation of the Earth's spin axis, which results in a 100,000-year cycle; (B) changes in the volume of the Earth's ice sheets, giving a 41,000-year cycle; and (C) the effect of the intensity of summer sunshine at northern latitudes, yielding a 23,000-year cycle.

second, between 3.5 and 2.5 million years ago, was associated with the first major buildup of Arctic ice and substantial expansion of Antarctic ice. The modern Sahara's roots lie at this point, too. This beginning of the modern Ice Age may have been initiated by a change in circulation patterns in the atmosphere and oceans as a result of the rise of the Panamanian Isthmus, which joined North and South America some 3.5 million years ago. The third event occurred nearly 1.7 million years ago. The fourth, arising approximately 0.9 million years ago, was possibly caused by uplift in western North America and of the Himalayan range and the Tibetan Plateau. Of the four events, the second was largest in extent. The overall pattern of climate change is therefore extremely complicated, driven by several different forcing agents.

Inevitably, species and the ecosystems of which they are a part do not remain immune to climate change of this

magnitude. The temperature extremes of the Milankovitch cycles exceed the habitat tolerances of virtually all species, turning a once suitable habitat into an inhospitable one; the larger shifts have an even more dramatic impact. The average lifespan of a terrestrial mammal species, for instance, is several million years; the periodicity of the cycles is just a fraction of that average. Thus, it is obvious that most species are able to survive these repeated climatic fluctuations. The principal response of species to climate is dispersal, tracking the change so as to remain in hospitable habitats. During global cooling, dispersal moves toward lower latitudes; during warm periods, it takes the reverse direction. Because different species have different tolerance limits, ecosystems do not migrate en masse, but rather become fragmented, eventually forming new communities.

Other biotic responses to climate change are possible as well, particularly when a threshold of tolerance is exceeded—namely extinction and speciation. These trends are central to the **habitat hypothesis**, which has been promoted principally by Yale University biologist Elisabeth Vrba. Although it has many components, the habitat hypothesis can be stated simply: species' responses to climate change represent the principal engine of evolutionary change. The major mechanism of such change is **vicariance**, or the creation of allopatric populations from once continuous populations, either by the establishment of physical barriers or the dispersal of populations across such barriers. After such populations become established, they are both vulnerable to extinction and have an opportunity for speciation (see unit 4).

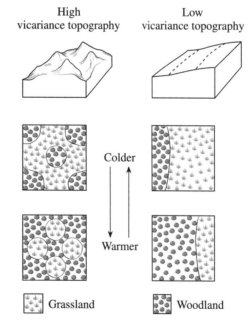

High
vicariance topography

Low
vicariance topography

Colder

Warmer

Grassland   Woodland

CLIMATE CHANGE AND HIGH TOPOGRAPHIC DIVERSITY: During times of climate cooling, regions of high topographic diversity will host many vicariant populations, which become isolated through the inability of organisms to track congenial habitats through dispersal. (Courtesy of E. Vrba.)

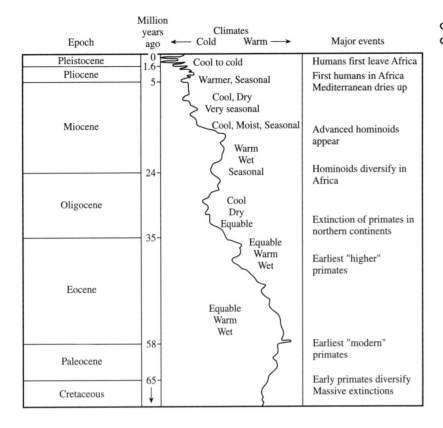

Epoch	Million years ago	Climates ← Cold Warm →	Major events
Pleistocene	0 / 1.6	Cool to cold	Humans first leave Africa
Pliocene	5	Warmer, Seasonal	First humans in Africa Mediterranean dries up
Miocene		Cool, Dry Very seasonal	
		Cool, Moist, Seasonal	Advanced hominoids appear
	24	Warm Wet Seasonal	Hominoids diversify in Africa
Oligocene		Cool Dry Equable	Extinction of primates in northern continents
	35	Equable Warm Wet	Earliest "higher" primates
Eocene		Equable Warm Wet	
	58		Earliest "modern" primates
Paleocene	65		Early primates diversify Massive extinctions
Cretaceous			

CLIMATE PATTERNS SINCE THE END-CRETACEOUS: An overall cooling trend with local fluctuations marks the Cenozoic period, which culminates in the Pleistocene Ice Age. Major events of primate evolution are shown in the right-hand column. (Courtesy of I. Tattersall.)

Because of their variable adaptations, different types of species exhibit different vulnerabilities to climate change. Warm-adapted species, such as tropical forests and the animals living there, cluster around the equator and will be extensive in warm times. Temperate forests and grasslands become increasingly dominant at higher latitudes. A fall in global temperature will produce a general equatorward migration, drastically reducing the area available for tropical forest, which responds by becoming reduced in extent and fragmented. In their equatorward migration, grasslands may be able to occupy an area similar to that in previous climes, leaving behind patches of vicariant habitat encroached upon by tundra. During such climatic times, therefore, warm-adapted species are likely to undergo higher rates of extinction and speciation than cold-adapted species. The reverse should be true during times of global warming. Because of the general cooling trend of the past 20 million years, the former pattern will have been predominant.

Differences are observed among warm-adapted and cold-adapted species, of course. Some species are habitat specialists, while others are generalists (these terms refer to the availability of required food resources, not just the breadth of diet). Anteaters, for instance, are food specialists; because their food is plentiful in many different ecosystems, however, they can tolerate significant habitat change. Food generalists, such as large carnivores and omnivores, can also tolerate habitat change because of their breadth of diet. Species that can survive in different kinds of habitats, or biomes, are known as **eurybiomic;** those with narrow biomic tolerance are deemed **stenobiomic.** Not surprisingly, stenobiomic species are more vulnerable to climate change than are eurybiomes—a pattern that is seen in the evolutionary history of African mammals, for instance. All clades of exclusive grazers and all clades of exclusive browsers consistently show higher speciation and extinction rates than species that can both graze and browse. As a result, biome generalist species are less numerous than biome specialists. ❀

KEY QUESTIONS

- What kind of fossil evidence would support the Red Queen hypothesis?
- How could the relative contributions of competition and climate change to speciation be tested?
- What is the most important component of the physical context of evolution?
- What changes in the physical environment might have been important in human evolution?

KEY REFERENCES

Broecker WS, Denton GH. What drives glacial cycles? *Sci Am* Jan 1990:48–56.

Carson HL. The processes whereby species originate. *BioScience* 1987;37:715–720.

deMenocal PB. Plio-Pleistocene African climate. *Science* 1995;270:53–59.

Foley RA. Speciation extinction and climate change in hominid evolution. *J Hum Evol* 1994;26:275–289.

———. The evolutionary geography of Pliocene hominids in African biogeography, climate change, and human evolution. In: Bromage TG, Schrenk R, eds. African biogeography, climate change, and early hominid evolution. New York: Oxford University Press, 1997.

Partridge TC, *et al*. The influence of global climate change and regional uplift on large-mammalian evolution in East and Southern Africa. In: Vrba ES, *et al.*, eds. Paleoclimate and evolution. New Haven, CT: Yale University Press, 1995:331–354.

Potts R. Humanity's descent: the consequences of ecological instability. New York: William Morrow, 1996.

———. Evolution and climate variability. *Science* 1996;273:922–923.

Ridley M. The Red Queen. New York: Macmillan, 1993.

Shackleton NJ. New data on the evolution of Pliocene climatic variability. In: Vrba ES, *et al.*, eds. Paleoclimate and evolution. New Haven, CT: Yale University Press, 1995:242–248.

Shreeve J. Sunset on the savannah. *Discover* July 1996:42–48.

Vrba ES. Habitat theory in relation to the evolution in African Neogene biota and hominids. In: Bromage TG, Schrenk F, eds. African biogeography, climate change, and early hominid evolution. New York: Oxford University Press, 1997.

White TD. African omnivores: global climate change and Plio-Pleistocene hominids and suids. In: Vrba ES, *et al.*, eds. Paleoclimate and evolution. New Haven, CT: Yale University Press, 1995:369–384.

EXTINCTION AND PATTERNS OF EVOLUTION

6

Life first evolved on Earth almost 4 billion years ago, in the form of simple, single-celled organisms. Not until half a billion years ago did complex, multicellular organisms evolve, in an event biologists call the **Cambrian explosion**. An estimated 100 **phyla** (taxonomic groupings based on body plans) arose in that geologically brief instant, with few, if any, new phyla arising later. The products of this initial, intensely creative moment in the history of life included all of the 30 or so animal phyla that exist today. The remaining 70 or so phyla disappeared within a few tens of millions of years of their origin.

In the 530 million years since the Cambrian explosion, 30 billion species have evolved. Some represented slight variants on existing themes, while others heralded major adaptive innovations, like the invention of jaws, the amniote egg, and the capacity of flight. Given that an estimated 30 million species exist today, it's clear that 99.9 percent of species that have ever lived are now extinct. Some extinctions occur at a steady, background rate of approximately one species every four years; others are part of **mass extinction** events, during which a great proportion of extant species disappear in a geologically brief period, measuring from a few hundred to a few million years. Although extinction—and particularly mass extinc-

tion—is an important fact of life, until recently evolutionary biologists have virtually ignored the topic, choosing instead to focus on mechanisms by which new species arise.

As a result of the recent burst of research into extinction processes, biologists' assumptions about mass extinction—about its causes and, more important, its effects—have been overturned. Mass extinctions were initially viewed as mere interruptions in the slow, steady increase in biological diversity that began after the Cambrian explosion. Now, however, they are recognized as playing a major role in guiding evolutionary change.

THE INFLUENCE OF CATASTROPHISM

In his *Origin of Species*, Darwin essentially denied the fact of mass extinction, stating that extinction is a slow, steady process, with no occasional surges in rate. He also argued that species become extinct because they prove adaptively inferior to their competitors. Darwin's equation of extinction with adaptive inferiority clearly derives from his theory of natural selection, and it powerfully shaped biologists' thinking.

The fact of extinction had been demonstrated before Darwin's time, by the French anatomist, Baron Georges Cuvier, in the late eighteenth century. Cuvier definitively showed that mammoth bones differ from those of the modern elephant. The inescapable conclusion was that the mammoth species no longer existed. Through his extensive study of fossil deposits in the Paris Basin, Cuvier went on to identify what he thought were periods of mass extinctions, or catastrophes, in Earth history when large numbers of species went extinct in very short periods of time.

Cuvier's observations inspired a great volume of geological work in the early part of the nineteenth century.

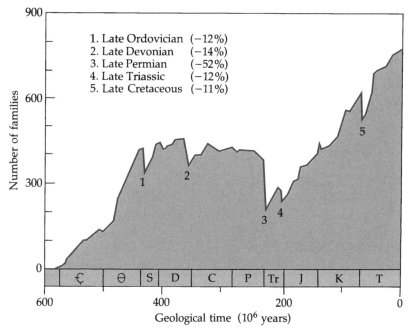

1. Late Ordovician (−12%)
2. Late Devonian (−14%)
3. Late Permian (−52%)
4. Late Triassic (−12%)
5. Late Cretaceous (−11%)

EPISODIC NATURE OF LIFE'S HISTORY: Since the origin of multicellular organisms in the early Cambrian, life's history has documented a steady rise in diversity, as recorded here by the increase through time in the number of families of marine vertebrates and invertebrates. Interrupting this rise, however, have been a series of mass extinction events (numbered 1–5), which have reduced diversity of families by the figures shown in parentheses. (The percentage loss of species is much higher.) Each extinction was followed by rapid radiations that quickly restored species diversity to pre-extinction levels. Typically, the groups that became dominant after the extinction differed from those before it. (Courtesy of David Raup.)

This research identified intervals of apparent major change in the history of life, which formed boundaries between geological **periods** that were given the following names: Cambrian, Ordovician, Silurian, Carboniferous, Permian, Triassic, Jurassic, Cretaceous, and Tertiary (which comprises the epochs Paleocene, Eocene, Oligocene, Miocene, Pliocene, Pleistocene, and Holocene).

Two particularly devastating catastrophes divided the history of multicellular life, known as the **Phanerozoic**, or visible life, into three **eras**: the **Paleozoic** (ancient life), from 530 to 225 million years ago; the **Mesozoic** (middle life), from 225 to 65 million years ago; and the **Cenozoic** (modern life), from 65 million years ago to the present. Cuvier lived in pre-Darwinian evolutionary theory times, of course, and he therefore saw the catastrophes as individual events (some 30 in all) that wiped out all of existing life, setting the stage for new waves of creation. This world view was known as **catastrophism**.

THE TRIUMPH OF UNIFORMITARIANISM

Even before Darwinian theory emerged, catastrophism came under attack, principally by the Scottish geologist Charles Lyell, who was following arguments made earlier by his fellow countryman, James Hutton. In his *Principles of Geology*, published in three volumes in the 1830s, Lyell argued that the geological processes we observe today—such as erosion by wind and rain, earthquakes and volcanoes, and so on—are responsible for *all* geological changes that have occurred throughout Earth history. He also denied the existence of mass extinctions of species.

Lyell's scheme came to be known as **uniformitarianism**. For a while, an intellectual battle pitted it against catastrophism. Uniformitarianism won decisively, and catastrophism was banished from the intellectual arena as a relic of earlier thinking. Catastrophism may have been

defeated as an idea, but paleontologists continued to find evidence of mass dyings in the fossil record. Earth history evidently is not one of gradualistic progression, as Lyell and Darwin averred, but instead a litany of sporadic and spasmodic convulsions. Some of these events have moderate impact, with 15 to 40 percent of marine animal species disappearing, but a few others are of much larger extent, constituting the mass extinctions.

This last group—known as the Big Five—comprises biotic crises in which at least 75 percent of species became extinct in a brief geological instant. In one such event, which brought the Permian period and the Paleozoic era to a close, more than 95 percent of marine animal species are calculated to have vanished. This handful of major events, from oldest to most recent, includes the following: the end-Ordovician (440 million years ago), the Late Devonian (365 million years ago), the end-Permian (250 million years ago), the end-Triassic (210 million years ago), and the end-Cretaceous (65 million years ago).

CAUSES OF MASS EXTINCTIONS

Numerous causative agents of mass extinction events have been suggested over the decades. These putative sources include a drastic fall in sea levels (sea-level regression), global cooling, predation, and interspecies competition. Of these, sea-level regression and global cooling have traditionally been held as most important.

In 1979, however, Luis Alvarez, a physicist at the University of California, Berkeley, and several colleagues suggested that the end-Cretaceous extinction, which marked the end of the dinosaurs' reign, was the outcome of Earth's collision with a giant asteroid. They based their conclusion on the presence of the element iridium in the layer that marks the Cretaceous/Tertiary boundary. Irid-

Era	Period	Age mya
Paleozoic	Permian	251
		290
	Carboniferous	
		353.7
	Devonian	
		408.5
	Silurian	
		439
	Ordovician	
		500
	Cambrian	
		543

Era	Period		Epoch	Age mya
Cenozoic	Quaternary		Holocene	0.01
			Pleistocene	1.8
	Tertiary	Neogene	Pliocene	5.2
			Miocene	23.8
		Paleogene	Oligocene	33.5
			Eocene	55.6
			Paleocene	65.0
Mesozoic	Cretaceous			144.0
	Jurassic			206.0
	Triassic			251.0

THE GEOLOGICAL TIME SCALE: Divisions in the time scale are based on major changes of biota in the fossil record. (Courtesy of David Jablonski.)

ium is rare in crustal and continental rock, but common in asteroids. The impact, striking with the force of a billion nuclear bombs, was postulated to have raised a dust cloud high into the atmosphere, effectively blocking out the sun for at least several months. The ensuing catastrophic results affected plant life first and then the animals that depended on it.

This idea was not well received initially, not least because it sounded too much like a return to catastrophism. In the years since its proposal, a large body of evidence has been gathered in its support, including evidence of an impact crater at the pertinent time, 65 million years ago. The dinosaur extinction, and several other mass extinctions, is therefore now more widely accepted as resulting from extraterrestrial impacts. Such impacts might not be the sole cause of extinction, however; the meteors might have struck a biota that was already fragile for other reasons, including those mentioned earlier, or they might have weakened the biota, making it vulnerable to secondary mechanisms of extinction.

BIOTIC RESPONSES TO MASS EXTINCTIONS

Whatever the cause of mass extinctions, the next question becomes, How do Earth's biota respond? And what determines which species survive through these crises and which do not?

One striking feature of the biota's response is that, following a rapid collapse, species diversity rebounds quickly. Within 5 to 10 million years of the event, the diversity equals and often exceeds pre-extinction levels. During this brief period, the rate of speciation greatly exceeds the rate of extinction. Typically, the groups of species that come to dominate the marine and terrestrial ecosystems differ from those that dominated prior to the collapse. Consider, for example, the end-Cretaceous extinction, which saw the disappearance of the dinosaurs as the major terrestrial animal species and their replacement by mammals. Mammals had coexisted with dinosaurs for more than 100 million years, but they were small and probably few in number. Modern orders of mammals may have originated 100 million years ago (see unit 5), but not until after the extinction did larger species evolve and become more numerous; these mammals eventually came to occupy the niches previously occupied by large reptiles.

This concept raises questions about what makes some groups of species vulnerable to extinction, or partial extinction, while others fare better. As the University of Chicago paleontologist David Raup has so succinctly put it, Was it bad genes or bad luck that consigned the losers to evolutionary oblivion? Most biologists agree that the prevailing force in times of background extinction is natural selection, in which competition plays an important

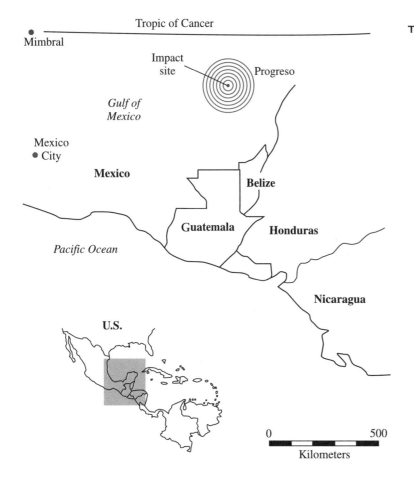

THE SMOKING GUN: The suggestion that the end-Cretaceous extinction might have been caused by asteroid impact was first made in 1979, based on the discovery of the rare element iridium at the Cretaceous/Tertiary boundary. Since then, much evidence has been amassed in support of the proposal, including the recent discovery of a huge impact crater in the Yucatan Peninsula, dated at 65 million years.

Phanerozoic Sea Level Curve

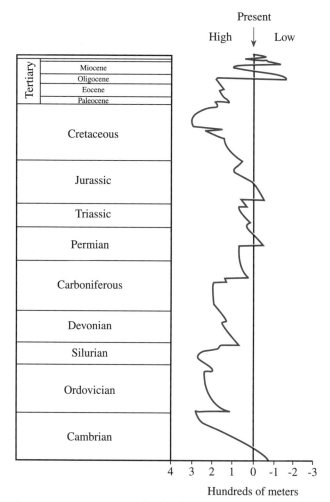

SEA-LEVEL CHANGES: Sea-level regression is a probable factor in some extinctions, and is associated with many of them. Many falls do not coincide with extinction, however, indicating that some mass extinctions are complex events, involving the interplay of several agencies.

part. But what about the bursts of higher rates of extinction? Is mass extinction merely background extinction writ large? Do marine regressions, climate cooling, and the effects of asteroid or comet impact merely tighten the screws of competition as times get tough? Until recently, the answer to these questions would have been an unequivocal "yes."

Counterintuitively, random processes can produce patterns. Raup and several colleagues tested the hypothesis that mass extinction events might represent such a pattern. In computer simulations of species communities over long periods of time, in which speciation and extinction were allowed to happen randomly with no external force operating, they found patterns similar in *form*, but not in *magnitude*, to the contents of the fossil record. In other words, species numbers fluctuated significantly with no

external driving force, but only rarely crashed in a way that could be termed a mass extinction. Thus, bad luck cannot be the sole cause of a species' demise in a mass extinction event. This research also partly inspired the realization that bad genes could not provide the sole explanation of the pattern of life. Instead, some combination of selection and bad luck operated in tandem.

The University of Chicago paleontologist David Jablonski has investigated the nature of that selection by comparing the pattern in background and mass extinction periods. During background extinction, several factors conspire to protect a species from extinction. Species that are geographically widespread resist extinction, for instance. Likewise, marine species that send their larvae far and wide (drifting with the currents) resist extinction, for similar reasons. A group of related species, a clade, resists extinction if it contains many species rather than only a few. Thus, the chance disappearance of a few species is more likely to threaten the survival of a clade that includes only three species, for example, than one that has 20.

When Jablonski examined the fate of mollusc species and species' clades across the end-Cretaceous extinction, he saw a very different picture. None of the above rules applied. The only rule he could discern was valid for groups of related species, clades. Once again, geographic distribution played a part in survival. If a group of species occurred over a wide geographic range, then they fared better in the biotic crisis than those that were geographically restricted, no matter how many species made up the clade. "During mass extinctions, quality of adaptation or fitness values . . . are far less important than membership in the particular communities, provinces, or distributional categories that suffer minimal disturbance during mass extinction events," wrote Jablonski. This finding was a landmark result, because it was the first to clearly indicate that the rules changed between background and mass extinction. Biotic crises are *not* simply background extinctions writ large.

This idea makes sense because, in the history of life, many successful species or groups of species have met abrupt ends in mass extinctions. The dinosaurs dominated their realms for more than 100 million years and were as diverse as they had ever been when they were vanished at the end-Cretaceous extinction. No evidence suggests that the mammals were better adapted in any way than dinosaurs, which they subsequently replaced as the major terrestrial tetrapod group.

Natural selection operates cogently at the level of the individual, in relation to local conditions, reflecting the impact of competitors and prevailing physical conditions. It is a powerful response to immediate biological experience, but it cannot anticipate future events. And it certainly cannot anticipate rare events. The average longevity of an animal species is about 2 million years, and extinction bursts occur on average no more than every 27 million years or so. Consequently, most species never experience such bursts. The mass extinction episodes are rarer still, making them invisible to natural selection. Species cannot adapt to conditions they do not experience. The Darwinian view that the history of life is one of continual improvement through adaptation led by natural selection is therefore incomplete.

Mass extinctions, then, restructure the biosphere, with an unpredictable set of survivors finding themselves in a world of greatly reduced biological diversity. With at least 15 percent and as much as 95 percent of species wiped out, ecological niches are opened or at least made much less crowded. This time provides an evolutionary opportunity offered to a lucky few.

Homo sapiens evolved amid a high point of global biodiversity. We are but one of millions of species here on Earth, the product of half a billion years of life's flow, lucky survivors of at least 20 biotic crises, including the catastrophic Big Five. If the ancestral primate species had been among the mammalian lineages that became extinct at the end-Cretaceous event, there would be no prosimians, no monkeys, no apes, and no *Homo sapiens* today. Its survival, and our subsequent existence, was largely a matter of factors having nothing to do with adaptive qualities. ✳

KEY QUESTIONS

- Why did uniformitarianism become so powerful a force in late nineteenth and twentieth century scientific thinking?
- How might mass extinction be explained as a consequence of natural selection?
- How does mass extinction influence the history of life?
- How can the hypothesis of asteroid impact be tested?

KEY REFERENCES

Alvarez W, Asaro F. An extraterrestrial impact. *Sci Am* Oct 1990:78–84.

Courtillot VE. What caused the mass extinction? A volcanic eruption. *Sci Am* Oct 1990:85–92.

Gould SJ. Jove's thunderbolts. *Nat Hist* Oct 1994: 6–12.

Hart MB, ed. Biotic recovery from mass extinction. London: Geological Society of London, 1996.

Hsü KJ. Uniformitarianism vs. Catastrophism in the extinction debate. In: Glen W, ed. The mass extinction debates. Palo Alto, CA: Stanford University Press, 1994:217–229.

Jablonski D. Mass extinctions: new answers, new questions. In: Kaufman L, Mallory K, eds. The last extinction, 2nd ed. Cambridge, MA: The MIT Press, 1993:47–68.

Lawton JH, May RM, eds. Extinction rates. Oxford: Oxford University Press, 1995.

McLaren DJ. Impacts and extinctions: science or dogma? In: Glen W, ed. The mass extinction debates. Palo Alto, CA: Stanford University Press, 1994:121–131.

Raup DM. Extinction: bad genes or bad luck? New York: Norton, 1991.

———. The role of extinction in evolution. In: Fitch WM, Ayala FJ, eds. Tempo and mode in evolution. Washington, DC: National Academy Press, 1995: 109–124.

van Valen LM. Concepts and the nature of selection by extinction. In: Glen W, ed. The mass extinction debates. Palo Alto, CA: Stanford University Press, 1994:200–217.

PART 2

BACKGROUND TO HUMAN EVOLUTION

Dating Methods
Systematics: Morphological and Molecular
Science of Burial
Primate Heritage

DATING METHODS

7

An accurate time scale is a crucial aspect of reconstructing the pattern of evolution of the anatomical and behavioral characteristics of early hominines. At least half a dozen methods of dating are now available that have the potential to cover events from 1000 years ago to many billions of years, albeit with some frustrating gaps. Paleoanthropologists' focus is on the last 10 million years or so, which includes some of those gaps.

Researchers who want to know the age of particular hominine fossils and/or artifacts in principle have two options for dating them: direct methods and indirect methods.

Direct methods apply the dating techniques to the objects. Two types of problem arise with this approach, however. First, for most objects of interest, no methods are as yet available for direct dating. Ancient fossils and most stone tools, for example, remain inaccessible to direct dating. Some methods, such as **carbon-14 dating** and **electron spin resonance**, may be applied directly to teeth or young fossils, and indeed to the pigments of rock shelter and cave paintings; in addition, **thermoluminescence dating** may be applied directly to ancient pots, flint, and sand grains. Second, fossils and artifacts are often too precious to risk destroying any part of them in the dating process.

In practice, indirect dating methods represent the typical approach. Here, an age for the fossil or artifact is obtained by dating something that is associated with them. This strategy may involve direct dating on nonhuman fossil teeth that occur in the same **stratigraphic layer**, by electron spin resonance, for instance, or by thermoluminescence dating of flints associated with human fossils. Both these approaches have been applied in recent years to fossils relating to the origin of modern humans (see unit 27). Fossils or artifacts may be attributed an age through information about the evolutionary stage of nonhuman fossils associated with them, a technique known as faunal correlation.

The most common indirect approach, where feasible, is to date stratigraphic layers that lie below and above the object in question. Stratigraphic layers accumulate from the bottom up, so that the lower layers are oldest and the upper layers youngest. The two dates, taken from below and above the object, provide brackets that include the date at which the object became buried in the stratigraphic system.

This unit will survey briefly the principal techniques available and identify where they are best applicable. The techniques may be classified into two types: those that provide relative dates and those that provide absolute dates. **Relative dating** techniques give information about the site in question by referring to what is known at other sites or other sources of information. **Absolute dating** techniques provide information by some kind of physical measurement of the age of material at the site in question.

RELATIVE DATING TECHNIQUES

Relative dating techniques include **faunal correlation** and **paleomagnetism**. Geologists and paleontologists have long used fossils to structure prehistory. For instance, the geological time scale for the history of life on Earth is built upon major changes in fossil populations, such as appearances and disappearances of groups. Because they are interested in a finer-scale approach, archeologists and anthropologists often look for evolutionary changes *within* groups. Among the most important species for paleoanthropologists are elephants, pigs, and horses.

The principle behind the faunal correlation is simple. If a hominine fossil is found in sedimentary layers that

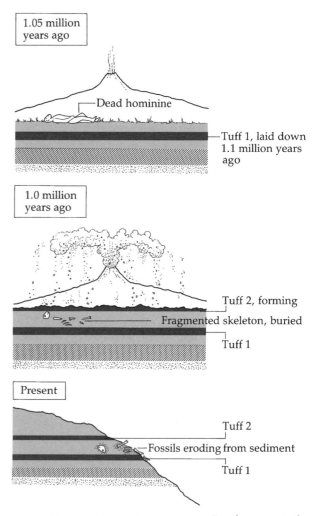

THE LIFE AND DATE OF A FOSSIL: Fossils cannot be dated directly. A date may be produced by dating volcanic ash layers that lie just below and just above the fossil, formed as shown.

Volcanic rocks

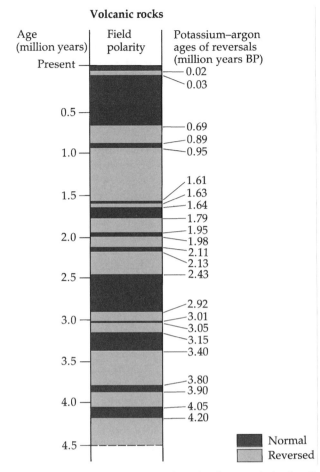

Age (million years)	Field polarity	Potassium–argon ages of reversals (million years BP)
Present		0.02
		0.03
0.5		
		0.69
		0.89
1.0		0.95
		1.61
1.5		1.63
		1.64
		1.79
		1.95
2.0		1.98
		2.11
		2.13
2.5		2.43
		2.92
3.0		3.01
		3.05
		3.15
3.5		3.40
		3.80
		3.90
4.0		4.05
		4.20
4.5		

■ Normal
▨ Reversed

PALEOMAGNETIC DATING: Occasional reversals in the direction of the Earth's magnetic field leave an imprint on iron-containing rocks as they form. The stripe pattern seen here represents the main reversals, and reference to it can help date sites.

also include fossil pigs that are known to have lived, for instance, between 2 million and 1.6 million years ago (as assessed, say, by tooth size or morphology), then this provides a bracket for the date of the hominine.

The principle behind paleomagnetism is based on the fact that the Earth's magnetic axis reverses periodically. We are currently in what is known as "normal" polarity, where magnetic north coincides with geographical north. During reversals, which occur every few hundred thousand or million years, a magnetic needle would point south. As rocks form, particularly after volcanic eruptions or during deposition of fine-grained material, the direction of the magnetic field is recorded in the orientation of iron-containing particles. Geologists have accumulated much information about past polarities and have constructed a chart showing the dates of reversals.

In paleomagnetic dating, a single piece of volcanic rock or certain types of sedimentary rock taken from a site can be tested for its polarity. By itself this information is insufficient to date a site, because the knowledge that a

particular layer has reversed or normal polarity leaves many options open. A series of layers that reveal a relatively large section of the overall pattern is sometimes sufficient to provide a more secure date. In general, however, paleomagnetic dating is rather imprecise and is used in combination with other methods, particularly **radiometric dating.**

ABSOLUTE DATING TECHNIQUES: RADIOPOTASSIUM DATING

The majority of absolute dating methods are radiometric, which depends on radioactive change in certain minerals. All methods share the same two principles. First, some action sets a radiometric "clock" to zero, such as the heating that rock experiences during volcanic eruption or burial in the earth. Second, the consequences of radioactive decay steadily accumulate, thus recording the passage of time.

The most important radiometric technique that has been applied in paleoanthropology is **radiopotassium (potassium/argon) dating.** This technique is based on the fact that potassium-40, a radioactive isotope of potassium that makes up 0.01 percent of all naturally occurring potassium, slowly decays to argon-40, an inert gas. Rocks that contain potassium, such as volcanic rocks, slowly accumulate argon-40 in their crystal lattices. The high temperature experienced during eruption drives out the argon (and other gases) from the mineral, and the clock is set to zero—the time of the eruption. As time passes, argon-40 builds up, with the amount in any particular rock depending on the initial potassium concentration and the time since the eruption. The age calculation is based on measurements of the potassium concentration and the accumulated argon-40 in potassium-rich minerals, such as feldspar.

A common problem is that a sample may be contaminated with older rock, which may happen when ash is erupting from a volcano, for instance, or mixing with other minerals as it accumulates on the landscape. Even a few crystals of, for example, Cambrian-age rock in a gram of 2-million-year-old ash can produce an erroneously old date.

The first major application of the potassium/argon technique to paleoanthropology occurred in 1960, in an assessment of ash layers at Olduvai Gorge. In 1959, Mary Leakey found the famous *Zinjanthropus* fossil (see unit 20), the first early hominine discovered in East Africa, at this site. The date produced for the fossil, 1.75 million years, was double the age inferred by indirect means. Both the discovery of the fossil and the application of the dating technique represented major milestones for paleoanthropology.

Since that time two important advances have taken place with radiopotassium-based dating. The first, developed in the 1960s, allows measurements to be taken in one sample rather than in two separate samples (one to measure potassium, the second to measure argon-40). The rock is initially irradiated with neutrons, which transforms the stable potassium-39 into argon-39; when the

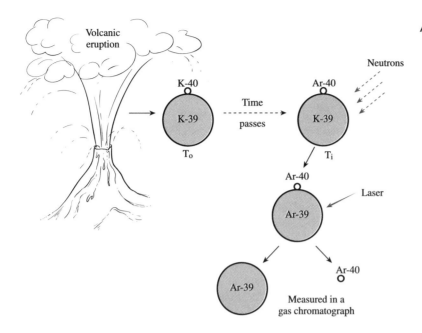

ARGON-39/ARGON-40 DATING: Volcanic ash contains potassium-rich minerals, such as feldspar. A small percentage of the potassium exists as a radioisotope, potassium-40, which has argon-39 as one of its decay products. In the laboratory, crystals of feldspar are irradiated with neutrons, which converts the stable potassium-39 isotope to argon-40. The crystals can then be individually heated by laser beam, and the emitted argon-39 and argon-40 measured in a gas chromatograph. The argon-40 represents a measure of the total amount of potassium that was present in the crystal, and the argon-39 provides a measure of the time since eruption.

rock is then heated, the two argon isotopes, 39 and 40, are released together and can be measured simultaneously on a gas chromatograph. The potassium-39 level provides a vicarious measure of the potassium originally in the rock, and the argon-40 measures the decay of potassium-40 since the rock was ejected from the volcano. This technique is known as argon-40/argon-39 dating.

The second advance, developed during the 1980s, allows the technique to be applied to single crystals taken from volcanic ash, compared with the several grams required for the conventional technique. The advantages of the new technique, known as single-crystal laser fusion, are several, including avoiding the problem of contamination. Until recently the youngest rocks that could be dated with radiopotassium techniques were approximately 0.5 million years old. Recent work, however, has shown that rocks containing potassium-rich minerals can be accurately dated with ages as young as 10,000 years—a range that overlaps with the limits of radiocarbon dating. There is no effective upper limit of age estimation.

ABSOLUTE DATING TECHNIQUES: FISSION TRACK, RADIOCARBON, URANIUM SERIES

The second radiometric technique is fission track dating, which is often used in combination with radiopotassium methods. Naturally occurring glass often contains the isotope uranium-238, which decays through powerful fission. This event effectively burns a tiny track in the glass, which represents the ticking of the clock. Once again, the clock is set to zero during volcanic eruption, which expunges existing tracks. The longer the time after eruption, the more tracks that will accumulate, depending on the concentration of uranium in the glass.

The preparation of glass for the technique is tedious, however, and the counting of tracks not always reliable.

In principle, this dating method can be applied to rocks as young as a few thousand years; in practice, the older the material, the more reliable the counting procedure.

Radiocarbon dating is the best known of all radiometric techniques, but because of its short time depth has limited applications in paleoanthropology. Most of the carbon dioxide in the atmosphere exists as a stable isotope, carbon-12. Some small percentage consists of carbon-14, a radioactive isotope that decays relatively rapidly. As plants incorporate carbon into their tissues, the ratio of the two isotopes in the tissues mirrors that found in the atmosphere. The same ratio applies for animal tissues, which effectively are built from plant tissues. Once an organism dies, however, the equilibrium between the isotopes in the air and in the tissues begins to change as carbon-14 continues to decay and is not replenished. As time passes, the ratio of carbon-14 to carbon-12 becomes increasingly smaller, a decline that forms the basis of the clock. Researchers can measure the proportions of the two isotopes in the organism's tissues and calculate when it died.

In principle, any organic material can be dated by the carbon-14 technique; in practice, many tissues decay too quickly to use this approach. The preferred material for dating by this technique is charcoal, as has recently been done on pigments in rock paintings in Europe and the United States. In Australia, rock paintings have recently been dated from blood that formed part of the pigment.

Contamination can represent a serious problem with radiocarbon dating (only a small amount of young material can substantially reduce the apparent age of older material). With the recent application of accelerator mass spectrometry to increase the sensitivity of measuring carbon-14, the useful range of the technique can be from a few hundred years to perhaps 60,000 years or a little more.

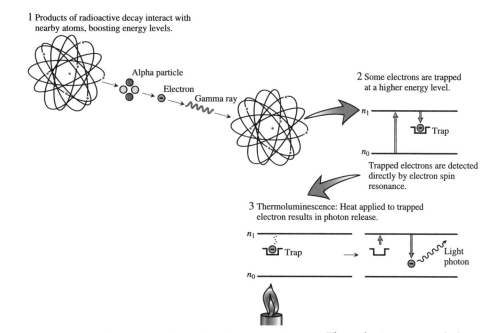

1 Products of radioactive decay interact with nearby atoms, boosting energy levels.

Alpha particle

Electron

Gamma ray

2 Some electrons are trapped at a higher energy level.

n_1

Trap

n_0

Trapped electrons are detected directly by electron spin resonance.

3 Thermoluminescence: Heat applied to trapped electron results in photon release.

n_1

Trap

n_0

Light photon

THERMOLUMINESCENCE AND ELECTRON SPIN RESONANCE DATING: Thermoluminescence and electron spin resonance techniques depend on the same natural process. Ionizing radiation from radioactive isotopes in the soil excites electrons in the crystal lattice of certain minerals (such as flint or tooth enamel) to rise from the ground state to an excited state, some of which remain there. As time passes, more electrons accumulate in the excited state. In the thermoluminescence technique, the quantity of accumulated, excited electrons (and therefore the age of the sample) may be measured by heating the sample, causing photons to be released as the electrons return to the ground state. In electron spin resonance, the excited electrons are measured more directly, by applying a magnetic field and microwave irradiation. (Courtesy of R. Lewin/Scientific American Library.)

Other methods of absolute dating include the uranium series technique, which relies on the decay of the radioisotopes uranium-238, uranium-235, and thorium-232, all of which decay ultimately to stable isotopes of lead. In addition, amino acid racemization has been used to date materials. This method depends on the slow transformation of the conformation of amino acid molecules used in living organisms (left-handed forms) to a nonliving mixture (right- and left-handed forms). Neither the uranium series technique nor amino acid racemization is as powerful or as applicable to paleoanthropology as the other absolute dating techniques.

THERMOLUMINESCENCE AND ELECTRON SPIN RESONANCE

Two relatively new dating techniques depend on the principle that electrons in certain minerals become excited to higher energy levels when irradiated by radio-

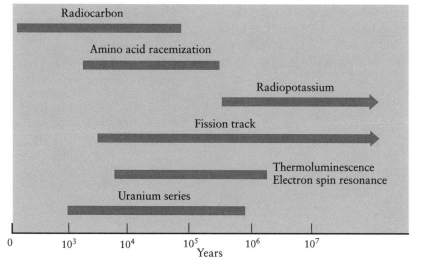

Radiocarbon

Amino acid racemization

Radiopotassium

Fission track

Thermoluminescence
Electron spin resonance

Uranium series

0   10^3 10^4 10^5 10^6 10^7
Years

RANGE OF DATING METHODS: The full range of dating methods available to paleoanthropologists begins at a few hundred years and extends to many millions of years. The recent development of thermoluminescence and electron spin resonance dating filled an important gap, between 50,000 and 0.5 million years.

isotopes of uranium, thorium, and potassium, which occur naturally in the ground and in cosmic rays. The radioactive rays knock off the negatively charged electrons from atoms, leaving positively charged "holes." These electrons diffuse through the crystal lattice and usually recombine with other holes, returning to the ground state. But all minerals contain impurities, such as lattice defects and atoms that can "trap" roving electrons, keeping them at an intermediate energy level. Exposure to heat, such as fire (for example, burned flint or fired pottery) or even sunlight in the case of sand grains, dislodges trapped electrons; these particles then return to nearby holes, setting the clock to zero as in radiopotassium dating. The number of trapped electrons in a newly unearthed mineral therefore provides a measure of the time that has passed since the mineral was last exposed to heat. These dating techniques, known as thermoluminescence and electron spin resonance, measure these trapped electrons by different means—the former indirectly, and the latter directly.

In the thermoluminescence technique, the artifacts are heated under controlled conditions to release the electrons. As they return to the ground state the electrons release photons (light), which can be detected by sensitive instruments. Electron spin resonance detects the trapped electrons in situ, where they act as minute magnets that become oriented when exposed to a strong magnetic field. Microwave energy flips the orientation of the electrons, yielding a characteristic signal. The strength of the signal provides a measure of the number of trapped electrons. The electron spin resonance technique can be applied to tooth enamel, but not, as yet, to bone.

In principle, both thermoluminescence and electron spin resonance techniques can reveal dates between a few thousand and 1 million years ago. This application range is particularly useful in paleoanthropology, because it fills a gap for material that is too old for radiocarbon dating and too young for radiopotassium dating.

The suite of dating techniques available to paleoanthropologists in principle covers the past 5 million years (the period of primary interest) completely. Unfortunately, many important fossil and archeological sites lack material suitable for dating, are embedded in a stratigraphy too complex to unravel, or both. The cave sites in South Africa are examples of a too-complicated stratigraphy. ❋

KEY QUESTIONS

- In what ways are accurate dating techniques important to paleoanthropology?
- Could a new date for an existing fossil specimen alter the interpretation of the species to which it belongs?
- Which currently available dating technique is most useful to paleoanthropology?
- What new development could revolutionize dating in paleoanthropology?

KEY REFERENCES

Aitken MJ, Valladas H. Luminescence dating relevant to human origins. *Phil Trans Roy Soc B* 1992;337: 139–148.

Brown FH, *et al.* An integrated Plio-Pleistocene chronology for the Turkana Basin. In: Delson E, ed. Ancestors: the hard evidence. New York: Alan R. Liss, 1985:82–90.

Chen Y, *et al.* The edge of time: dating young volcanic ash layers with the argon-40/argon-39 laser probe. *Science* 1996;274:1176–1178.

Deino A, *et al.* Argon-40/argon-39 dating in paleoanthropology and archeology. *Evol Anthropol* 1998; 6:63–75.

Feathers JK. Luminescence dating and modern human origins. *Evol Anthropol* 1996;5:25–36.

Grün R. Electron spin resonance dating in paleoanthropology. *Evol Anthropol* 1993; 2:172–181.

Lewin R. Rock of ages—cleft by laser. *New Scientist* Sept 28, 1991:35–40.

Schwarcz HP. Uranium series dating in paleoanthropology. *Evol Anthropol* 1992;1:56–61.

York D. The earliest history of the Earth. *Sci Am* Jan 1993:90–96.

SYSTEMATICS: MORPHOLOGICAL AND MOLECULAR

8

Systematics is the study of the diversity of life and the relationships among **taxa** at all levels in the hierarchy of life, from species to genus to family to order, and so on up to kingdom. A **taxon** (singular of "taxa") is a category of organisms at any level in that hierarchy: a species is a taxon, as is a genus, family, order, and so on. Conventionally, taxa above the level of genus are referred to as higher taxa.

To communicate unambiguously about the diversity of life and relationships within it, biologists require a consistent method of classifying the taxa of interest. Traditionally, classification has been based on anatomical characters. More recently, molecular data have been used. The advantages and disadvantages of both approaches will be discussed.

The Linnaean system of classification is hierarchical, as illustrated in the accompanying diagram for the grey wolf, *Canis lupus*. Species are grouped into genera; in this case the grey wolf appears in the same genus as the golden jackal, *Canis aureus*. Genera are grouped into families; here, the wolf and the jackal are in the same family as foxes (genus *Vuples*), with the family name Canidae. Several families constitute the order Carnivora; and the Carnivora combine with other mammalian orders (including primates) to form the class Mammalia. The class Mammalia joins with other vertebrate classes (such as Carnivora and Insectivora) to form the phylum Chordata, which is one of approximately 30 animal phyla that constitute the kingdom Animalia.

The basic unit of Linnaean classification is the species, whose identification includes two parts: the genus name and the specific name, termed a **binomen**. Different species may share the same specific name but are linked to a different genus names, such as *Proconsul africanus* (a fossil ape; see unit 16) and *Australopithecus africanus* (an early hominine; see unit 20). The laws governing the naming of species are quite strict under the Code of Zoological Nomenclature, so that if a species is reclassified (based on new discoveries, for instance), the genus name may be changed but the specific name must remain the same.

PHILOSOPHIES OF CLASSIFICATION AND SYSTEMATICS

How is classification arrived at? For Linnaeus, in the mid-eighteenth century, the criterion was simply anatomical similarity and, naturally, had nothing to do with evolution. After 1859 and the publication of Darwin's *Origin of Species*, however, biologists could approach classification with evolution explicitly in mind. Darwin argued that because species are related by common descent, genealogy represented the only logical basis for classification. Recent years have witnessed surprisingly heated debate over precisely how classification should be performed. Should it emphasize the results of evolution, in terms of adaptation? Or should it reflect relatedness, or **phylogeny**, as Darwin argued? This issue is particularly pertinent when classifying the great apes and humans (discussed in this unit and unit 15).

Currently three major schools of classification address the hierarchies of living things: **phenetics** (also called numerical taxonomy), which emphasizes overall anatomical similarity, and is therefore rooted in adaptation and does not necessarily reflect phylogeny; **cladistics** (also called phylogenetic systematics), which emphasizes only phylogeny; and **evolutionary systematics**, which is somewhat intermediate between the other two approaches in its philosophy.

If evolution proceeded at regular rates, so that after branching two lineages diverged steadily in terms of morphological adaptations, then the phenetic pattern would

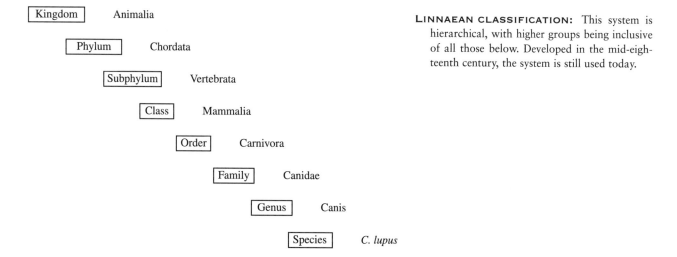

Kingdom — Animalia

Phylum — Chordata

Subphylum — Vertebrata

Class — Mammalia

Order — Carnivora

Family — Canidae

Genus — Canis

Species — *C. lupus*

LINNAEAN CLASSIFICATION: This system is hierarchical, with higher groups being inclusive of all those below. Developed in the mid-eighteenth century, the system is still used today.

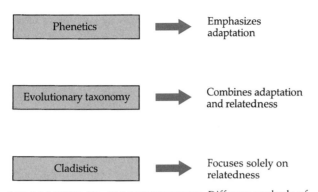

APPROACHES TO CLASSIFICATION: Different methods of describing relationships among organisms effectively emphasize different aspects of the world. For instance, by concentrating on characters that reflect genetic relatedness, cladistics produces an evolutionary tree. In contrast, phenetics measures all aspects of similarity among organisms and therefore emphasizes similarities in adaptation. Evolutionary taxonomy steers a middle path between the two.

be identical to the phylogenetic pattern. This generally does not happen, however. Sometimes a new lineage will diverge quickly, accumulating many evolutionary novelties that put a great morphological distance between it and its sister species; sometimes a new lineage will remain almost identical to its sister species over vast periods of time, with the morphological distance remaining minimal while genetic distance increases. As a consequence of these different tempos of evolution, phenetics will sometimes yield a different pattern from that produced by cladistic analysis.

The choice of a classification system therefore becomes a matter of philosophy: Should the grouping be developed according to overall morphological similarity, which emphasizes adaptation? Or should it reflect relatedness? Which is the more "natural" system? Proponents of phenetics claim that their analysis is completely objective and completely repeatable, and therefore will reflect meaningful patterns in nature. Cladists argue that the phylogenetic hierarchy is the only important reality, whether we discover it or not. Only one pattern of phylogenetic branching exists—the path that evolution actually followed. The challenge is being able to infer that pattern from the morphology and other evidence, such as genetics.

RELATIVE IMPORTANCE OF HOMOLOGY

A vertebrate species' morphology is composed of a large suite of anatomical characters: shapes of bones, patterns of muscular attachments, skin color, and so on. Phenetics compares as wide a range of characters as possible between a group of species to produce multivariate cluster statistics, which is effectively an average of all such comparisons. The more characters that are included, the more objective the technique is said to be, automatically spitting out a phenetic hierarchy from the assembled cluster statistics. In fact, practitioners frequently must choose among several possible patterns, betraying the fact that the method is less objective than is often claimed.

Unlike pheneticists, biologists who wish to infer evolutionary relatedness among species will not usually include all available characters. Although many characters that are shared among species are the result of common descent—that is, homology—some will reflect convergent, or parallel, evolution—that is, analogy. Only homologous characters can be used to reconstruct phylogenies, because they are what link evolutionarily related species

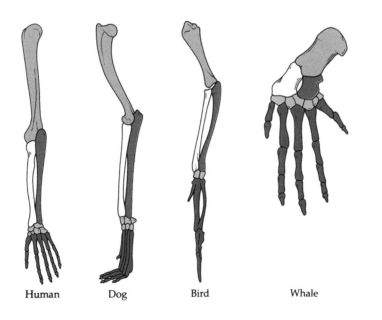

Human   Dog   Bird   Whale

THE PRINCIPLE OF HOMOLOGY: The biological derivation relationship (shown by colors) of the various bones in the forelimbs of four vertebrates is known as homology and was one of Darwin's arguments in favor of evolution. By contrast, the wing of a bird and the wing of a butterfly, although they perform the same task, are not derived from the same structures: they are examples of analogy.

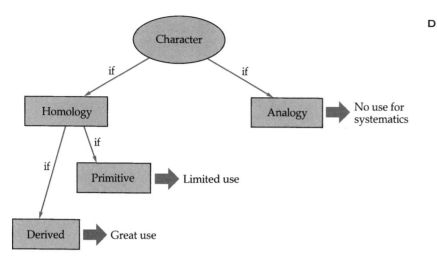

DEDUCING RELATIONSHIPS: A genetic relationship can be deduced between two species only if homologous—not analogous—characters are used. Homologous characters come in two forms: primitive and derived. Primitive characters have limited use in deducing relationships because they occur in the ancestor of the group and therefore give no information about species within the group. Derived characters are the key to relationships because they occur in only some of the species under study and therefore can be used to differentiate within the group.

together (see unit 4). Moreover, whereas pheneticists sometimes deal with a quantitative measure of a character, such as the dimension of a character (the size of a cusp on a tooth, for example), cladists prefer to deal with the form of character (the particular shape or number of cusps, for example). This latter attribute is called **character state**.

Even when characters have been reliably identified as homologous rather than analogous, they are not all equally valuable in inferring evolutionary relatedness. In any group of species under comparison, some homologous characters will be considered primitive and some derived; it is the derived characters that uniquely link species.

Primitive characters are those inherited from the ancestral stock for that group. For instance, baboons, chimpanzees, and humans all have nails on the ends of their fingers. These species are not uniquely linked by this character, however, because New World monkeys and all prosimians have fingernails as well. Fingernails are a characteristic feature of all primates. For baboons, chimpanzees, and humans, the possession of fingernails is therefore a primitive character with respect to primates. Some dozen or so characters are found uniquely among baboons, chimpanzees, and humans that are absent from New World monkeys and prosimians; these attributes represent **shared derived characters** for the Catarrhini (the infraorder that encompasses the Old World monkeys, apes, and humans as a group) with respect to primates.

Obviously, the classification of homologous characters into primitive and derived is always relative to the level of the hierarchy being considered. For instance, although the possession of fingernails is a primitive character within the Catarrhini with respect to other primates, it is a derived character for primates as a whole: it distinguishes them from other mammals. Generally, derived characters at one level will become primitive at the next level up (moving in the species-to-kingdom direction). Deciding whether a character state is primitive or derived in a particular species comparison is known as deciding its **polarity**.

To infer a unique phylogenetic relationship among a group of species, one must identify derived characters—the evolutionary novelties that separate the species from their common ancestor. This idea, simply stated, is the principle behind cladistics. A collection of all species with shared derived characters that emerged from a single ancestral species is said to be a **monophyletic** group, or clade; a diagram indicating relationships is a **cladogram**. Cladists reject **paraphyletic** groups and **polyphyletic** groups as unnatural groups. A paraphyletic group contains a subset of descendants from a single ancestor. If, for instance, only some descendants of the common ancestor diverged significantly away from the original adaptation, the phenetic approach would recognize two groups; those that remain most similar to the ancestral state form the paraphyletic group. Polyphyletic groups arise when members of different lineages converge on a similar adaptation; pheneticists would recognize these species as a natural group. Paraphyletic groups include the common ancestor, while polyphyletic groups do not.

CLADISTIC PRACTICE AND HOMININE CLASSIFICATION

The cladistic approach was originally developed by the German systematist Willi Hennig in 1950, and in recent years it has become the approach of choice for many researchers in paleoanthropology. As a result, the literature is becoming littered with cladistic analyses and cladistic terminology, which unfortunately includes real tongue twisters. For instance, shared derived characters are **synapomorphies**. Shared primitive characters are **symplesiomorphies**. A derived character not shared with other species is an **autapomorphy**. Convergent characters are **homoplasies**.

Determining relationships between species involves two steps. First, homologies must be separated from homoplasies, which requires careful attention to the traps of functional convergence. Second, polarities of homologous character states must be selected: Are they primitive (plesiomorphic) or derived (apomorphic)? How is polarity determined?

Suppose, for example, one is assessing the bony ridge above the eyes, which is found in chimpanzees, gorillas, and the human lineage, but not in orangutans. Is this brow ridge a synapomorphy (shared derived character) linking the three as a clade? Or could it be a symplesiomorphy (shared primitive character) for hominoids that was lost in the orangutan? The answer is obtained by looking further down the hierarchy, at more distantly related species. This process is known as an **outgroup comparison**. In this case, one would also examine a gibbon and an Old World monkey, for instance. The brow ridge happens to be absent in Old World monkeys, which implies that indeed it a synapomorphy for the African apes and humans. By these criteria, then, the African apes and humans form a monophyletic group, or clade.

No one, however, likes to base such a judgment on a single character. Most analyses therefore survey many characters. The importance of multicharacter comparison becomes evident as one often finds that one subset of characters might imply one pattern of relationship while a second subset points to another. Cladistic analysis of hominoids is no exception. The conclusion from this apparent confusion is that anatomical characters are often extremely difficult to assess and interpret.

For instance, one cladistic analysis of the hominoids in recent years ranked chimpanzees, gorillas, and orangutans as a monophyletic group, leaving humans as a separate clade. A second analysis showed humans and orangutans as a clade, with chimpanzees and gorillas as a second clade. Most current cladistic analyses favor chimpanzees, gorillas, and humans as a monophyletic group, leaving the orangutan separate, although the preference is not particularly strong.

Now, suppose that this last-mentioned phylogenetic pattern is correct—and molecular data support this classification (see unit 15). Surely it should be reflected in the formal classification, one might think. Traditionally, humans and their direct ancestors have been assigned to the family Hominidae, while the African apes and the orangutan occupy a separate family, the Pongidae. Such a grouping reflects overall morphological similarity, because humans have diverged dramatically from the apes; it ignores strict phylogeny, however, which groups humans with African apes and puts the orangutan separate.

If phylogeny is to be accurately reflected in classification, then one possibility is as follows. Hominidae would include the African apes and humans, with orangutans occupying the family Pongidae. Humans would be the sole occupant of the subfamily Homininae—hence the more general term "hominine" rather than the previously

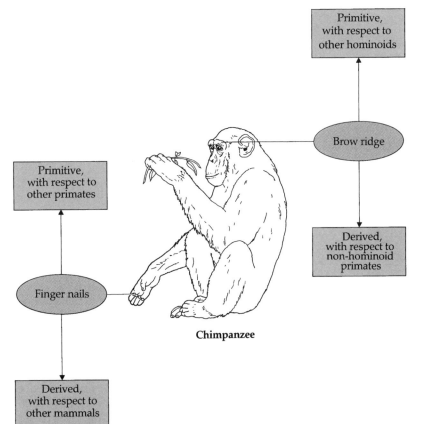

Chimpanzee

Relative status of characters: The state of a character depends on the reference point. For instance, for an ape, fingernails are considered primitive in relation to other primates because all other primates have fingernails. Thus, fingernails would not serve to distinguish apes from, for example, monkeys. Fingernails are a derived character for primates as a whole, however, because no other mammals have them. Thus, fingernails serve to distinguish primates from other mammals. The second character illustrated here—brow ridges—is found only in hominoids, not in other primates, and is therefore derived for hominoids. This character distinguishes apes from monkeys. In a chimpanzee, however, brow ridges would be considered to be primitive with respect to other hominoids; that is, the character would not distinguish a chimpanzee from, for example, a gorilla.

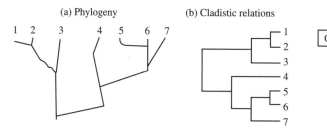

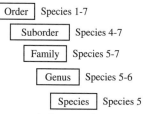

RELATION BETWEEN PHYLOGENY, CLADISTIC CLASSIFICATION, AND LINNAEAN CLASSIFICATION: (a) The phylogeny of seven species. (b) A cladistic classification. (c) The Linnaean classification for species 5.

used "hominid." The gorilla and chimpanzee would occupy the subfamily Gorillinae.

Accurate in cladistic terms though this grouping may be, pheneticists and evolutionary taxonomists would demur. Classification should also reflect the very drastic ecological shift that has occurred in the hominine line compared with its ape cousins, they contend. According to this argument, maintaining family status for the apes but separate family status for humans is therefore appropriate.

MOLECULAR SYSTEMATICS

Genetic evidence has recently taken its place alongside morphology, creating the approach known as **molecular systematics**. Various kinds of data are relevant here, including DNA sequences, comparison of immunological reactions of proteins, comparison of electrical properties of proteins (gel electrophoresis), and DNA-DNA hybridization, which effectively compares the entire genetic complement of one species with that of another. All of these methods, with the exception of DNA sequence data, provide a measure of **genetic distance** between the species being compared, which is equivalent to the phenetic measure of overall similarity. Consequently, cladists reject the use of these methods. Only those techniques that produce information about DNA sequence are accessible to cladistic analysis, because the sequence data are equivalent to characters whose state can be determined directly (that is, the presence or absence of particular nucleotides).

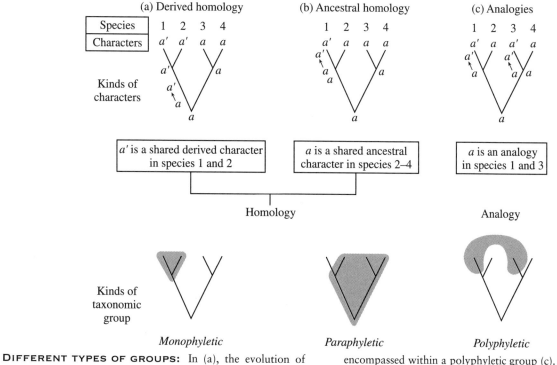

DIFFERENT TYPES OF GROUPS: In (a), the evolution of shared-derived characters leads to the formation of a monophyletic group. When a species (or several species) diverges significantly from the ancestral status, it may be excluded, leaving a paraphyletic group (b). Convergent evolution may yield species with similar adaptations; these species may be encompassed within a polyphyletic group (c). Cladists recognize only monophyletic groups as being natural groups because they truly reflect phylogeny. Pheneticists accept the reality of both paraphyletic and polyphyletic groups because they reflect the results of evolution, or adaptation.

Molecular systematics relies on the fact that when two species diverge, mutations will accumulate independently in the DNA of the daughter lineages. Scrutiny of similarities and differences among species' DNA therefore permits their evolutionary relationships to be inferred. In its early days, molecular systematics was perceived (by molecular biologists) as being inherently superior to traditional methods, for several reasons.

First, because molecular data are derived from the genes of a species, they were envisioned as carrying the fundamental record of evolutionary change. Second, molecular data were considered to be immune to the problem of convergence, for the following reason. Natural selection produces convergence, through adaptation to similar environmental conditions. Because the majority of mutations that accumulate are selectively neutral, they remain invisible to natural selection. Convergence toward similar mutations in different lineages is therefore highly unlikely, except by chance. Third, molecular and morphological evolution were thought to proceed at very different tempos (the former always regular, the latter always erratic), which was assumed to imply that molecular data were more reliable.

Moreover, because genetic differences between lineages were suspected to proceed in a regular manner, the notion of a **molecular evolutionary clock** was developed. Not only would it be possible to reliably determine the *branching order* of related species with genetic data, but one could also calculate when the lineages diverged from one another—that is, the *branch length*. Last, morphological features express complex and mostly unknown sets of genes and regulatory interactions among genes. In

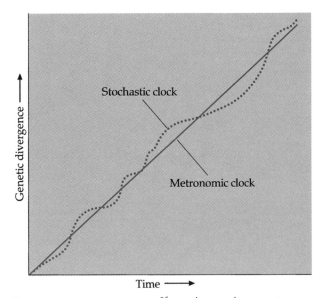

The molecular clock: If genetic mutation were to occur at a constant rate, then biologists would have access to a completely reliable, "metronomic" molecular clock. In fact, the rate of mutation for any particular region of DNA is likely to fluctuate through time, giving a "stochastic" molecular clock. By bringing together data on genetic divergence from different regions of DNA, it is possible in principle to average out these fluctuations, thereby providing a good, average clock. Because the technique of DNA hybridization effectively compares the entire DNA complement of two related species, fluctuations in mutation rate in different parts of the genome are automatically averaged out.

contrast, molecular data relate to much smaller and strictly defined sets of genes. According to proponents of molecular systematics, simplicity yields reliability.

LIMITATIONS OF MOLECULAR SYSTEMATICS

Today, molecular approaches to systematics are recognized as less simple, and therefore less immediately reliable, than previously supposed for several reasons. For instance, it is now recognized that the dynamics of mutation are highly complex, including the fact that not all regions of a gene or other regions of DNA are equally susceptible to change; indeed, some regions are highly susceptible to similar kinds of change. For this reason, convergence can and does occur in DNA sequences. Moreover, some mutation events may become hidden through "multiple hits." Imagine that a particular nucleotide position in a gene mutates early in the lineage's history. As time passes, other mutations will accumulate as well. If all subsequent mutations occur at different sites, a count of the mutations present will give an accurate record of the lineage's mutational history. If a later mutation occurs at a previously mutated site, however, then the count will be too low, giving an erroneous conclusion. The longer the time period under investigation, the greater a problem that multiple hits become. Statistical methods are being developed to try to accommodate this factor.

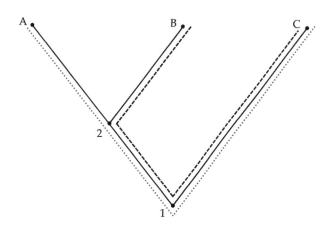

The relative rate test: The diagram represents two evolutionary events. At 1, a split occurred, leading to species C and a second lineage. The second lineage then split at node 2, leading to species A and B. According to the rate test, if the average rate of genetic divergence is the same in all lineages, then the genetic distance from species A to species C (dotted line) should be the same as the genetic distance from species B to species C (dashed line). If gene mutation slowed down in lineage B, then the B-to-C genetic distance would be shorter than the distance from A to C.

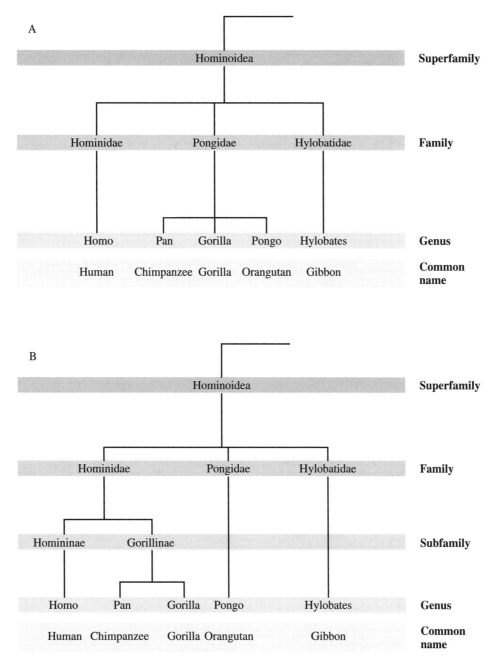

**TWO VIEWS OF HOMINOID (HUMANS AND APES) CLAS-
SIFICATION:** Classification A, the traditional scheme, emphasizes adaptation, putting the African and Asian great apes in
one family, the Pongidae, with humans being the sole occupant of the family Hominidae. Until relatively recently, this
classification was also considered to reflect the evolutionary
history of hominoids. Classification B is based on a phylogenetic perspective, particularly on genetic evidence, and groups
humans and the African apes in the family Hominidae. Humans are then assigned to the subfamily Homininae, and the
chimpanzees and gorillas are assigned to the subfamily
Gorillinae. Strict adherence to the most recent genetic evidence would alter the classification further (see unit 15).

Another potential confounding problem is that the
degree of sequence divergence between the same gene in
different lineages might not accurately measure the point
at which the lineages diverged. The issue here relates to
the potential difference between the **species tree** and the
gene tree. A species tree describes the evolutionary history
of the species—that is, the true phylogeny. If all genes in
two daughter species begin to diverge only when the
populations diverged, then the gene tree would be the
same as the species tree. This scenario, however, is not
always the case. Genes often develop variants (**polymorphisms**) within a population, so that some individuals

may possess one variant while other individuals carry the other. Once such variants become established, they begin to accumulate mutations independently.

Suppose a polymorphism of a gene X arose in a species A some 4 million years ago, giving variant X in some individuals and variant X1 in others. Suppose, too, that allopatric populations became established 2 million years later, with X remaining in the parent population while X1 appeared exclusively in the newly isolated population. Such a situation can lead to speciation (see unit 4). Now suppose that the modern populations of the descendant species are subjected to molecular systematics analysis, using gene X. Calculations based on the nucleotide differences between X and X1 would indicate that the two daughter species diverged 4 million years ago, when their separate sequences would have begun to diverge (this is the gene tree). In fact, the species did not begin to diverge until 2 million years ago (as the species tree reveals). In general, therefore, when the gene tree/species tree problem arises, the divergence date inferred from the molecular data will be too old.

This example assumes, of course, that molecular data can be used to calculate time since divergence, based on the molecular clock concept. It was once assumed that mutations accumulated at a regular rate in all genes, in all lineages, and at all times in lineages' histories. Some genes might mutate at higher rates than others because of functional constraints—globin genes mutate at a higher rate than histone genes, for example. In fact, a series of clocks would operate, each ticking regularly but at different rates.

Again, however, this assumption turns out to be too simplistic. It has now been established that some genes in some lineages at some points in their evolutionary history do indeed accumulate mutations in a clocklike manner. Differences arise, however, in mutation rates in the same gene *between* lineages, as well as differences in rates in a single gene *within* a single lineage at different points in its history. The notion of a *global clock* is therefore no longer tenable. The existence of *local clocks* is, nevertheless, a reality, and they have great utility. Researchers must determine whether their gene of interest is behaving in a clocklike manner, using the relative rate test (see diagram), before they can proceed to measure branch lengths in phylogenies.

MORPHOLOGY AND MOLECULES COMPARED

Paradoxically, one of the advantages of morphological systematics stems from the erratic nature of the tempo of morphological evolution. An important feature of evolution is adaptive radiation, which occurs when a new group diversifies at its establishment, yielding many lineages with unique features that subsequently may change little. If such a radiation occurred deep in evolutionary history, a clocklike accumulation of genetic mutations would be unable to track the details of the brief burst of change, for the following reasons. A slow rate of mutation in DNA sequences would leave the event unrecorded. DNA sequences that change rapidly, on the other hand, would capture such change, but this information would

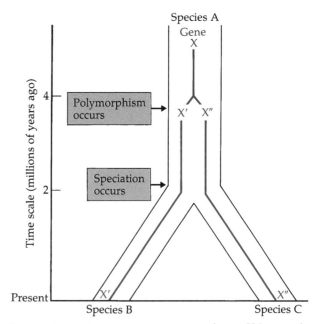

SPECIES TREES AND GENE TREES: A gene X in a species A undergoes polymorphism, producing variants X and X1, which then continue to accumulate differences between them. A speciation event occurs later, producing species B and C. Through various circumstances, gene variant X1 predominates in species B while variant X predominates in species C. A comparison of the differences between X1 and X would overestimate the time at which the daughter species B and C diverged. In other words, the gene tree is older than the species tree.

be overwritten to the point of illegibility by subsequent mutation. By contrast, the morphological changes that accompany the radiation would, in principle, persist in the lineages' subsequent history, preserving the event for comparative morphologists to discern. The rapid radiation of placental mammals near the end-Cretaceous extinction, 100 million years ago, is a good example of this type of development.

A major advantage of molecular phylogenetics is the potential extent of information it can evaluate, which at the limit is equal to the entire genome (in humans, for example, the genome includes 3 billion nucleotides). Morphological characters necessarily represent only a subset of this information. Moreover, because different sectors of the genome accumulate mutations at variable rates, genetic methods offer access to both ancient divergences (with slow-changing DNA, such as ribosomal DNA) and recent events (with fast-changing DNA, such as mitochondrial DNA). Morphological information cannot encompass this range of evolutionary history. It is also powerless to discern evolutionary history in cases involving limited morphology, such as in the early divergence of microorganisms nearly 3 billion years ago.

Molecular systematics has been important in three areas of human prehistory. Its application to the issue of the origin of the hominine clade has already been mentioned

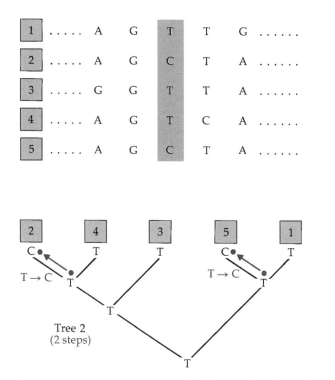

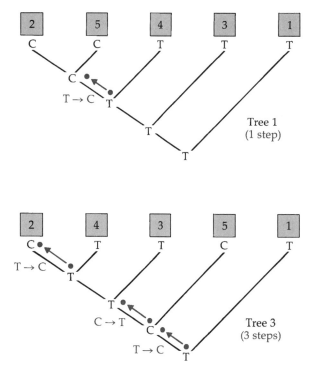

THE PARSIMONY TECHNIQUE: In this example of five individuals (1–5), we see part of the DNA sequence. By concentrating on position 3 in this instance, the parsimony technique seeks to find the tree with the lowest number of mutational steps to link all population members. Three trees are drawn here, with one, two, and three steps taken to link the five individuals. The parsimony technique would select tree 1 as the most likely relationship among the five individuals.

(see unit 15). A second area is in the origin of modern humans (see unit 28), while the third relates to the timing of human colonization of the Americas (see unit 35).

METHODS OF PHYLOGENETIC ANALYSIS

The raw data, whether molecular or morphological, are just the starting point for phylogenetic reconstruction. Half a dozen analytical methods have been developed, some of which work with distance data (such as from DNA-DNA hybridization and immunological measures) and some that rely on character state (such as protein or DNA sequence). Whatever the method, the task is formidable. Even with just a handful of species, the number of possible evolutionary trees is vast. With a mere 50 species, for example, 2.8×10^{74} trees are possible—or some 10,000 times as many trees as there are atoms in the universe. A computer that could scrutinize a trillion trees per second (no computer even approaches this speed as yet) would take 8.9×10^{54} years to complete the job—that is, 2×10^{45} times the age of the Earth. Analytical methods therefore must negotiate this challenge by rapidly seeking the most likely tree. In reality, many trees are produced, each with equal or nearly equal probability of being correct. Statistical methods are then required to narrow down the list of possibilities.

Methods employing the **parsimony** principle are currently the most popular and powerful for phylogenetic analysis. Briefly put, parsimony seeks the simplest explanation, with the belief that this path is the most likely to have been followed. Evolutionary change is inherently of low probability, so simple paths going from character state A to character state B are themselves inherently likely to be simple rather than complex (involving, for example, reversals of evolutionary direction). In the context of phylogenetic analysis, the parsimony method looks for the tree (or trees) that uses the fewest changes to link the given species in an evolutionary hierarchy. ✳

KEY QUESTIONS

- Why do phenetic and phylogenetic patterns often differ?
- What are the advantages of classifications based on adaptation relative to those based on phylogeny?
- How does molecular systematics compare in efficacy to morphological systematics?
- How can inferences from molecular systematics be strengthened?

KEY REFERENCES

Britten RJ. Rates of DNA sequence evolution differ between taxonomic groups. *Science* 1986;31: 1293–1298.

Hillis DM, *et al.* Application and accuracy of molecular phylogenies. *Science* 1994;264:671–667.

Kimura M. Molecular evolutionary clock and the neutral theory. *J Molec Evol* 1987;26:24–33.

King MC, Wilson AC. Evolution at two levels in humans and chimpanzees. *Science* 1975;188:107–116.

Lewin R. Patterns in evolution: a molecular view. New York: W. H. Freeman, Scientific American Library, 1996.

Molecular evolutionary clock. *J Molec Evol* 1987;26: 1–171.

Patterson C, ed. Molecules and morphology in evolution. Cambridge: Cambridge University Press, 1987.

Patterson C, *et al.* Congruence between molecular and morphological phylogenies. *Annu Rev Ecol Systematics* 1993;24:153–188.

Stewart CB. The powers and pitfalls of parsimony. *Nature* 1993;361:603–607.

SCIENCE OF BURIAL

9

The fossil and archeological records serve as the principal sources of evidence upon which human prehistory is reconstructed. Unless that evidence can be interpreted with some confidence, the reconstruction—however convincing—may not be valid. In recent years, a tremendous emphasis has been placed on understanding the multifarious processes that impinge on bones and stone artifacts that become part of the record. This science of **taphonomy** has revealed that the prehistoric record is littered with snares and traps for the unwary.

Death is a bewildering, dynamic process in the wild. First, many animals meet their end in the jaws of a predator rather than passing away peacefully in their sleep. Once the primary predator has eaten its fill, scavengers, which in modern Africa would include hyenas, jackals, vultures, and the like, move in. The carcass is soon stripped of meat and flesh and the softer parts of the skeleton, such as vertebrae and digits, are crushed between the devourers' powerful jaws. The remaining bones dry rapidly under the sun. Even in this initial phase the skeleton is probably partially disarticulated, with hyenas having torn off limbs and other body parts to be consumed in the crepuscular peace of their dens. Passing herds of grazing animals bring a new phase of disarticulation and disintegration as hundreds of hooves kick and crush the increasingly fragile bones.

Thus, within a few months of a kill, the remains of a zebra, for example, might be scattered over an area of several hundred square meters, and a large proportion of the skeleton will apparently be missing. Some of the skeleton may indeed be miles away, lying among the cache of bones in a hyena's den. Some bones will have been shattered and disintegrated into minuscule pieces. Others will have been compressed into the ground by the pressure of passing hooves, often being splintered in the process. Only the toughest skeletal parts, such as the lower jaw and the teeth, remain intact.

Given that such a fate awaits most animals in the wild, it is perhaps unsurprising that the fanfared announcements of ancient hominine discoveries typically mean an interesting tooth jaw, arm bone, or, rarely, a complete cranium. The most complete specimen found to date is the famous "Turkana boy," whose virtually complete skeleton was found in deposits on the west side of Lake Turkana in 1984 (see unit 24). Dated at approximately 1.5 million years old, this *Homo erectus* specimen lacks only a few limb bones and most of the bones of the hands and feet. The individual, who was about nine years old when he died, came to rest in the shallows of a small lagoon. Even this case is marred by evidence of passing animals, in the form of a limb bone that was snapped in two as a hoof stood on it, pressing it into the soft sand.

DYNAMICS OF BURIAL

To become fossilized, a bone must first be buried, preferably in fine alkaline deposits and preferably soon after death. Rapidity of burial following death is surely the key factor in determining whether a bone will enter the fossil record. Most hominine specimens have been found near ancient lakes and rivers, partly because our ancestors (like most mammals) were highly dependent on water, and partly because these sites provide the depositional environments favoring fossil formation.

As it happens, the forces that can bury a bone—for example, layers of silt from a gently flooding river—can later unearth it as the river "migrates" back and forth across the flood plain through many thousands of years. When this removal occurs, the bones become subject once again to sorting forces. Light bones will be transported some distance by the river, perhaps to be dumped where flow is slowed, while heavier bones are shifted only short distances. Anna K. Behrensmeyer, a leading taphonomist at the Smithsonian Institution (Washington, D.C.), identifies transport and sorting by moving water as one of the most important taphonomic influences. Abrasions caused when a bone rolls along the bottom of a river or stream provide tell-tale signs of such activity, as do the characteristic size profiles and accumulations in slow-velocity areas of an ancient channel. For hominine remains, this activity often results in accumulation of hundreds of teeth and little else, as the researchers working along the lower Omo River in Ethiopia know only too well.

Large numbers of hominine fossils have been recovered from the rock-hard breccia of a number of important caves in South Africa. At one time, hominines were thought to live in these caves, and the bones of other animals found with them were suspected to represent remains of food brought there to be consumed in safety. In addition, the fractures and holes present in virtually all hominine remains were considered to be the outcome of hominine setting upon hominine with violent intent. In many ways, the South African caves present one of the most severe taphonomic problems possible, but with years of patience a group of workers (in particular, C. K. Brain) has cut through the first impressions and progressed a little closer to the truth.

Most of the bone assemblages in the caves were almost certainly the remains of carnivore meals accumulated over very long periods of time. The profile of skeletal parts present matches what would be expected after carnivores had eaten the softer parts. In addition, the damage recorded in the hominine crania found at these sites simply reflected the compression of rocks and bones into them as the cave deposits mounted. Exactly how much time is represented in these fascinating accumulations, and when they occurred, is difficult to determine. But the question, as in many taphonomic investigations, is a key one.

One area of investigation in which taphonomic analysis has been particularly crucial in recent years is in the

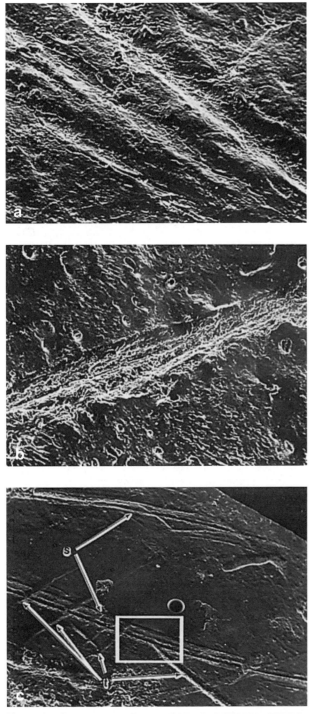

**BONE SURFACES UNDER THE ELECTRON MICRO-
SCOPE:** (a) The surface shows the round-bottomed groove
made by a hyena gnawing at a modern bone. (b) A sharp
stone flake makes a V-shaped groove in a bone surface (modern). (c) This fossil bone from the Olduvai Gorge carries
carnivore tooth marks (t) and stone flake grooves (*left at
center*); the scavenger activity followed the hominine's activity
on this occasion. (Courtesy of Pat Shipman and Richard
Potts.)

study of ancient assemblies of bones and stones—in other
words, putative living sites. Some of the best known and
oldest of these sites occur in the lowest layers of Olduvai
Gorge, Tanzania, and are dated to almost 2 million years
ago. These concentrations of broken bones and chipped
stones have long been assumed to be the product of
hunting and gathering activity such as that seen among
surviving foraging peoples. The occurrence of such sites
appears to increase in frequency through time, giving the
impression of an unbroken trail of litter connecting people ancient and modern who shared a common lifeway
(see unit 26).

In some cases, however, careful taphonomic analysis of
the geological setting and the composition of the bone
and stone assembly has shown such "sites" to result from
water flow, with the material having been dumped by a
stream in an area of low energy—in other words, the
assembly is not an archeological site, but a hydrological
jumble. Even when a collection of bones and stones can
be shown not to be produced by water flow, there remains
the task of deciding how the various materials reached the
site, and whether they were related. For example, did
early hominines use the stones to butcher carcasses?

Taphonomists have determined the stages through
which bones go as they lie exposed to the elements—this
process, known as weathering, can be calibrated. By looking at the degree of weathering evident in a fossil bone, it
is therefore possible to determine how long the bone lay
on the surface before its burial. Applying this technique
to the sites at Olduvai reveals that in many cases bones
accumulated over periods of 5 to 10 years, which would
be unheard of in modern hunter-gatherer sites, which are
occupied only briefly.

CLUES FROM MARKS ON BONES

In the late 1970s and early 1980s, several researchers
discovered on the surface of a small percentage of the
Olduvai bones what appeared to be marks made by stone
tools. Thus, although the sites might not have been typical
hunter-gatherer home bases, it did appear that a connection existed between the bones and the stones: the hominines almost certainly were eating meat. By looking at the
pattern of distribution of cut marks over a bone—on the
shaft as compared with the articular ends, for example—investigators can obtain some idea of whether the
marks were made during the disarticulation of a carcass or
during the removal of meat or skin from the bone.

Determining the identity of marks on the surface of
fossil bones is an important taphonomic activity: gnawing
carnivores and nibbling porcupines can all leave their
signatures. Likewise, sand grains can leave behind tell-tale
signs. In 1986, Behrensmeyer and two colleagues from the
Smithsonian Institution reported that bones trampled in
sandy sediment can sustain abrasions that are virtually
indistinguishable from genuine stone-tool cutmarks. "Microscopic features of individual marks alone provide insufficient evidence for tool use versus trampling," warn
Behrensmeyer and her colleagues. "If such evidence is
combined with criteria based on context, pattern of mul-

tiple marks and placement on bones, however, it should be possible to distinguish the two processes in at least some cases bearing on early human behavior."

Not all taphonomists agree about the difficulty of distinguishing between the effects of trampling and genuine cut marks, however. For instance, Sandra Olsen and Pat Shipman have examined the problem experimentally and stated: "Macroscopic and microscopic comparison of experimentally trampled bones and those which have had soft tissue removed with a flint tool demonstrate significant differences between the surface modifications produced by the two processes." ❋

KEY QUESTIONS

- What is implied by the fact that the great majority of hominine fossil remains have been recovered from sediments layed down near sources of water, such as streams and lakes?
- Why is the fossil record of the African great apes virtually nonexistent for the past 5 million years—during which time the hominine record is relatively good?
- Fossil fragments from almost 500 hominine individuals representing perhaps four species over a period of 4 million years ago to 1 million years ago have been recovered from the Lake Turkana region of Kenya. What percentage does this amount represent of the original populations?
- What is the single most important factor in shaping the life history of a fossil?

KEY REFERENCES

Behrensmeyer AK. Taphonomy and the fossil record. *Am Scientist* 1984;72:558–566.

Behrensmeyer AK, Hill AP. Fossils in the making. Chicago: University of Chicago Press, 1980.

Behrensmeyer AK, Hook RW. Paleoenvironmental contexts and taphonomic modes. In: Behrensmeyer AK, *et al*. Terrestrial ecosystems through time. Chicago: University of Chicago Press, 1992:15–136.

Olsen SL, Shipman P. Surface modification on bone. *J Archeol Sci* 1988;15:535–553.

Shipman P. Life history of a fossil. Cambridge, MA: Harvard University Press, 1981.

Tappen M. Savannah ecology and natural bone deposition. *Curr Anthropol* 1995;36:223–260.

PRIMATE HERITAGE

10

Homo sapiens is one of approximately 200 species of living primate, which collectively constitute the order Primates. (There are 22 living orders in the class Mammalia, which includes the bats, rodents, carnivores, elephants, and marsupials.) Just as we, as individuals, inherit many resemblances from our parents but also are shaped by our own experiences, so it is with species within an order. Each species inherits a set of anatomical and behavioral features that characterize the order as a whole, but each species is also unique, reflecting its own evolutionary history.

Matt Cartmill, of Duke University, says of anthropology that: "Providing a historical account of how and why human beings got to be the way they are is probably the most important service to humanity that our profession can perform." An understanding of our primate heritage provides the starting point for writing that historical account. In this unit we will consider what it is to be a primate, in terms of anatomy and behavior.

The study of primates—primatology—has undergone important changes in recent years for two reasons. First, ecological research has been thoroughly incorporated into primate studies. As a result, primate biology can be interpreted within a more complete ecological context. Second, the science of sociobiology has enabled a keener insight into the evolution of social behavior (see unit 13). And primates, if nothing else, are highly social animals. Modern primatology therefore promises to serve as the focus of some of the most serious intellectual challenges of behavioral ecology.

Modern primates can be classified into four groups:

- The prosimians, which include lemurs, lorises, and bushbabies;
- New World monkeys, such as the marmosets, spider monkeys, and howler monkeys (prosimians are also known as strepsirhines);
- Old World monkeys, such as macaques, baboons, and colobus monkeys; and
- The hominoids, which comprise apes and humans.

Monkeys and apes are known collectively as anthropoids. Twenty-eight of the 200 modern primate species live in Madagascar (the lemurs), with approximately 50 species each found in Central and South America, Africa, and Asia. There are no native, modern primate species in Europe, North America, or Australia.

Modern primate species constitute an extraordinarily varied order, in terms of both morphology and behavior. Some species are among the most generalized and primitive of all mammals, while others display specializations not seen in other mammalian orders. Nevertheless, primate bodies are generally primitive. True, some have lost tails and others have developed large brains. None, however, has turned hands into wings (as bats have), or reduced fingers and toes to single digits (as horses have), or lost limbs altogether (as baleen whales have, being without hindlimbs), or transformed its dentition into something that no self-respecting primate would put into its mouth (as the baleen whales have, with their hairlike combs designed for filtering tiny prey water).

Modern primates vary enormously in size, ranging from the diminutive mouse lemur, which weighs in at 80 grams, to the male gorilla, at more than 2000 times the mouse lemur's size. Whatever their size, primates are quintessentially animals of the tropics. Although different primate species occupy every major type of tropical environment—from rainforest, to woodland, shrubland, sa-

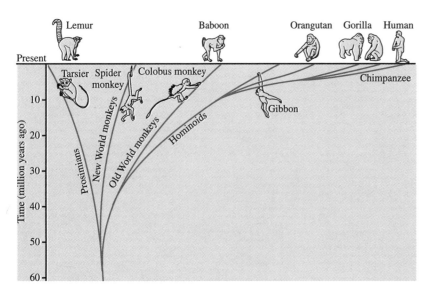

PRIMATE FAMILY TREE

vannah, and semidesert scrub—80 percent of them are creatures of the rainforest. Several Old World monkeys and one ape—the mountain gorilla—live in temperate and even subalpine zones. Among primates, *Homo sapiens* is unique in ranging so wide geographically and in tolerating so extreme a variety of environments.

DEFINITION OF PRIMATE

Although humans have clearly departed from our primate roots in colonizing so broad a range of habitats, many of the characteristics that we often envision as separating us from other primates—such as habitual upright walking, great intelligence, and more complex forms of social organization and behavior sociality—are actually extensions of, rather than discontinuities with, what it means to be a primate. We should therefore ask, What is it to be a primate?

Surprisingly, this question, which essentially asks for a definition of "primate," has proved difficult to answer concisely. "It has, in fact, been a common theme throughout the literature on primate evolution that primates lack any clear-cut diagnostic features of the kind found in other species of placental mammals," notes Robert Martin, of the Anthropological Institute, Zurich. The difficulty, he suggests, stems from an overemphasis on "skeletal features identifiable in the fossil record." If one looks at living primate species instead, encompassing all aspects of their anatomy and behavior, a definition constructed from universal or near-universal characteristics is possible, says Martin.

"Primates are typically arboreal inhabitants of tropical and sub-tropical forest ecosystems," begins Martin's definition. It goes on to describe features of hand and foot anatomy, overall style of locomotion, visual abilities, intelligence, aspects of reproductive anatomy, life-history factors (such as longevity and reproductive strategy), and dental architecture. The definition generally depicts species that have a rather special niche in the world. As Rockefeller University anthropologist Alison Jolly recently noted, "If there is an essence of being a primate, it is the progressive evolution of intelligence as a way of life."

Some of the key components of Martin's definition are described below.

Primate hands and feet have the ability to grasp and are therefore equipped with opposable thumbs and opposable great toes. Humans are an exception, as the human foot has lost its grasping function in favor of forming a "platform" adapted to habitual upright walking. In higher primates, fingers and toes have nails, not claws; and finger and toe pads are broad and ridged, which aids in preventing slippage on arboreal supports and in enhancing touch sensitivity. But some primates have retained claws on certain digits.

Primate locomotion is hindlimb-dominated, whether it consists of vertical clinging and leaping (various small species), quadrupedal walking (monkeys and the African great apes), or brachiation (apes). In each case, the center of gravity of the body is located near the hindlimbs, which produces the typical diagonal gait (forefoot preceding hindfoot on each side). It also means that the body is frequently held in a relatively vertical position, making the transition to habitual bipedalism in humans a less dramatic anatomical shift than is often imagined.

Vision is greatly emphasized in primates, while the olfactory (smell) sense is diminished. In all primates, the two eyes have come to the front of the head, producing stereoscopic vision, to a greater extent than in other mammals. Although some primates (the diurnal species) have color vision, this character does not discriminate the order from many other vertebrate groups. The shifting of the eyes from the side of the head to the front, combined with the diminution of olfaction, produces a shorter snout; this character is accompanied by a reduction in the number of incisor and premolar teeth from the ancestral condition of 3 incisors, 1 canine, 4 premolars, and 3 molars (denoted 3.1.4.3) to a maximum of 2.1.3.3. (Prosimians and New World monkeys demonstrate this latter pattern, whereas Old World monkeys and hominoids have one fewer premolar.)

Partly because of the emphasis on vision, primate brains are larger than those found in other mammalian orders. This increase also reflects a greater "intelligence." In this character, the lemurs, lorises, and other prosimians are, however, less well endowed than monkeys and apes. Tied to this enhanced encephalization is a shift in a series of life-history factors: animals with large brains for their body size tend to have a greater longevity and a low potential reproductive output. For instance, primate gestation is long relative to maternal body size, litters are small (usually one), and offspring precocious; age at first reproduction is late, and interbirth interval is long. "Primates are, in short, adapted for slow reproductive turnover," observes Martin.

If we think of humans as animals with particular physical and behavioral habits, this discursive definition describes us as well, apart from the fact that we do not live in trees. For instance, a quarterback would not be able to stand behind his offensive line and accurately throw a deep pass, unless he were a primate. Hindlimb-dominated locomotion, grasping and touch-sensitive hands, stereoscopic vision, and intelligence—all are required in that activity, and all are general characteristics of primates. More historically, when hominines first began making stone tools, they were not "inventing culture" in the sense that is often used, but merely applying primate manipulative skills to a new task. Although it is true that even by primate standards *Homo sapiens* is particularly well endowed mentally, our generous encephalization merely represents an extension of just another primate trait.

Later we will return to some of these and other themes, particularly the issue of life-history strategy and brain size (see units 12 and 31). In this unit, we will address the question of how primates arrived at their current form—that is, how a small, ancestral, arboreal mammal species developed the above suite of characteristics.

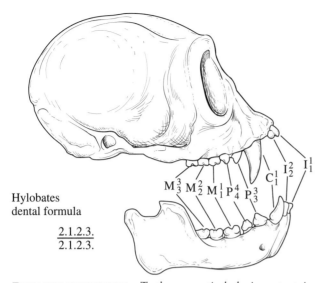

Hylobates
dental formula

$$\frac{2.1.2.3.}{2.1.2.3.}$$

PRIMATE DENTITION: Teeth are particularly important in the reconstruction of primate phylogeny, for two reasons. First, their extreme hardness means that they are the most common item recovered from the fossil record, and hence provide a disproportionate amount of information about fossil species. Second, teeth give very clear information about dietary habits because the shape is strongly influenced by the type of food eaten. By convention, dental formula is written as shown in the diagram. This species (a siamang) possesses two incisors, one canine, two premolars, and three molars (a common scheme in higher primates). (Courtesy of John Fleagle.)

THEORIES OF THE ORIGIN OF THE PRIMATE ADAPTATION

The first systematic attempt to account for the differences between primates and other mammals was made by T. H. Huxley, in his 1863 book, *Evidence as to Man's Place in Nature*. In the early twentieth century, the British anatomists Grafton Elliot Smith and Frederic Wood Jones continued this quest. Ancestral primates and, by extrapolation, humans were different from other mammals, they argued, because of adaptation to life in the trees—hence the **arboreal hypothesis** of primate origins. Grasping hands and feet provided a superior mode of locomotion, according to these scientists, while vision was a more acute sensory system than olfaction in among the leaves and branches.

As Cartmill noted, however, "The arboreal theory was open to the most obvious objection that most arboreal mammals—opossums, tree shrews, palm civets, squirrels, and so on—lack the short face, close-set eyes, reduced olfactory apparatus, and large brains that arboreal life supposedly favored."

The British anthropologists valiantly defended their theory, invoking ingenious and often inconsistent lines of argument. In any case, the arboreal theory was modified and extended in the 1950s by another British researcher, the eminent Sir Wilfrid Le Gros Clark. It continued to thrive for another two decades, until Cartmill felled it in 1972.

In reassessing the arboreal theory in the early 1970s, Cartmill applied biologists' most powerful tool—comparative analysis. "If progressive adaptation to living in trees transformed a tree shrew-like ancestor into a higher primate, then primate-like traits must be better adapted to arboreal locomotion and foraging than their antecedents," reasoned Cartmill. In other words, if primates are truly the ultimate in adaptation to arboreal life, you would expect that they would be more skillful aloft than other arboreal creatures. "This expectation is not borne out by studies of arboreal nonprimates," he noted. Squirrels, for instance, do exceedingly well with divergent eyes, a long snout, and no grasping hands and feet, often displaying superior arboreal skills to those of primates. "Clearly, successful arboreal existence is possible without primate-like adaptations," concluded Cartmill.

If the close-set eyes and grasping hands and feet were an adaptation to something other than arboreality, what was it? Once again Cartmill used the comparative approach to find an answer that formed the basis of the **visual predation hypothesis.** Boldly put, the hypothesis states that the suite of primate characteristics represents an adaptation by a small arboreal mammal to stalking insect prey, which are captured in the hands.

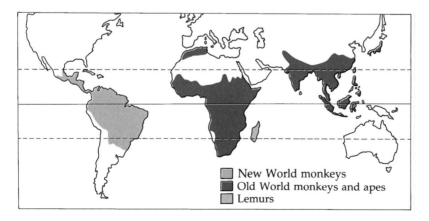

GEOGRAPHICAL DISTRIBUTION OF LIVING PRIMATES

New World monkeys
Old World monkeys and apes
Lemurs

MODES OF PRIMATE LOCOMOTION: The monkey (*top right*) walks quadrupedally, while the gibbon (*top left*) is an adept brachiator (it swings from branch to branch like a pendulum). The orangutan (*mid-left*) is also adept in the trees, but as a four-handed climber. The gorilla (*bottom left*), like the chimpanzee, is a knuckle walker (it supports its weight through the forelimbs on the knuckles of the hand rather than using a flat hand as the monkey does). The tarsier (*foreground*) moves by vertical clinging and leaping. The hominine (*right*) is a fully committed biped. Note the grasping hands and forward-pointing eyes characteristic of primates. (Courtesy of John Gurche/Maitland Edey.)

Cartmill sought individual elements of the primate suite in a range of other species. For instance, chameleons have grasping hindfeet, which they use to steady themselves when approaching insect prey on slender branches. Some South American opossums show similar behavior, capturing their prey by hand or mouth. And, of course, the convergence of the eyes is found in many predatory animals that need to be able to judge distance accurately, such as cats, owls, and hawks.

"Most of the distinctive primate characteristics can thus be explained as convergence with chameleons and small bush-dwelling marsupials (in the hands and feet) or with cats (in the visual apparatus)," concluded Cartmill. "This implies that the last common ancestor would have subsisted much as modern tarsiers, the mouse lemur, and some lorises do today." These species should not be considered "living fossils" because, like humans, they are also the products of 60 million years of evolution. It is simply that their ecological niche resembles the niche occupied by their ancestors.

Cartmill's visual predation hypothesis has recently been challenged by American primatologist Robert Sussman. He points out that many primate species locate their prey by smell or hearing, so that visual predation by itself is not sufficient to explain this suite of primate adaptations. He also argues that the earliest primates evolved at a time when flowering plants were in the midst of an evolutionary diversification. Grasping hands and feet would have enabled small primate species to move with agility in terminal branches rich with fruit; keen visual acuity would allow fine discrimination of small food items. Sussman's hypothesis is obviously similar in some ways to the earlier arboreal hypothesis. Cartmill's hypothesis remains the most cogent explanation of primate adaptations.

Living primates do not follow a single "primate diet." Insects, gums, fruit, leaves, eggs, and even other pri-

mates—all are found on the menu of one primate species or another, and most species regularly consume items from two or more of these categories. The key factor that determines what any individual species will principally subsist on is body size. Small species have high energy requirements per unit of body weight (because of a high relative metabolic rate), and they therefore require food in small, rich packets. Leaves, for instance, are simply too bulky and require too much digestive processing to satisfy small primates. Because of their reduced relative energy demands, large species have the luxury of being able to subsist on bulky, low-quality resources, which are usually more abundant. From the small to the large species, the preferred foods shift, roughly speaking, from insects and gums, to fruit, to leaves.

A good deal of variation upon this basic equation exists, however. As Yale primatologist Alison Richard points out, "Almost all primates, regardless of size, meet part of their energy requirements with fruit, which provides a ready source of simple sugars." What sets the basic equation, she says, is "how they make up the difference in energy and how they meet their protein requirements." This issue is where body size is crucial, and why, for instance, the bushbaby's staple is insects and the gorilla's is leaves.

THE ORIGIN AND EVOLUTION OF PRIMATES

The overall evolutionary pattern of primates remains still unsettled. Some kind (or kinds) of species ancestral to all primates survived the mass extinction 65 million years ago that spelled the end of the Age of Reptiles, with the dinosaurs being the most notorious of the extinctions. Soon into the subsequent Age of Mammals, "primates of modern aspect" appeared approximately 50 million years ago, beginning an adaptive radiation that included an increase in range of body size and a concomitant broadening of diet. The 200 modern species represent the remains of that adaptive radiation, which, in total, probably gave rise to some 6000 species.

The known fossil record provides only the briefest of glimpses of this radiation, a sketchy outline at best; somewhere between 60 and 180 fossil primate species can be recognized. Some researchers consider the earliest primate group to be the plesiadapiformes, the best-known specimen of which was *Purgatorious*, which was found a century ago in Montana and later at several other sites. The plesiadapiformes constituted a successful group living in the Paleocene and early Eocene (55 to 65 million years ago) of North America and Europe, amounting to some 25 genera and 75 species. The range of body size was considerable, stretching from 20 grams to more than 3 kilograms. Most members of the group were probably insectivores. Their supposed phylogenetic link with later primates is somewhat limited, resting on the primatelike structure of the cheek teeth and ear structure. In other respects the plesiadapiformes are somewhat specialized, including the possession of large anterior teeth and three or fewer premolars (many of the earliest prosimians have four premolars). For these reasons, the plesiadapiformes were probably not ancestral to prosimians, but possibly formed a sister group in the primate clade. Some researchers contend, however, that the plesiadapiformes were not primates at all, but instead are linked with the modern colugo (also misleadingly called flying lemurs).

The 1990s have witnessed a flurry of discoveries related to early primates. These advances are helping to resolve the early history of the group, extend its known geographic range, and root its origins and diversification deeper in the past, perhaps even before the end-Cretaceous extinction.

For instance, in 1990, French researchers announced the discovery in Morocco of a collection of 10 undoubtedly primate cheek teeth, which were described as a new species, *Altiatlasius koulchii*. The species, which is estimated to have weighed less than 100 grams, is thought to belong to the family Omomyidae, one of two major groups of early, true primates. A North American discovery, consisting of a relatively rare cache of fossil skulls, is also said to be an omomyid, of the species *Shoshonius cooperi*, which lived a little more than 50 million years ago. The omomyids—tiny, nocturnal, fruit-eating species—are considered to be ancestral to tarsiers. The second major group of early primates, the Adapidae—diurnal folivores, frugivores, and insectivores—were larger than omomyids and are putative ancestors of lemurs. An adapid specimen was found early in the nineteenth century. Although these two large and geographically widely dispersed families now seem well accepted as the earliest known primates, the question of their origin persists, if they are not derived from the plesiadapiformes.

One of the most spectacular discoveries, announced in 1994, included five new types of early primate, of both omomyid and adapid affinities at the Shanhuang site in southeastern China. The diversity of species at this site

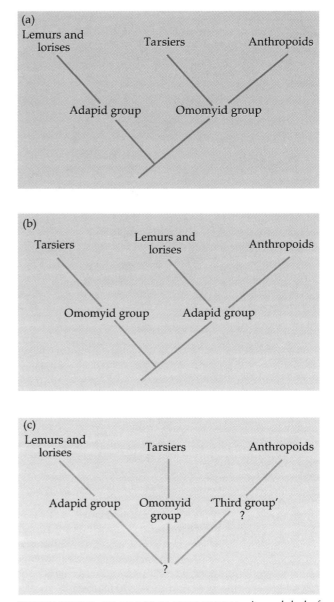

THREE VIEWS OF PRIMATE EVOLUTION: A good deal of uncertainty exists over the pattern of primate evolution. Until recently most opinion was divided between schemes (a) and (b), which show differences over the origin of anthropoids. A third view (c) has also been proposed, which postulates a third, early group of primates that was ancestral to modern anthropoids. Based on the most recently discovered fossil evidence, however, scheme (a) is now most strongly supported.

exceeds that found in all of the rest of Asia and in well-documented sites in Europe and North America. One of the most interesting finds involved teeth that are virtually identical to those of modern tarsiers. Huxley speculated that the anatomical range of the lower-to-higher primates in today's world gives a window into the group's evolutionary history. The Chinese find indeed implies the mod-

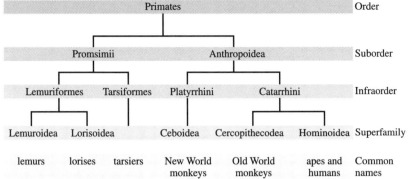

PRIMATE CLASSIFICATION

ern tarsier might be a "living fossil." (Not literally, of course, but the group simply has not changed much since its origin.)

Uncertainty has long swirled around the evolutionary root of the suborder Anthropoidea (monkeys, apes, and humans). Some anthropologists have argued that its origin lies within the adapids; others have favored the omomyids. Both schemes put the origin of anthropoids close to 35 million years ago. A recently developed argument suggests that neither group is ancestral to anthropoids, but that a third group existed *Algeripithecus minutus*, discovered in Algeria and reported in May 1992, is suggested to be a specimen of the latter group. The Shanhuang fossils provide support for the omomyid affin-

ity with anthropoids, however. The dental formula of one specimen, *Eosimias*, is what would be expected of an ancestral anthropoid (hence its name). *Eosimias* is more closely linked to omomyids than to adapids, thus forging a link with the ancestral tarsier group. It now seems likely that modern tarsiers and modern anthropoids shared a specific common ancestor. If correct, this pattern of primate evolution would put the origin of anthropoids closer to 50 million years rather than the 35 million years ago that was previously believed.

The earliest known fossil of the superfamily Hominoidea, which includes all living and extinct species of humans and apes, is some 20 million years old; it was found in Africa (see unit 16).

KEY QUESTIONS

- What general trends did the primate order follow through evolutionary time that are common in other mammalian orders?
- What are the most important problems in trying to reconstruct the phylogeny of primates?
- What key adaptations do humans share with non-human primates?
- How great a departure is bipedalism from the mode of locomotion of monkeys and apes?

KEY REFERENCES

Cartmill M. New views on primate origins. *Evol Anthropol* 1992;1:105–111.

Culotta E. A new take on anthropoid origins. *Science* 1992;256:1516–1517.

Fleagle JG. Primate adaptation and evolution. New York: Academic Press, 1997.

Fleagle JG, Kay RF. Anthropoid origins. New York: Plenum Press, 1994.

Kay RF, *et al*. Anthropoid origins. *Science* 1997;275:797–804.

Martin RD. Primates: a definition. In: Wood B, Martin L, Andrews P, eds. Major topics in primate and human evolution. London: Academic Press, 1986.

———. Primate origins: plugging the gaps. *Nature* 1993;363:223–234.

Sussman RW. Primate origins and the evolution of angioserms. *Am J Primatol* 1991;23:209–223.

HUMANS AS ANIMALS

BODIES, SIZE, AND SHAPE

11

It is striking that human populations in different parts of the world vary significantly in their body form, suggesting, among other things, an adaptation to different climates. An understanding of anatomical adaptation of many animal species to different climates has a long history, with two specific "rules" relating to this issue. **Bergmann's rule**, published in 1847, states that in a geographically widespread species, populations in warmer parts of the range will be smaller-bodied than those in colder parts of the range. **Allen's rule**, published in 1877, states that populations of a geographically widespread species living in warm regions will have longer extremities than those living in cold regions.

PRINCIPLES OF CLIMATIC ADAPTATION IN HUMAN POPULATIONS

Despite the long pedigree of Bergmann's and Allen's ecogeographical rules, anthropologists were slow to apply them to human variation. Interest in this relationship emerged in the 1950s and 1960s, when climate began to

be recognized as an important influence in determining anatomical differences among different geographical populations. For instance, the bodily differences between the tall, thin Nilotics at the equator and the short, bulky Eskimos in the Arctic came to be viewed as a direct reflection of optimal strategies for balancing heat production and dissipation at different latitudes with different prevailing climates.

In recent years, Christopher Ruff, of Johns Hopkins University, has been bringing together the study of ancient and modern human variation in relation to climate. For his analyses, Ruff views the human body as a cylinder, the diameter of which represents the width of the body, or, more specifically, the width of the pelvis; the length of the cylinder represents trunk length. The link between anatomy and climate relates to thermoregulation, or the balance between heat produced and the ability to dissipate it. This relationship translates to the ratio of the surface area to the volume of the cylinder, or body mass. In hot climates, a high ratio—that is, a large surface area relative to body mass—facilitates heat loss. In cold climates, a low ratio—that is, a small surface area relative to body mass—allows heat retention. Simple geometry shows that the ratio of surface area to body mass is high when the cylinder is narrow, and low when it is wide. This finding forms the basis of Bergmann's rule.

A strong prediction flows from this analysis: people living at low latitudes will have narrow bodies and a

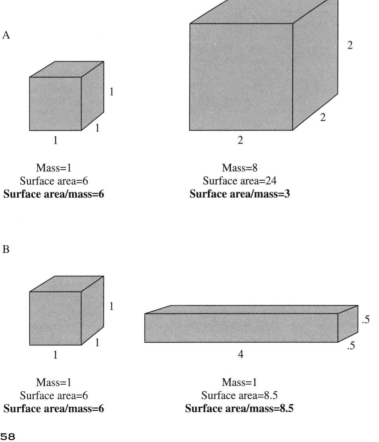

A

Mass=1
Surface area=6
Surface area/mass=6

Mass=8
Surface area=24
Surface area/mass=3

B

Mass=1
Surface area=6
Surface area/mass=6

Mass=1
Surface area=8.5
Surface area/mass=8.5

(A) GEOMETRIC BASIS OF BERGMANN'S RULE: An increase in size decreases the ratio of surface area to mass; in humans, this relationship is reflected in the breadth of the trunk.

(B) GEOMETRIC BASIS OF ALLEN'S RULE: An elongated shape increases the ratio of surface area to mass; in humans, this relationship is reflected in limb length.

linear stature, while those at high latitudes will have wide bodies and a relatively bulky stature. When Ruff surveyed 71 populations around the globe, he found that the prediction was sustained very well. He also discovered that Allen's rule applies convincingly, with tropical people having longer, thinner limbs, which maximizes heat loss, while people at high latitudes have shorter limbs. A comparison of the tall Nilotic people of Africa with the relatively stocky Eskimos in the northern-most latitudes of North America illustrates this difference very clearly.

Body width represents the key variable, even though tropical people also tend to be linear. A further step of simple geometry shows that linearity is not a necessary feature of low-latitude populations. The ratio of the surface area to body mass in a cylindrical model of a certain width is not altered by changing its length, as the accompanying diagram shows. Peoples who live in similar climatic zones will have the same body width, no matter how tall or short they are, because they have the same surface area to body mass ratios. This fact is revealed in a comparison of Nilotic people, whose average height is more than six feet, and Mbuti Pygmies, who are two feet shorter on average. Why the difference in stature?

The answer is related to efficiency of heat dissipation. Humans rely heavily on sweating to cool their bodies. Nilotics live in open environments, where sweating is efficient; in contrast, Mbuti Pygmies, like most Pygmy populations, live in moist, humid forests, where the air is still and sweating is an inefficient cooling mechanism. Under these environmental conditions, the best strategy is to limit the amount of heat generated during physical

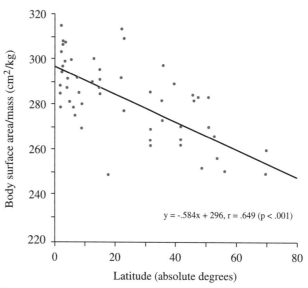

$y = -.584x + 296, r = .649 \ (p < .001)$

RELATIONSHIP BETWEEN BODY BREADTH AND LATITUDE: People living at high latitudes have broad bodies, as measured by the bi-iliac (pelvic) breadth; those residing at low latitudes have narrow bodies. This relationship is a consequence of Bergmann's rule. (Courtesy of C. B. Ruff.)

exertion, which is achieved by reducing the volume of the cylinder. With the width of the cylinder remaining constant, this requirement implies a reduction of its length—in other words, reduced stature.

This insight may have implications for the lifestyles of both Lucy (an *Australopithecus afarensis*) and the Turkana boy (*Homo ergaster*, tall in stature), whose differences in stature are similar to the differences observed between

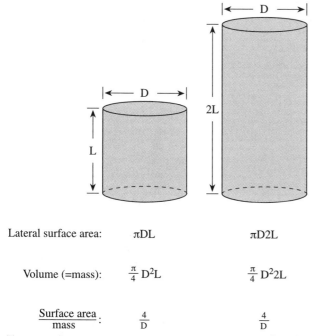

Lateral surface area:	πDL	πD2L
Volume (=mass):	$\frac{\pi}{4} D^2L$	$\frac{\pi}{4} D^2 2L$
$\frac{\text{Surface area}}{\text{mass}}$:	$\frac{4}{D}$	$\frac{4}{D}$

THE CYLINDRICAL MODEL OF BODY SHAPE: An increase in the length (*L*) of the trunk has no effect on the ratio of surface area to body mass. (Courtesy of C. B. Ruff.)

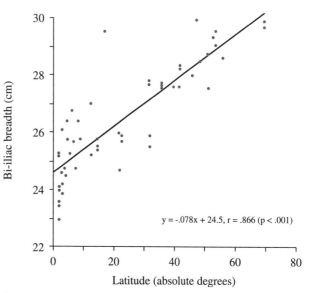

$y = -.078x + 24.5, r = .866 \ (p < .001)$

RELATIONSHIP BETWEEN THE RATIO OF SURFACE AREA TO BODY MASS AND LATITUDE: People living at high latitudes have a low ratio as a consequence of Bergmann's rule. (Courtesy of C. B. Ruff.)

Nilotics and Pygmies. Despite their varying statures, Lucy and the Turkana boy had very similar body widths, comparable with the width of modern tropical populations. This observation makes sense because, living in East Africa as they did, they were exposed to a tropical climate (albeit more than a million years apart). Ruff speculates that, like the Nilotics of today, the Turkana boy and his fellow *Homo ergaster* people lived an active life in open environments. Lucy and her companions, by contrast, inhabited more closed, forested environments, comparable with the environment of modern Pygmies.

Climatic adaptation of body form can also be seen in Neanderthals, who lived in Europe between 250,000 and 27,000 years ago—a time when, for the most part, the Pleistocene Ice Age still held the continent in its grip (see unit 30). The frigid conditions under which the Neanderthals evolved is reflected in their wide bodies and their relatively short limbs, characteristics comparable to those seen in modern Eskimos.

CHANGES IN RECENT HUMAN POPULATIONS

We now turn to changes in body form of humans through time. Many anthropologists agree that from early in the *Homo* lineage, some 2 million years ago, to the appearance of archaic *Homo sapiens*, 300,000 or 400,000 years ago, robusticity steadily increases before finally reaching a plateau. In this case, we are talking about people having thick skulls and heavily muscled limbs. (Brain size increased from approximately 900 ccs to more than 1400 ccs in this time.) These people were immensely strong, reflecting their arduous subsistence pattern. Early anatomically modern humans, who appeared 200,000 years ago, were significantly less robust than archaic *sapiens*, but much more so than people today. (As mentioned earlier, the early anatomically modern people in Europe were also more linear, because of their African origin.) The robusticity of early moderns decreased gradually over a long period, and then dramati-

cally so after the end of the Ice Age, 10,000 years ago, but not in all populations. Australian Aborigines, Patagonians, and Fuegans, for instance, are still relatively robust in their skull and skeletal anatomy. Where it occurred, the loss of robusticity occurred principally between 10,000 and 5000 years ago, then halted. Reductions in brain size (to 1300 ccs), size of teeth and jaws, and overall stature followed similar patterns, but to different degrees.

For instance, in his studies of Australian populations, Peter Brown, of the University of New England, Armidale, found the following changes in the five millennia after the Ice Age: tooth reduction, 4.5 percent; facial size reduction, 6–12 percent; brain size reduction, 9.5 percent; and stature reduction, 7 percent. Where data exist in other parts of the world, such as in Europe and Southeast and West Asia, similar changes are observed, although paleoanthropologists disagree on whether, for instance, significant brain shrinkage began as early as 30,000 years ago or only 10,000 years ago. Whatever the details of the timing of events in these later stages, it seems irrefutable that, until the nutritional effects of the last century or so kicked in, modern people were comparative midgets on the human evolutionary stage.

What overall pattern held, beginning with the increase in robusticity until archaic *sapiens* arrived? Subsistence was strenuous in those days, as our ancestors plied a life of hunting and gathering with only rudimentary technology to aid them. Muscles—not missiles—were their weapons. Other explanations for this trend have been suggested, too. For instance, Robert Foley, of Cambridge University, speculates that people became stronger because they were embroiled in increasing conflict between neighboring groups. The conflicts arose, he says, because the groups were dominated by bands of males, probably closely related, who sought to appropriate the plentiful resources in their area, including females from other groups.

Why, then, did robusticity decline with the origin of anatomically modern humans, and continue to diminish

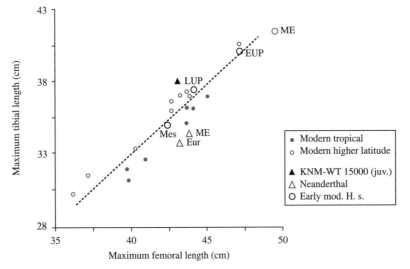

TIBIA LENGTH RELATIVE TO FEMUR LENGTH IN MODERN AND PREHISTORIC POPULATIONS: A short tibia relative to the femur indicates a relatively short leg, a consequence of Allen's rule. The dotted line represents the division between modern tropical and high-latitude populations. Tropical populations fall above the line, indicating a long leg; high-latitude populations fall below it, indicating a short leg. European (open triangle, Eur) and Middle Eastern (open triangle, ME) Neanderthals had relatively short legs; early African *Homo* (closed triangle, KNM-WT 15000, the Turkana body) had relatively long legs. The early modern human populations (open circles) are intermediate between tropical and high-latitude populations. (Courtesy of C. B. Ruff.)

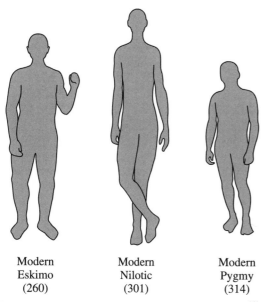

BODY OUTLINES OF MODERN POPULATIONS: Figures below the outlines give the surface area to body mass ratio (cm²/kg). Note the broad body and short stature of the Eskimo, and a low ratio; the Nilotic body is narrow and linear, with a high ratio. The Pygmy has the same body breadth as the Nilotic and a similar ratio. (Courtesy of C. B. Ruff.)

for tens of millennia? Not because these humans changed their social structure and became more peaceable, says Foley, but because technological inventions usurped the role previously played by sheer strength. One key invention involved projectiles, spears in which stone points were hafted onto wooden shafts. Stone tools became more versatile, which perhaps buffered people from some bare-hands contact with their environment. And people were smarter, too, indicating that guile rather than brawn might have filled the larder. Loring Brace, of the University of Michigan, has long been a proponent of technology, or culture, as an important force in diminishing human robusticity. Eventually, food preparation, through cooking, took the pressure off teeth, which became smaller as a result. This development emerged at different times in different parts of the world. But, as Peter Brown observes of the Australians, tooth reduction can occur even in the absence of food preparation, so other forces must be operating as well.

Many dramatic changes transpired with the end of the Ice Age, not least of which was the disappearance of plentiful game, some of it very large. Gone were the mammoth and mastodon, for instance. Foley suggests that this reduction of resources forced recourse to one of two subsistence strategies. The first was food production, or agriculture (see unit 35); the second was a shift to a different kind of social structure in hunting and gathering bands. Although we now think of agriculture as producing plentiful food, early food production was a hazardous venture, with many lean times. The archeological record shows that nutritional stress was rife for early farmers—a

sure way of keeping body size small. In the second strategy, because male hunters were unable to monopolize food resources to the same degree as their ancestors had done, they were unable to monopolize many females as mates (a practice known as polygyny). As a result, less aggressive competition for females occurred among males, and therefore less of a premium was placed on raw strength. Thus, males became smaller because they didn't need to fight as much.

Nutritional stress is a popular explanation of human shrinkage for many anthropologists, not because of the loss of the megafauna of the Ice Age but because of a booming human population. Limitations on resources often lead to reduced body size, says Christopher Stringer, of the Natural History Museum, London, as is seen in the dwarfing of species on islands. According to Robert Martin, of the Anthropological Institute, Zurich, the stress results from a shift to early weaning, a strategy that boosts reproductive output in the face of the competition associated with increased populations. Early weaning inevitably leads to a reduction in brain size, though not,

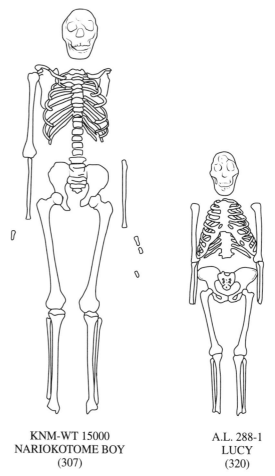

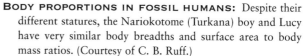

BODY PROPORTIONS IN FOSSIL HUMANS: Despite their different statures, the Nariokotome (Turkana) boy and Lucy have very similar body breadths and surface area to body mass ratios. (Courtesy of C. B. Ruff.)

says Martin, necessarily to a reduction in body size. This viewpoint separates Martin from most anthropologists, who argue that reductions of brain size in recent history simply followed in the path of body size reductions.

The advent of agriculture was once viewed as the universal change in human culture that produced a universal change in human physique. As more was learned about the shift from foraging to food production, however, this notion appeared less likely to be the answer. Agriculture was developed at different times in different parts of the world, and in some places not at all. In Australia and parts of the Americas, for instance, people were still hunting and gathering as they had been for tens of millennia, and yet the pattern of body size reduction still applied. The one change that applies everywhere, of course, is the increase in global temperature associated with the end of the Pleistocene. That fact alone is sufficient to force serious consideration of climate as the causative agent. Moreover, body size reduction has occurred in many nonhuman animals in this same period, in Australia, Israel, and Indonesia, for example. Wherever researchers look (and so far not many places have been analyzed), the same phenomenon is found in nonhuman animals. ✻

KEY QUESTIONS

- Why is local climate such a strong predictor of body form for human populations in different parts of the world?
- How is subsistence pattern superimposed on climatic adaptation in relation to body form?
- What problems are encountered in reconstructing the body form of extinct human species?
- What is the most likely explanation of the reduction of robusticity in recent human populations?

KEY REFERENCES

Aiello LC. Allometry and the analysis of size and shape in human evolution. *J Hum Evol* 1992;22: 127–147.

Holliday TW, Falsetti AB. Lower limb length of European early modern humans in relation to mobility and climate. *J Hum Evol* 1995;9:141–153.

Kappelman J. The evolution of body mass and relative brain size in fossil hominids. *J Hum Evol* 1996;30: 243–276.

McHenry HM. Behavioral ecological implications of early hominid body size. *J Hum Evol* 1994;27: 77–87.

Ruff CB. Morphological adaptation to climate in modern and fossil hominids. *Yearbook Physical Anthropol* 1994;37:65–107.

Ruff CB, Walker A. Body size and body shape. In: Walker A, Leakey R, eds. The Nariokotome *Homo erectus* skeleton. Cambridge, MA: Harvard University Press, 1993:234–265.

Ruff CB, *et al*. Body mass and encephalization in Pleistocene *Homo*. *Nature* 1997;387:173–176.

Smith RJ. Biology and body size in human evolution. *Curr Anthropol* 1996;37:451–481.

BODIES, BRAINS, AND ENERGY

12

This unit will explore the impact of size—of both brains and bodies—on **life-history variables** and behavioral ecology. Life-history variables are those factors that describe how individuals of a species proceed from infancy through maturity to death, and the strategies involved in producing offspring. We will see why hominines, with their large body size, have many more options open to them in terms of diet, foraging range, sociality, expanded brain capacity, and so on, than, for example, the diminutive mouse lemur.

In 1978, Princeton ecologist Henry Horn encapsulated the range of potential ecological options by posing the following set of questions: "In the game of life an animal stakes its offspring against a more or less capricious environment. The game is won if offspring live to play another round. What is an appropriate tactical strategy for winning this game? How many offspring are needed? At what age should they be born? Should they be born in one large batch or spread out over a long lifespan? Should the offspring in a particular batch be few and tough or many and flimsy? Should parents lavish care on their offspring? Should parents lavish care on themselves to survive and breed again? Should the young grow up as a family, or should they be broadcast over the landscape at an early age to seek their fortunes independently?"

In responding to these challenges, the animal kingdom as a whole has come up with a vast spectrum of strategies, ranging from species (oysters, for instance) that produce millions of offspring in a lifetime, upon which no parental care is lavished, to species (such as elephants) that produce just a handful of offspring in a lifetime, each born singly and becoming the object of intense and extensive parental care. In the first case, the potential reproductive output of a single individual is enormous, though typically curtailed by environmental attrition; in the second case, it is small.

PRIMATES AS LARGE MAMMALS

By their nature, mammals are constrained in the range of life-history patterns open to them: mammalian mothers are limited in the number of offspring that can be carried successfully through gestation and suckling. Nevertheless, potential reproductive output can be relatively high if more than one litter is raised each year and the lifetime lasts several years.

In the order Primates, potential reproductive output is low compared with that of mammals as a whole, with litters being restricted in the vast majority of species to a single offspring. In the parlance of population biology, primates are said to be *K*-selected. (Species with a high potential reproductive output are described as being

r-selected.) Of all primates, humans are the most extremely *K*-selected species. (See Table 12.1 for correlates of *r*- and *K*-selection.)

Within the overall Primate order, however, a wide range of life-history patterns exists, as biologists Paul Harvey, Robert Martin, and Tim Clutton-Brock have pointed out. "Adult female mouse lemurs [the smallest species of primate] can probably produce one or two litters of two or three offspring each year, and the young can be parents themselves within a year of their own birth. On the other hand, adult female gorillas [the largest species of primate] produce a single offspring every 4 or 5 years, and the young do not breed until they are about 10 years old."

In terms of potential reproductive output, the female mouse lemur (which weighs 80 grams) can leave 10 million descendants in the time it takes the female gorilla (weighing 93 kilograms) to produce just one. "Such differences between species have presumably evolved as adaptations for exploiting different ecological niches," note Harvey and his colleagues. "Each niche is associated with a particular optimum body size, dictated in part by an animal's ability to garner and process available food supplies."

Success in simple Darwinian terms is often measured in the currency of reproductive output, which is determined by a series of interrelated life-history factors. These factors include age at maturity, length of gestation, litter size, duration of lactation period, interbirth interval, and lifespan.

A DIFFERENCE IN BODY SIZES: The gorilla and the mouse lemur represent the largest and the smallest of the primates, with the females of the species weighing 93 kilograms and 80 grams, respectively. Such differences in body size have many implications for a species' social and behavioral ecology. One of the most dramatic involves potential reproductive output: the female mouse lemur can grow to maturity and, theoretically, leave 10 million descendants in the time it takes a female gorilla to produce a single offspring.

TABLE 12.1 Characteristics of *r*- and *K*-selection: *r*-selected species (such as oysters) live high-risk lives and are more affected by external factors than by competition from within the population. *K*-selected species pursue low-risk strategies in which intraspecies competition is an important factor in success. Primates as a whole, and apes and humans in particular, are *K*-selected.

	r-Selection	*K*-Selection
Climate	Variable and/or unpredictable; uncertain	Fairly constant and/or predictable; more certain
Mortality	Often catastrophic, nondirected, density independent	More directed, density dependent
Survivorship	High juvenile mortality	More constant mortality
Population size	Variable in time, nonequilibrium; usually well below carrying capacity of the environment; unsaturated communities or portions thereof; ecological vacuums; recolonization each year	Fairly constant in time, equilibrium, at or near carrying capacity of the environment; saturated communities; no recolonization necessary
Intra- and interspecific competition	Variable, often lax	Usually keen
Selection favors	• Rapid development • High maximal rate of increase, r_{max} • Early reproduction • Small body size • Single reproduction • Many small offspring	• Slower development • Greater competitive ability • Delayed reproduction • Larger body size • Repeated reproduction • Fewer larger progeny
Length of life	Short, usually less than 1 year	Longer, usually more than 1 year
Leads to	Productivity	Efficiency

Some species live "fast" lives—during their short lifespan, they mature early, produce large litters after a short gestation period, and wean early. The result is a large potential reproductive output. Other species live "slow" lives—during their long lifespan, they mature late, produce small litters (a single offspring) after a long gestation period, and wean late. Thus, their potential reproductive output is small.

As it happens, the best predictor as to whether a species lives "fast" or "slow" is its body size. Small species live fast lives, while large species live slow lives. As potential reproductive output is highest in species that experience fast lives, it might seem that all species would be small. That some species are large implies that a bigger body size provides some benefits that offset the reduced potential reproductive output.

Such benefits might include (for a carnivore) a different spectrum of prey species or (for a potential prey) better antipredator defenses. Another potential benefit of increased body size is the ability to subsist on poorer-quality food resources. Basal energy demands increase as the 0.75 power of body weight; in other words, as body weight increases, the basal energy requirement *per kilogram of body weight* decreases, a relationship known as the Kleiber curve. This concept explains why mouse lemurs must feed on energy-rich insects and gums, for instance, while gorillas can subsist on energy-poor foliage. A further potential benefit of increased body size is improved thermoregulatory efficiency.

The generally close relationship between body size and the value of the various life-history factors is the outcome of certain basic geometric and bioenergetic constraints. Any particular body size increase is associated with a more or less predictable change in, for example, gestation length and age at maturity. For each life-history variable, therefore, a log/log plot against body size produces a straight line, with a particular exponent that describes the relationship [0.75 for basal energy needs, 0.37 for interbirth interval (in primates), 0.56 for weaning age, and so on]. In effect, such plots *remove* body size from species

Life-history factors

↓ Metabolic needs ↑	
↑ Age at maturity ↓	
↑ Gestation length ↓	
↓ Litter size ↑	
↑ Interbirth interval ↓	
↑ Lactation period ↓	
↑ Longevity ↓	

Large body size

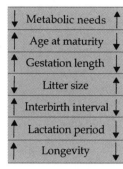

Small body size

LIFE-HISTORY FACTORS: Body size affects a broad range of life-history factors, as illustrated here. For instance, a large primate will have a long lifespan, mature late, have a long gestation time and lactation period, have a long period between litters, but litters will be small (usually one) and basal metabolic requirements will be low.

comparisons and allow us to assess the significance of particular organs—for example, brain size—after body size is taken into account. This examination amounts to analyzing how far particular features depart from predictions based on body size.

If basic engineering constraints were all that underpinned life-history factors, then every species would be directly equivalent with every other species *when body weight is taken into account.* That is, all figures for each life-history variable would fall on the appropriate straight lines. In fact, individual figures often fall above or below the line, indicating a good deal of life-history variation. This variation reveals an individual species' (or, more usually, a group of related species') adaptive strategy.

Researchers now know that, in addition to body size, brain size is also highly correlated with certain life-history factors, in some cases much more so than is body size.

ALTRICIAL AND PRECOCIAL STRATEGIES

Among mammals as a whole, a key dichotomy exists in developmental strategy that has important implications for life-history measures: the altricial/precocial dichotomy. **Altricial** species produce extremely immature young that are unable to feed or care for themselves. The young of **precocial** species, on the other hand, are relatively mature and can fend for themselves to a certain degree.

Life-history factors critically associated with altriciality and precociality include gestation length. In altricial species, gestation is short and neonatal brain size is small. Gestation in precocial species is relatively long, and neonatal brain size is large. There is, however, no consistent difference in *adult* brain size between altricial and precocial species. Primates as a group are precocial with the exception of *Homo sapiens,* which has developed a secondary altriciality and an unusually large brain (see unit 31).

In addition to the distinction between fast and slow lives based on absolute body size, some species' lives may be fast or slow *for their body sizes.* Such deviations have traditionally been explained in terms of classic *r-* and *K*-selection theory. According to this theory, environments that are unstable in terms of food supply (that is,

are subject to booms and busts) encourage *r*-selection: fast lives, with high potential reproductive output. Alternatively, stable environments (which are close to carrying capacity and in which competition is therefore keen) favor *K*-selection: slow lives with low potential reproductive output and high competitive efficiency.

As mentioned earlier, primates are close to the *K*-selection end of the spectrum among mammals as a whole, but some primates are less *K*-selected than others. For instance, Caroline Ross has shown that, when body size is taken into account, primate species that live in unpredictable environments have higher potential reproductive output than species residing in more stable environments.

A second factor that influences whether a species might live relatively fast or slow for its body size has been

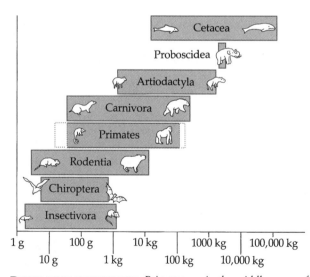

BODY SIZE COMPARED: Primates are in the middle range of mammalian body sizes. Nevertheless, the biology of hominoids is the biology of large mammals. Most mammalian species are concentrated in Rodentia, Chiroptera, and Insectivora, and are therefore small.

identified by Paul Harvey and Daniel Promislow. In a survey of 48 mammal species, the two found that "those species with higher rates of mortality than expected had shorter gestation lengths, smaller neonates, larger litters, as well as earlier ages at weaning and maturity." In other words, species that suffer high natural rates of mortality live fast. "The reason is that species with higher rates of mortality are less likely to survive to the following breeding season and will therefore be selected to pay the higher costs associated with the earlier reproduction."

Again, does the very slow life lived by *Homo sapiens* imply evolution from an ancestor that experienced very low levels of mortality?

Given that most mammals measure less than 32 centimeters in length, hominines—even the early, small species—must be classified as large mammals. One of the earliest known hominine species, *Australopithecus afarensis*, stood 1 meter (females) to 1.7 meters (males) tall, and weighed some 30 to 65 kilograms (see unit 19). These general proportions persisted until approximately 1.5 million years ago and the evolution of *Homo ergaster*, which stood close to 1.8 meters (with a much reduced difference between males and females). (See unit 24.)

PREDICTIONS FOR EARLY HOMININE SPECIES

With a knowledge of these general body proportions and the estimates of brain size, it becomes possible to make estimates of various life-history factors for the early hominine species, given what is known of the only extant hominine, *Homo sapiens*. Surely, hominines lived slow lives in the terms of life-history variables, with a vastly increased brain capacity eventually distorting some of them.

In addition, we can identify several behavioral ecology traits that would be associated with large body size, as Cambridge University anthropologist Robert Foley has done. For instance, dietary scope could be broad; day and home ranges could be large; mobility could be high; predator-prey relations would be shifted from that of smaller primates; thermoregulatory efficiency would be improved; sociality would be extended; and enhanced encephalization would be energetically possible.

In sum, studies of life-history strategies have identified body size, brain size, environmental variability, and mortality rates as being crucial to the rate at which a species will live. Much of human evolution may therefore be explained in terms of a large hominoid exploiting a relatively stable food supply, its stability perhaps being enhanced by virtue of its breadth. Technology may eventually have contributed to this stability by permitting more efficient exploitation of meat and certain plant foods, thus broadening the diet even further. A reduction in mortality, perhaps through improved antipredator defense, would further encourage a "slow" life-history strategy. The selection pressure leading to increased body size remains to be identified. ✸

KEY QUESTIONS

- What are the limitations of a simple Darwinian measure of reproductive success?
- At any particular body size, which is the riskier strategy: living fast or living slow?
- Primates as a group are twice as encephalized as other mammals. How might this character have arisen?
- Could the first hominines have originated in tropical rainforests?

KEY REFERENCES

Bogin B. The evolution of human childhood. *BioScience* 1990;40:16–25.

Charnov EL, Berrigan D. Why do female primates have such long lifespans and so few babies? Or life in the slow lane. *Evol Anthropol* 1993;1:191–194.

Foley R. Humans before humanity. Oxford: Blackwell, 1995

Harvey P, Krebs JR. Comparing brains. *Science* 1990;249:140–146.

Harvey P, Martin R, Clutton-Brock T. Life histories in comparative perspective. In: Smuts BB, Cheney DL, Seyfarth RM, Wrangham RW, Struthsaker TT, eds. Primate societies. Chicago: University of Chicago Press, 1986:181–196.

Harvey P, Nee S. How to live like a mammal. *Nature* 1991;350:23–24.

Hill K. Life history theory and evolutionary anthropology. *Evol Anthropol* 1993;1:78–88.

Martin RD, Harvey P. Human bodies of evidence. *Nature* 1987;330:697–698.

Promislow D, Harvey P. Living fast and dying young. *J Zool Soc London* 1990;220:417–437.

Ross C. Environmental correlates of the intrinsic rate of natural increase in primates. *Oecologia* 1992;90:383–390.

———. Primate life histories. *Evol Anthropol* 1998;6:54–62.

Smith BH. Life history and the evolution of human maturation. *Evol Anthropol* 1992;1:134–142.

Stearns SC. The evolution of life histories. Oxford: Oxford University Press, 1992.

BODIES, BEHAVIOR, AND SOCIAL STRUCTURE

13

The great majority of primate species are social animals, living in groups that range from 2 to 200 individuals. Whatever the size of the group, it serves as the focus of many important biological activities, including foraging for food, raising offspring, and defending against predators. The group is also the center of intense social interaction that has little apparent direct bearing on the practicalities of life: in the human sphere we would call it socializing, the making and breaking of friendships and alliances (see unit 31). The size, composition, and activity of a group defines what is usually meant by a species' social organization.

Animal behavior is a far more variable characteristic than, for instance, anatomy or physiology. Consequently, an order such as the Primates will display an astonishingly wide range of social organization, in which even closely related species may carry out their daily social lives in very different ways. We saw in unit 12 that body size can have a powerful influence on many aspects of a species' way of life, but social organization is not one of them. Even if we consider just the apes—the largest of the nonhuman primates—the array of social organization found in the species is as great as among the primates as a whole.

As highly social creatures, we may find it odd to ask, Why should animals live in groups? This problem is, in fact, a very good biological question because gregariousness carries many costs. For instance, a lone individual doesn't have to share its food with another individual, but competition for all resources characterizes a group. A lone individual is not exposed to diseases that flourish in communities, which provide a viable host pool for pathogens. A lone individual is much less conspicuous to predators than is a group of individuals, and so on. Clearly, as most primates do live in groups, the benefits must outweigh the costs.

This unit will discuss current thinking about the benefits—that is, the causes—of living in groups. It will also examine some of the consequences of group living—not the costs mentioned, but the ways in which individuals might adapt behaviorally and anatomically to different types of social structures.

SOCIAL ORGANIZATION IN APES

To obtain a feel for some of the details of social organization and the range to be found among primates, we will first survey the social lives of the apes: gibbon (and siamang), orangutan, chimpanzee, and gorilla.

Gibbons and siamangs, the smallest of the apes (sometimes called the lesser apes), live in forests in south-east Asia. The basic social structure of these highly acrobatic, arboreal creatures is very similar, consisting of a monogamous mating pair plus as many as three dependent offspring. Gibbons are territorial, and eat a diet of fruit and leaves. On reaching maturity, the offspring leave the natal group and eventually establish one of their own by pairing with another young adult of the opposite sex. Mature males and females have essentially the same body size. Gibbons provide a good example of life-long monogamy.

The other Asian ape, the orangutan, is much larger than the gibbon and pursues a very different lifeway, although it is also highly arboreal. The core of its social organization is a single mature female and her dependent offspring. The mother and offspring occupy a fairly well-defined home range, which usually overlaps with that of one or more other mature females and their offspring. In contrast, males are rather solitary creatures, with each occupying a large territory that usually contains the home ranges of several mature females with whom he will mate. Males, which are about twice the size of females, actively defend their territories against incursion by other males. The mating system is therefore one of a loosely organized harem, with one male mating with several females (technically known as unimale polygyny).

Gorillas, the largest of the apes, live in the forests of central and west Africa. These animals follow a mating system similar to that of the orangutan—unimale polygyny—although their ecology and organization are distinctly different. Predominantly terrestrial animals that live on low-quality herbage found in abundant but widely dispersed patches, gorillas live in close groups composed of from 2 to 20 individuals. The adult male—the silverback—has sole mating access to the mature females, whose immature offspring also live in the group. Mature males compete for control of the group. Nevertheless, a female, usually a young adult, will sometimes transfer from one group to another, seemingly as a matter of free choice. New groups are established when a lone silverback begins to attract transferring females. As with orangutans, male gorillas are twice the size of females.

Chimpanzees, which are terrestrial and arboreal omnivores, live in rather loose communities composed of between 15 and 80 individuals, representing a mixture of mature males and mature females and their offspring. Unlike savannah baboons, which live in close, cohesive troops of mature males, related females, and their offspring, sometimes numbering 200 individuals in total, chimpanzee communities are maintained by occasional contact between males and females. The core of chimpanzee social life is a female with her offspring; these units are often found by themselves but sometimes link up with other females and their offspring. Each female maintains a core area, which usually overlaps with that of one or more other females. By contrast with orangutans, single chimpanzee males do not maintain exclusive control of a group of female home ranges. Instead, a group of males defends the community range against the males of neigh-

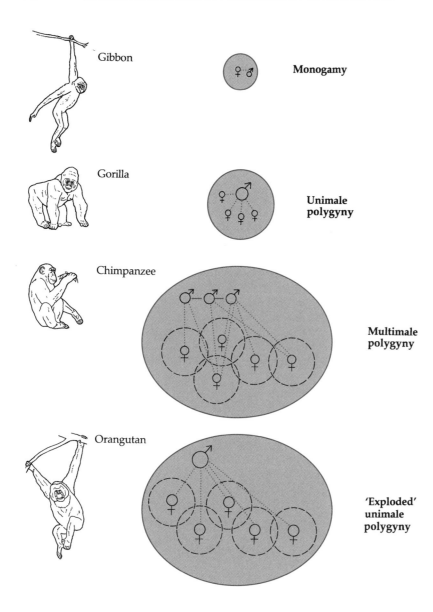

Gibbon

Monogamy

Gorilla

Unimale polygyny

Chimpanzee

Multimale polygyny

Orangutan

'Exploded' unimale polygyny

HOMINOID SOCIAL ORGANIZATION:
The range of social organizations among the apes matches that found among anthropoids as a whole. Gibbons are monogamous, with no size difference between males and females. In gorillas, a single male exerts control over a group of females (and their offspring); this system is known as unimale polygyny. Single male orangutans also defend a group of females (and their offspring), but the females are distributed over a large area; this organization is sometimes known as exploded polygyny. In chimpanzees, several related males cooperate to defend a group of widely distributed females (and their offspring); this system provides an example of multimale polygyny.

boring communities. Mating in chimpanzee communities is promiscuous, with each estrus female copulating with several males. The social organization is therefore known as multimale polygyny.

A key feature of chimpanzee social organization is that, unlike in the general pattern of multimale societies among primates, males remain in their natal group while young adult females transfer (or are sometimes kidnapped) to other communities. As a result, the adult males who are cooperating to defend their community are usually closely related to one another. Adult male chimpanzees are typically 25 to 30 percent larger than females.

Among the apes, then, one finds monogamy, unimale polygyny, and multimale polygyny. (Polyandry—one female having exclusive access to several mature males—which is rare in mammals generally, is absent here.) This spectrum of social organization raises questions about

several aspects of group living. For example, how big will a social group be? What is the ratio of adult females to adult males? Among which sex is there the greater degree of relatedness? What difference arises in the size of males and females?

CAUSES OF SOCIAL ORGANIZATION

The fact that such a rich array of social organizations exists among primates as a whole, and among the apes in particular, surely indicates that a rather complex set of processes underlies them. For each species, some kind of interaction must take place between its basic phylogenetic heritage—its anatomy and physiology—and key factors in the environment. Thus, different species will probably react differently to the same environmental factors, creating at least part of the observed diversity. What else plays a part?

"There is no consensus as to how primate social organization evolves," Richard Wrangham of the University of Michigan observed, "but a variety of reasons suggest that ecological pressures bear the principal responsibility for species differences in social behavior." Indeed, for more than two decades ecological influences have been a popular source of explanations. As Wrangham explains, the problem is that "we do not know exactly what the relevant ecological pressures are, or which aspects of social life they most directly affect, or how."

One of the most frequently advanced explanations of the benefits of group living has been defense against predation. Even though it may be more conspicuous than a lone individual, a group can be more vigilant (more pairs of eyes and ears) and more challenging (more sets of teeth). Effective defense against predators has been observed in many group-living species of primate.

It is certainly true that terrestrial species, which face greater risk from predators than arboreal animals, live in larger groups and commonly include more males in the group; in addition, the males in such species frequently are equipped with large, dangerous canine teeth. For each of these factors, however, one can advance equally plausible explanations of their origin that have nothing to do with protection against predation. So, it is possible that terrestrial primates evolved these characteristics for these other reasons; once evolved, the properties proved highly effective in mitigating the threat of predation. Protection against predation may to some degree be a consequence, not the primary cause, of group living.

Food distribution has also been suggested as a trigger of social organization. Groups might be more efficient than individuals at discovering discrete patches of food, for instance, or, where food patches are defensible by territorial species, the patch size will then influence the optimum group size. Wrangham has proposed a theory of social organization that includes food distribution as a key influence, but the focus of this model differs from that of earlier ideas.

Wrangham's model examines the evolutionary context of male and female behavior, and proposes that "it is selection pressures on female behavior which ultimately determine the effect of ecological variables on social systems." In other words, whatever ecological setting a species might occupy, the behavior of females is fundamental to the social system that evolves within it.

The reproductive success of female primates, as with all mammals, is determined by the number of offspring she can successfully raise. Access to mature males is not usually a limiting factor, whereas access to food resources most certainly is. Male primates, in company with 97 percent of all male mammals, bestow no parental care on their offspring. As a result, their reproductive success is determined by successful access to mature females.

In the great majority of primate societies, females remain in their natal group while males transfer. Any explanation of why primates should form social groups at

DIFFERENT REPRODUCTIVE STRATEGIES: For a female primate, the variable that determines ultimate reproductive success is access to food resources. By contrast, a male's reproductive success is limited by his access to mature females. This difference critically influences the overall social structure of primate societies.

all must also explain this asymmetry. Attempts to correlate different types of habitat with the tendency to form different types of social groups fail to satisfy this criterion. Wrangham's model does offer an explanation, as follows.

If food generally comes in patches that can support only one female and her offspring, then females will forage alone, as orangutan and chimpanzee females do much of the time. Food that comes in larger, defensible patches can, however, support several mature females and their offspring. Sharing a food resource also brings an element of competition into the group, which leads to loss of time and energy through aggressive encounters. Wrangham suggests that the costs of competition within a group are balanced against the benefits of cooperating with group members to outcompete other groups for access to food patches. Cooperation is most beneficial when it occurs

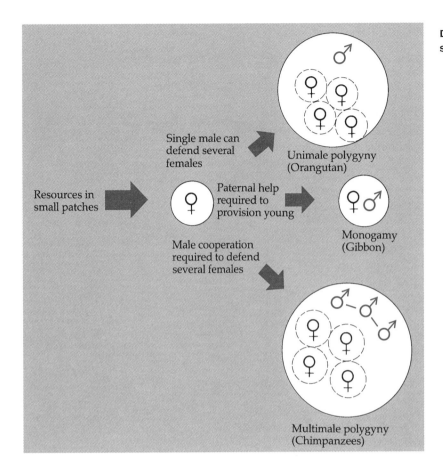

Distribution, with small resource patches: When food exists in patches too small to support more than one mature individual, females will forage singly (with their offspring). If a male can defend a "community" of lone females, unimale polygyny will result, as observed with the orangutan. If a male can defend only one female, or if paternal help is required in raising offspring, monogamy will result, as found with the gibbon. If a community of females can be defended only by several males, then a group of related males will defend a number of unrelated females, as observed in chimpanzees.

among relatives: helping kin is like helping yourself, because they share your genes.

Thus, when a species exploits food resources that come in discrete, defensible patches, multifemale social groups will evolve in which the females are closely related to one another. In anthropology, such groups are known as matrilocal; with nonhuman primates, a better term is female-bonded. Where do the males fit in? If patches of food resources are relatively densely distributed, allowing a group of females to defend them all and exercise territoriality, extra males are somewhat extraneous and a unimale social system usually forms. If, however, territoriality is not possible and increased group size does not create major problems, several adult males can be accommodated. Indeed, extra males can prove useful in the occasional competitive encounters with other groups. In such a situation, some kind of multimale system would form.

In non–female-bonded systems, such as the chimpanzee and orangutan, where food does not come in defensible patches and females are mostly alone, the distribution of males depends on whether they can defend a community range alone or need the cooperation of other males. For orangutans, community defense by a single male is feasible, but for chimpanzees, cooperation is essential. Again, cooperation is most effective among relatives.

Thus, chimpanzees have evolved a multimale social system in which females, not males, transfer to other groups on reaching maturity.

Consequences of social organization

Given these underlying influences, says Wrangham, several predictions can be made in terms of behaviors within and between groups. For instance, intense social interaction—grooming and so on—is expected among females in female-bonded groups, but is less frequent in non–female-bonded groups. Aggression within female-bonded groups should arise over access to food resources, and females should play a very active role in the encounters. By contrast, aggression within non–female-bonded groups should relate to access to females, and males should be the principal aggressors. These predictions appear to have some support.

Another possible consequence of primate sociality is body size—specifically, the difference between males and females, known as sexual dimorphism in body size. Male primates often must compete with other males for access to breeding females, and the bigger their body size, the more likely they are to succeed. Natural selection in species in which such male-male competition occurs is likely to lead to increased male body size. Other factors that

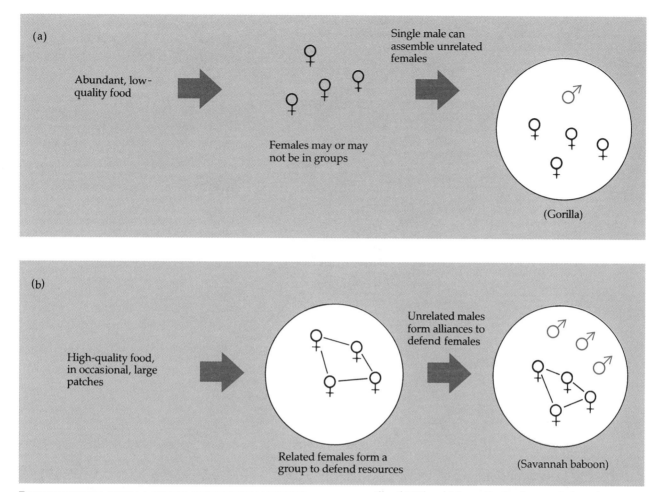

DISTRIBUTION, WITH LARGER RESOURCE PATCHES:
(a) When low-quality food is widely distributed, females may forage alone or in groups (in which the individuals are unrelated). A male may be able to assemble a harem, as does the gorilla. (b) When high-quality food occurs in large but scarce patches, related females will form a group to defend them. Alliances among unrelated males may form to defend the females from other males, as in savannah baboons.

might be important in such encounters—canine teeth, for example—may also become exaggerated in males.

In monogamous species, in which competition between males is low or absent, males and females are typically the same size. In addition, all species characterized by significant sexual dimorphism exhibit some degree of polygyny. Enlarged canines are also found in polygynous species. The equation is not simple, however, because no direct correlation exists between the degree of polygyny and the degree of body size dimorphism. Species in which males typically control harems of, for example, 10 females do not necessarily display greater body size dimorphism than species in which males control harems of two females.

Although the notion that body size dimorphism represents the outcome of competition among males for access to females is popular among biologists, other explanations are also possible. The simplest is that males are large and aggressively equipped so as to provide effective protection against predators. Once again, the problem of circularity arises here. Another suggestion is that males and females assume different sizes as a way of exploiting different food resources, thus avoiding direct resource competition.

Robert Martin of the Anthropological Institute in Zurich adds an important note of caution to this discussion, noting that perhaps our explanations have been too male-oriented in seeking to explain why the male size has increased. Instead, he suggests, perhaps the size difference reflects that the females have become smaller. "Smaller females may breed earlier," he notes, "selection for earlier breeding might explain the development of sexual dimorphism in at least some mammalian species."

Even though many aspects of the interaction of species and their different environments remain to be fully worked out, one thing is clear: the complete social behavior of a species is the outcome of a mix of causes and consequences of individuals coming together to coexist in groups. ✻

Polygynous social system
Dimorphic canines

Monogamous social system
Monomorphic canines

SEXUAL DIMORPHISM, TEETH, AND BODIES: In polygynous social systems, the males are typically larger than females, both in terms of body size and canine teeth, as illustrated here for baboons. By contrast, in monogamous species, body size and canine size are usually very similar between the sexes, as illustrated here for gibbons. (Courtesy of John Fleagle/Academic Press.)

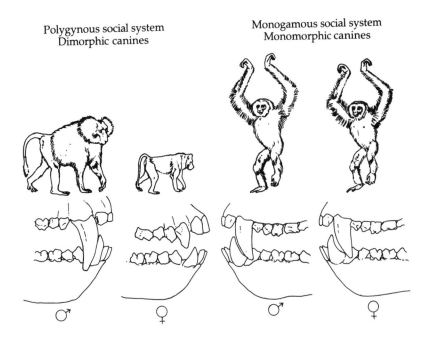

KEY QUESTIONS

- Under what circumstances would you expect an animal to live in a social group?
- Under what circumstances might male primates find themselves forced to contribute to the raising of offspring?
- If cooperating with one's kin indirectly benefits one's own genes, what advantages might be provided by cooperating with non-kin in a social group?
- In considering fossil species—such as early hominines—what anatomical clues might be available about the social system of the living animals?

KEY REFERENCES

Di Fiore A, Rendall D. Evolution of social organization. *Proc Natl Acad Sci USA* 1994;91:9941–9945.

Foley RA. Evolution and adaptive significance of hominid maternal behavior. In: Pryce CR, *et al.* Motherhood in human and nonhuman primates. Basel: Karger, 1994:27–36.

Foley RA, Lee PC. Finite social space, evolutionary pathways, and reconsidering hominid behavior. *Science* 1989;243:901–906.

Harvey PH, Bennett PM. Sexual dimorphism and reproductive strategies. In: Ghesquiere J, Martin RD, Newcombe F, eds. Human sexual dimorphism. London: Taylor and Francis, 1985.

Lee PC. Social structure and evolution. In: Slater PJB, Halliday TR, eds. Behavior and evolution. Cambridge: Cambridge University Press, 1994:266–303.

Leigh SR. Socioecology and the ontogeny of sexual dimorphism in anthropoid primates. *Am J Physical Anthropol* 1995;97:339–356.

Martin RD, *et al.* The evolution of sexual dimorphism in primates. In: Short RV, Balaban E, eds. The difference between the sexes. Cambridge: Cambridge University Press, 1994:159–200.

Rendall D, Di Fiore A. The road less travelled: phylogenetic perspectives in primatology. *Evol Anthropol* 1995;4:43–52.

Rodseth L, *et al.* The human community as a primate society. *Curr Anthropol* 1991;32:221–254.

Smith EA. Human behavioral ecology: II. *Evol Anthropol* 1992;1:56–61.

Wrangham RW. Evolution of social structure. In: Smuts BB, *et al.*, eds. Primate societies. Chicago: Chicago University Press, 1987.

NONHUMAN MODELS OF EARLY HOMININES

14

The behavioral ecology of living primates can give insight into behavior of our forebears. For instance, one can ask questions about the social organization of the last common ancestor between hominines and apes and of the early hominines themselves. Did they live in groups, and, if so, what was their size? What was the ratio of mature males to mature females? It is also important to remember that "early hominines" include at least half a dozen species. If the experience of observing the behavior of modern ape species teaches us a lesson, it is that we can expect different forms of social organization among different hominine species.

THREE APPROACHES TO MODELS

There are several ways in which modern primates can be used to model the lives of the extinct species. First, one can identify a living species that appears to match some basic hominine characteristics and then seek lessons about behavior that might transfer to hominines. Second, guided by phylogeny, one can consider only the living African apes and humans and seek commonalities of behavior that might therefore have been present in a common ape/hominine ancestor. Third, now that an understanding of behavioral ecology is beginning to develop (see unit 13), one can infer from basic principles the social organization of hominine forebears.

PRIMATE MODELS

The first of these three models—the specific primate model—is the longest-established approach. Several different species have been offered as *the* most appropriate model at certain times, including the savannah baboon, the common chimpanzee, and, most recently, the pygmy chimpanzee.

Although the baboon is a monkey, not an ape, and is therefore genetically related to hominines only rather distantly, baboons are attractive as models for some early hominines, because they share a similar habitat: bushland savanna. (The very earliest hominines appear to have lived in forest environments.) Living in troops with as many as 200 individuals, the savannah baboon offers a striking picture of the social life of our forebears. A troop consists of mature females (often related to one another) and their offspring, and many mature males (unrelated to one another). The males are larger than the females and are equipped with impressively threatening canines. In other words, baboons operate within a multimale, female-bonded social organization.

So powerful was this image, and so well studied were these animals, that Shirley Strum, a baboon-watcher herself, observed that "the 'baboon model' had a disproportionate impact on our ideas about primates."

The chimpanzee has also been proposed as a model for the last common ancestor and the early hominines, and for good reason: it is our closest genetic relative. One problem with the chimpanzee, as with all specific models, is the trap of the present: just as extinct species are likely to be unique anatomically and not represent some slight variant of a living species, so the behavior of extinct species is also likely to be unique. When, for instance, a chimpanzee model is proffered, "an ape-human dichotomy is created," says Richard Potts, an anthropologist at the Smithsonian Institution (Washington, D.C.). "The problem with placing early hominids along a chimp-human continuum is that it precludes considering unique adaptations off that continuum."

Potts points out that the dentition of the early hominine genus *Australopithecus*—large, thickly capped cheek teeth set in robust jaws—resembles that of neither chimpanzees nor humans. "Thus, in this aspect of dental anatomy *Australopithecus* did not fall on the proposed continuum," notes Potts.

The most recent entry into the primate model stakes is the pygmy chimpanzee, or bonobo, which was proposed in 1978 by Adrienne Zihlman, John Cronin, Vincent Sarich, and Douglas Cramer. Randall Susman, of New York University at Stony Brook, is also a proponent of the model. The pygmy chimpanzee, which is now found only in a small area in Zaire, is strikingly similar in overall body proportions to the early hominine species, *Australopithecus afarensis*. One characteristic of the bonobo is that its society is female-centered and egalitarian, in which sex is a substitute for aggression. Bonobos engage in sex in every possible partner combination, often face-to-face, usually as a way of reducing tension in the group. This species, *Pan paniscus,* may therefore represent an even better model of early hominines than the common chimp, *Pan troglodytes,* not least in its partial separation of sex from reproduction, as happens in our own species. Nevertheless, one must always remember that even closely related species may exhibit distinctly different social structures when they occupy different habitats.

In addition to primate models for hominine forebears, studying social carnivores has occasionally been said to be instructive. The case here is based upon strict analogy with a supposed behavior: cooperative hunting. As cooperative hunting among hominines may have been a rather late evolutionary development (see unit 26), this model may have limited utility.

The second approach—phylogenetic comparisons—is considerably more conservative, seeking only to identify basic shared behavioral characteristics among humans and African apes. The rationale, as explained recently by Richard Wrangham of the University of Michigan, is as follows: "If [a behavior] occurs in all four species, it is likely (though not certain) to have occurred in the common ancestor because otherwise it must have evolved independently at least twice. If the four species differ with

A CATALOG OF CANDIDATES: Several different species have been nominated as instructive models for early hominine evolution. Here we see the pygmy chimpanzee (*top left*), the common chimpanzee (*top right*), the savannah baboon, and the lion (a social carnivore).

respect to a particular behavior, nothing can be said about the common ancestor."

PHYLOGENETIC MODELS

Wrangham examined 14 different behavioral traits—such as social group structure, male-female interactions,

intergroup aggression, and so on. He found eight traits to be common to gorillas, the two chimpanzee species, and humans; six traits were not shared. On this basis Wrangham infers that the common ancestor of hominines and African apes "had closed social networks, hostile male-dominated intergroup relationships with stalk-and-attack interactions, female exogamy and no alliance bonds between females, and males having sexual relationships with more than one female."

This ancestral suite, as Wrangham calls it, is merely a foundation upon which past social behavior can be constructed. But, for instance, it does seem to preclude the suggestion made in 1981 by Owen Lovejoy that the then earliest known hominine, *Australopithecus afarensis,* was monogamous and nonhostile.

BEHAVIORAL ECOLOGY MODELS

The third approach—reconstructing social organization from first principles of behavioral ecology—is the newest and most promising. Developed most thoroughly by Robert Foley, the technique seeks to establish the range of social structures that might have been available to hominine ancestors, and then determine how these structures might be altered in the face of changing environments.

The basis of the analysis is the recognition of phylogenetic constraints in ecological context. Just as ancestral anatomy limits the paths of subsequent evolution, so too does ancestral social structure. For instance, evolving from a multimale, non–female-bonded organization to a multimale, female-bonded structure is highly unlikely, because the intermediate steps would be inappropriate under prevailing conditions. In other words, only certain evolutionary pathways are available for ecologically driven shifts in social organization. Thus, if you know where an ancestral species began among the many possible social structures, you can predict the nature of ecologically driven social change, because you know the available pathways.

The phylogenetic context for hominines is, of course, the apes—particularly the African apes. The social structures found among the apes vary greatly, ranging from solitary individuals among orangutans, through monogamous families in gibbons, to single-male units with small numbers of unrelated females among gorillas, to complex fission-fusion communities of chimpanzees (see unit 13). "Despite this variation, these social structures . . . constitute a limited set of social outcomes and are evolutionarily closely related," observed Foley recently, in conjunction with Phyllis Lee, also of Cambridge University. In marked contrast with common monkey social organization, none of the ape social structures involves a core of related females.

Foley and Lee suggest that the most likely social structure in species ancestral to African apes and hominines is relatively gorilla-like. Toward the end of the Miocene, approximately 10 million years ago, a steadily cooling climate was reducing forest cover. A drier, more diverse habitat developed, especially in East Africa, which created

TABLE 14.1   Ancestral social organization: Using different models, it is possible to determine those aspects of behavior that might have appeared in an ancestral species. In the phylogenetic comparison, each of the 14 questions asks if a particular aspect of behavior exists in *all* modern African apes. If it does, then this same behavior likely also appeared in the common ancestor with hominines. (Courtesy of Richard Wrangham.)

Method	Phylogenetic comparison	Chimpanzee model	Chimpanzee model	Behavioral ecology	Behavioral ecology
Species	Common ancestor	A late prehominine	The earliest hominine	The earliest hominine	An early hominine
1 Closed social network	Yes	—	Yes	Yes	—
2 Party composition	?	Unstable	Unstable	Stable	Unstable
3 Females sometimes alone	?	Yes	Yes	No	Yes
4 Males sometimes alone	Yes	Yes	Yes	No	Yes
5 Female exogamy	Yes	—	Yes	No	—
6 Female alliances	No	—	No	Yes	—
7 Male endogamy	?	Often	Yes	No	Yes
8 Male alliances	?	—	Yes	—	—
9 Males have single mates	No	No	No	No	Yes
10 Length of sexual relationships	?	Short	Short	Short	Long
11 Hostile relations between groups	Yes	No	Yes	—	—
12 Males active in intergroup encounters	Yes	—	Yes	—	—
13 Stalking and attacking	Yes	—	—	—	—
14 Territorial defense	?	—	?	—	—

a patchy distribution of food resources. Such an ecological shift would favor the evolution of a chimp-like social structure: communities of dispersed females and their offspring, with genetically related males defending the community against males from other groups.

"The emergence of the hominids can be seen as part of the African hominoid radiation, with this clade exhibiting increasingly strong male kin alliances under certain ecological conditions," say Foley and Lee. As long as the cooling persisted, the ecological shift would continue. "Such environments appear to promote, among other things, larger group size among primates—partly as a response to the greater threat of predation, partly due to the effects of resources being more patchily distributed." Given the evolutionary pathways available under the model, the larger social groups "are more likely to be built upon the male kin alliances rather than related females."

Given this background, Foley and Lee suggest that "the most probable social organization for the early australopithecines . . . consists of mixed sex groups, with males linked by a network of kinship. Females, forced to forage over larger areas to find dispersed and seasonally limited food and to aggregate in the face of some predation, would be expected to form stable associations with either specific males within the alliance, or with the entire alliance of males."

This social package, although apparently consistent with prevailing ecological conditions, is at odds with the extensive fossil record of one of the earliest known hominines (at least in the way it is currently interpreted). This hominine, *Australopithecus afarensis*, is said to exhibit a degree of body size dimorphism as extreme as that in orangutans. Dimorphism on this scale is usually taken to imply extreme competition between males for access to females. If, as Foley and Lee suggest, the males in these social groups are closely related to one another, then cooperation rather than fierce competition should emerge. Thus, either the model is incomplete, or the fossils currently interpreted as *Australopithecus afarensis* represent two less dimorphic species (see unit 19).

Within the hominine species of 3 to 1 million years ago there developed a degree of morphological diversity, presumably reflecting adaptation to different patterns of subsistence. At one extreme, the robust australopithecines apparently exploited a diet of coarse, low-quality plant

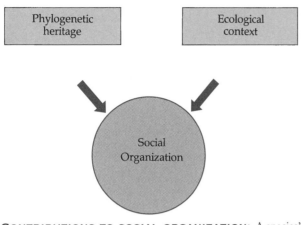

CONTRIBUTIONS TO SOCIAL ORGANIZATION: A species' social structure will be determined by the outcome of interaction between its phylogenetic heritage—body size, and so on—and the environment in which it lives. Species with different phylogenetic constraints may therefore exhibit different social structures under the same environmental conditions.

foods. Such foods tend to occur in large, widely dispersed patches. "The expected effect on the ancestral hominine socioecology . . . would have been to weaken male kin bonds within a less structured large or fluid group," note Foley and Lee. More competition between males would develop, presumably accompanied by dimorphism in body size.

At the other extreme, *Homo erectus/ergaster* produced adaptations including increased brain size and much reduced dental apparatus. Faced with the same problem of subsisting in tropical savannah environments—that is, maintenance of a constant food supply in the face of seasonality—this hominine species adopted a strategy different from that of the robust australopithecines. Instead of exploiting low-quality food resources, its members increased their consumption of meat, a patchily distributed but high-quality resource.

"The causes of meat-eating are ecological," note Foley and Lee. "The consequences for the hominines would have been distributional and social." One consequence would be greatly increased home and day ranges, which would complicate direct defense of females by males. "The predicted response would be resource defense through territorial exclusion, and given the size of the area involved, this would involve alliances of males rather than individual defense. Such a pattern of behavior would enhance male kin associations as a means of coping with high levels of intergroup competition and interactions."

In addition to changes wrought by this subsistence strategy, *Homo* would face another key change: the consequences of brain enlargement. Producing and rearing large-brained offspring are energetically expensive. At some point it would have become too expensive for the mother to provide for the offspring by herself, necessitating paternal involvement. "The effect . . . would be to increase the frequency, intensity, and stability of male-female associations," say Foley and Lee. Is this point the beginning of the nuclear family, so much a part of Western society? No, because the nuclear family is actually rather uncommon among human societies; an analysis of social structure variation among modern human societies shows that 74 percent are polygynous.

The 20 percent body size dimorphism in modern humans would indicate a degree of male-male competition in our recent past, not monogamy. And the fact that more resources are devoted to male fetuses than female fetuses, thus giving them a higher birth weight, is also consistent with male-male competition. One further factor is the size of the male's testis, an indicator of subtle competition among males in multimale groups. For instance, chimpanzees live in promiscuous, multimale groups. One way that an individual male might outcompete his fellows is to produce more sperm in his ejaculate. Gorillas and orangutans do not face this kind of competition, and consequently they have small testes.

What of *Homo sapiens*? Human testes are small as well, apparently ruling out competition in promiscuous, multimale groups. Monogamy also appears to be eliminated, leaving a form of unimale polygyny. But, as Robert Martin and Robert May commented recently, "these biological antecedents are today often overlain by extremely powerful socioeconomic determinants." ✽

KEY QUESTIONS

- Why are extinct species likely to have displayed unique behaviors?
- Why are "first principle" approaches to inferring ancestral social behavior likely to be more difficult, yet more informative than single-species models?
- How important are phylogenetic constraints likely to be in the face of sharply changing environmental conditions?
- What are the consequences of a male primate undertaking parental care through provisioning?

KEY REFERENCES

Boesch C. Hunting strategies of Gombe and Tai chimpanzees. In: McGrew WC, *et al.*, eds. Chimpanzee cultures. Cambridge, MA: Harvard University Press, 1994:77–91.

Foley RA, Lee PC. Finite social space, evolutionary

pathways and reconsidering hominid behavior. *Science* 1989;243:901–906.

Nishida T, *et al.* Meat-sharing as a coalition strategy by an alpha male chimpanzee. In: Nishida T, *et al.*, eds. Topics in primatology: human origins. Tokyo: University of Tokyo Press, 1992:159–174.

Potts R. Reconstructions of early hominid socioecology: a critique of primate models. In: Kinzey WG, ed. The evolution of human behavior: primate models. Stony Brook, NY: SUNY Press, 1987:28–50.

Rodseth L, *et al.* The human community as a primate society. *Curr Anthropol* 1991;32:221–241.

Stanford CB. Chimpanzee hunting behavior. *Am Scientist* 1995;83:256–261.

de Waal FBM. Bonobo sex and society. *Sci Am* March 1995:82–88.

Wrangham RW. The significance of African apes for reconstructing human social evolution. In: Kinzey WG, ed. The evolution of human behavior: primate models. Stony Brook, NY: SUNY Press, 1987: 51–71.

Zihlman AL, *et al.* Pygmy chimpanzee as a possible prototype for the common ancestor of humans, chimpanzees and gorilla. *Nature* 1978;275: 744–746.

PART 4

HOMININE BEGINNINGS

Ape and Human Relations: Morphological and Molecular Views

15

The superfamily **Hominoidea** (colloquially, hominoids) includes all living and extinct ape and human (hominine) species. This unit will address the relationships among living hominoids and their formal classification, the timing of the evolutionary divergence between the human and ape lineages, and the probable anatomical characteristics of the ancestor of humans common to both humans and apes. Unit 16 will examine our knowledge of extinct ape species and their possible relationship to living hominoids.

MORPHOLOGICAL INTERPRETATIONS

Since the time of Darwin and Huxley, anthropologists have recognized that humans' closest relatives are the African great apes, the chimpanzee and gorilla, with the Asian great ape, the orangutan, more distant. This conclusion is based principally on comparative anatomy of the hominoids. A continuing question is the evolutionary relationship between humans, chimpanzees, and gorillas.

For instance, the African apes share many anatomical similarities, particularly in their forelimbs, which show adaptations to their knuckle-walking mode of locomotion, and in their dentition, which has a thin layer of enamel on the cheek teeth. Modern humans and (most of) their extinct relatives have thick enamel (but see unit 19 for a qualification), as do many fossil apes. In several cladistic analyses of living hominoids (by, for example, Lawrence Martin of the State University of New York at Stony Brook and Peter Andrews of the Natural History Museum, London), the shared limb anatomy and dental features of African apes were judged to be derived characters that linked chimpanzees and gorillas as a separate

clade from humans. Under this scheme, humans were seen as having diverged first from the hominoid lineage, with gorillas and chimpanzees sharing a common ancestor in which knuckle-walking and thin tooth enamel evolved. A second scheme—a trichotomy in which African apes and humans diverged simultaneously from a common ancestor—was also said to be possible, though less likely.

The Martin/Andrews view of human/African ape affinity won wide support, although different views were expressed as well. For instance, one cladistic analysis grouped the orangutan with the African apes in a clade separate from humans, while another identified an African ape clade and a human/orangutan clade. In this plethora of morphological analyses, only one, published in 1986 by the Australian anatomist Colin Groves, concluded (weakly) that humans and chimpanzees are one another's closest relatives; this assessment was based on forelimb anatomy, particularly the wrist. That is, gorillas were suggested to have diverged first from the hominoid ancestor, with humans and chimpanzees sharing a common ancestor from which they later diverged. (A later, more detailed study, reached the same conclusion.) As we shall see, this counterintuitive view was also emerging from molecular studies of the time, and it became ever more strongly supported throughout the following decade.

Morphologists resisted this latter interpretation, because the many anatomical similarities between gorillas and chimpanzees were assumed to be shared derived characters. If the human/chimpanzee association was indeed correct, then morphologists faced awkward puzzles. For instance, the many striking anatomical similarities of gorillas and chimpanzees must be explained either as homoplasies (independent, parallel evolution), which seems unlikely, or as shared primitive characters that were present in the common ancestor of apes and humans (see below). Furthermore, why have the homologous features that reveal the human/chimpanzee link been so hard to find? Groves was the lone voice in identifying any at all. Recently, however, analyses of fossil and living hominoids have added further evidence related to this point.

For instance, David Begun, of the University of Toronto, compared cranial and dental features in the Miocene ape *Dryopithecus*, an early member of the hominine clade, and living hominoids. He concluded that

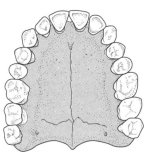

Ramapithecus   Human

***RAMAPITHECUS* RECONSTRUCTED:** In the original reconstruction of the two fragments of upper jaw (maxilla) of Lewis's *Ramapithecus* specimen, the shape appeared to be humanlike. This partly explains why the Miocene ape was thought to be an early hominine. The reconstruction was inaccurate, in part because of missing portions of the specimen.

Adaptations to bipedal locomotion

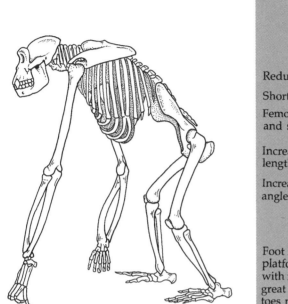

Head held vertically

Large bulbous cranium

Short face

Reduced anterior dentition, small canines, large cheek teeth covered with thick enamel

Reduced lower back

Short, wide pelvis

Femoral head angled and strengthened

Increased hindlimb length

Increased valgus angle of knee

Shortened forelimb

Hand with enlarged thumb, enhanced fingertip sensitivity, non-curved fingers (a manipulative, rather than a locomotor, structure)

Foot forms a platform structure with non-opposable great toe. Lateral toes not curved.

APE AND HUMAN ANATOMY: The ape (*left*) is adapted to a form of quadrupedalism known as knuckle-walking, which is seen only in chimpanzees and gorillas. Rather than support the forelimb on the palm of the hand (like most primates) or the palmar surface of the fingers (like baboons), the African apes support it on the dorsal surface of the third and fourth digits of their curled hands. The wrist and elbow anatomy is adapted so as to "lock," thus providing a firm support for the body weight. Human bipedalism (*right*) involves a number of anatomical differences from that seen in quadrupedalism, as indicated. Anthropologists are divided over whether the common ancestor of humans and African apes was a knuckle-walker.

many characters in gorillas once considered to be derived are actually primitive, and that humans, chimpanzees, and australopithecines share several characters that are derived for the group as a whole. This finding links humans and chimpanzees as one another's closest relatives.

MOLECULAR STUDIES

The term "molecular anthropology" was coined in 1962 by Emile Zuckerkandl, who, with Linus Pauling, invented the notion of using molecular evidence to uncover evolutionary histories (see unit 8). At the time, Zuckerkandl had already discerned a hint of what was to unfold in the science when he compared enzymic digests of proteins from humans, gorillas, chimpanzees, and orangutans. As mentioned earlier (see units 3), first Morris Goodman and then Allan Wilson and Vincent Sarich actually went on to establish the new field of research. They used immunological reactions of certain blood proteins to measure genetic distances among the living hominoids. In the early 1960s, Goodman established the human/African ape affinity, while in the late 1960s Wilson and Sarich used the genetic distances to identify times of divergence between the ape and human lineages.

As with all such calculations, Wilson and Sarich calibrated their molecular clock using known (or assumed) divergence times derived from the fossil record. They applied the then-accepted divergence time of Old World monkeys (superfamily Cercopithecoidea) and Hominoidea of 30 million years ago. According to their research, the genetic distance between humans and African apes was one-sixth of that between living African hominoids and Old World monkeys. This finding implied that African apes and humans diverged 5 million years ago (one-sixth of the 30 million years that anthropologists believed to be the case, based on fossil evidence, namely *Ramapithecus*).

In the three decades since this first calculation of human/ape divergence based on molecular data, many different techniques have been applied to the problem, including electrophoresis of proteins, amino acid sequencing of proteins, restriction enzyme mapping of various types of DNA, sequencing of mitochondrial and nuclear DNA, and DNA-DNA hybridization. Although their re-

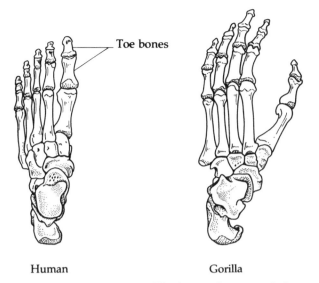

ANATOMY OF THE FEET: The human foot is a platform, built for bipedalism, while the gorilla foot is more of a grasping organ. A key difference, therefore, is in the relationship of the great toe to the other toes of the foot. In humans, the great toe is parallel with the other toes; in apes, it is opposable.

sults are by no means unanimous, the great majority of these techniques have supported the human/African ape linkage and have yielded a divergence time near 5 million years, and probably not greater than 6 million. This finding is in good accord with the known fossil record (see unit 19).

Much controversy surrounded this work, and not all disagreements pitted molecular biologists against morphologists. For instance, considerable debate surrounded the issue of the rate at which genetic change in the hominoid lineages accumulated. Supporters of the molecular clock (such as Wilson) argued that the rate was constant and universal. Others (such as Goodman) believed that accumulation rates could change over time and in different lineages. Indeed, Goodman initially attributed some of the surprisingly small genetic distance between humans and African apes to a slowdown in the clock. A slowdown could, of course, affect calculations of divergence times: a small genetic distance might disguise a long evolutionary separation. By now, fluctuations in the clock's rate in general have been accepted, and a slowdown among hominoids in particular. Nevertheless, as long as such fluctuations are taken into account, it remains possible to use genetic data for calculations of divergence times via local clocks (see unit 8). For instance, using extensive DNA sequences of certain globin genes, Goodman (previously a critic of the clock) and his colleagues recently calculated the human/chimpanzee divergence as 5.9 million years.

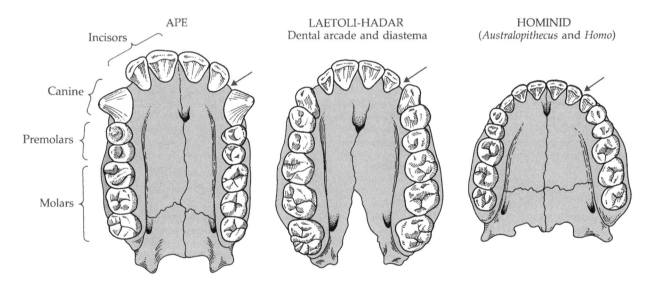

PALATE AND TOOTH ANATOMY: In apes, the jaw is U-shaped; in modern humans and later extinct hominines, it is parabolic. The jaws of early hominines such as *Australopithecus afarensis* are somewhat intermediate in shape. Ape incisors are large and spatulate; a gap, the diastema, separates the second incisors from the large canine; the premolars and molars have high cusps. In humans, the incisors are small; no diastema appears; the canines are small; the premolars and molars have low cusps. In *Australopithecus* species, the incisors are larger than in modern humans, as are the canines; a diastema is sometimes present in early species; the premolars and molars are large with low cusps. The very earliest hominine species are more chimplike in their dentition. (Courtesy of Luba Gudz.)

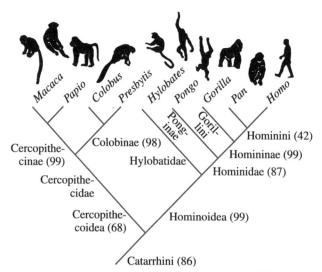

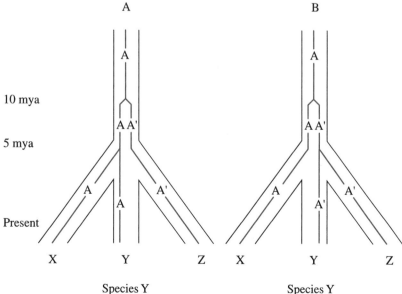

CLADOGRAM OF CATARRHINE RELATIONS: This analysis of 264 morphological characters leads to a chimpanzee/human association as the most parsimonious tree; a tree with a hominoid trichotomy is less parsimonious. This study is one of very few morphological analyses that identifies chimpanzees and humans as one another's closest relatives. (Adapted from Shoshani *et al.*)

During the first two decades of molecular anthropology, the vast majority of work agreed on two things: the reality of a human/African ape affinity and an inability to break the trichotomy. The latter factor implied that either the trichotomy was real or the techniques were not sensitive enough to detect what might be rather short branches in a tree with two divergence points. In the mid-1980s, evidence began to build in favor of a tree with two divergence points: the separation of the gorilla, followed later by a human/chimpanzee split. During the subsequent decade, most molecular data sets of various types supported the same pattern. Cladistic analysis requires specific characters (not genetic distance); in this context, it means gene sequences. Of 10 such independent data sets collected to date, eight support a human/chimpanzee link, two a chimpanzee/gorilla link, and none a human/gorilla link. (Humans are known to share 98.3 percent identity in nuclear, noncoding DNA sequence and more than 99.5 percent identity in nuclear coding sequences, or genes.)

Molecular phylogenetics involves several potentially confounding complications, in particular the gene tree/species tree problem (see unit 8). This can yield a phylogenetic pattern of the sort now heavily supported, even though the evolutionary reality is a simple trichotomy. A thought experiment will illuminate the point.

Imagine that an ancestral species possessed a gene A. Now imagine that a variant of the gene, A′, arose 10 million years ago, making the gene polymorphic. Individuals in the population of the common ancestor may now possess two copies of variant A (that is, homozygous for A), two copies of variant A′ (homozygous for A′), or one copy of each variant (heterozygous). Suppose that 5 million years ago the ancestral species split into three daughter species, X, Y, and Z. In the population that leads to X, the variant A′ is lost, leaving just A. In the population that leads to Z, variant A is lost, leaving just A′. A comparison of the sequences of this gene in species X and Z would indicate that they diverged 10 million years ago, despite the fact the speciation event occurred only 5 million years ago. This erroneous dating, based on conflation of so-called gene trees and species trees, would follow from the gene polymorphism.

GENE TREES VERSUS SPECIES TREES: Gene polymorphism in an ancestral species followed by differential sorting of variants can lead to erroneous conclusions, both in the timing of divergence and the relationship among descendant species. (A) Genetic analysis would make species Y look more closely related to species X than to species Z. (B) Y looks more closely related to Z than X. The reality is a trichotomy. (See text for details.)

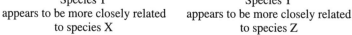

Species Y appears to be more closely related to species X

Species Y appears to be more closely related to species Z

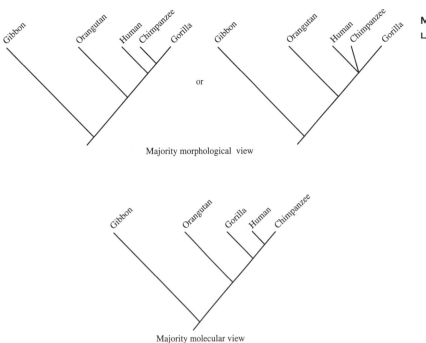

Majority morphological view

Majority molecular view

MORPHOLOGICAL VERSUS MOLECU-
LAR VIEWS: The cladograms show the
current views that most paleoanthropologists take on the two approaches. Most
morphological analyses favor either a
chimpanzee/gorilla clade or a trichotomy.
Most molecular analyses favor a hu-
man/chimpanzee clade.

What about species Y? If its population lost variant A, a comparison of all three species would imply that Y is more closely related to species Z than to species X; similarly, if Y lost variant A', it would appear to be more closely related to species X than to species Z. In fact, all three species are equally related.

As this model indicates, for ancestral species possessing many highly polymorphic genes, no simple, single picture will emerge in a comparison of its descendants' genes. This complexity, suggests Rogers, explains the mixed data for the hominoids, stating that a trichotomy is the most likely pattern.

It is true that the gene tree/species tree problem can lead to an erroneously old divergence date. It is also true that the problem can yield a pattern of two divergences apparently separated in time whereas the reality is a trichotomy. How is hominoid history to be assessed, given the data at hand?

The processes involved are stochastic, in terms of the timing of the origin of polymorphisms and the subsequent sorting of variants. As a result, many data sets are required to test hypotheses. The fact that so many data sets point to a similar divergence time for the inferred human/chimpanzee split provides some confidence in that date, unless all genes just happened to have produced polymorphisms at the same time in the ancestral species prior to speciation—an unlikely event.

The same principle can be applied to the putative two-divergence pattern, as Maryellen Ruvolo, of Harvard University, has argued. Given the stochastic nature of the sorting of variants, there is a one-third probability of genetic data implying a human/chimpanzee alliance and a two-thirds probability of seeing chimpanzee/gorilla or hu-

man/gorilla alliances. Statistically speaking, Ruvolo calculates, the probability of eight human/chimpanzee alliances emerging from ten data sets as a matter of chance is close to one in 3000. In other words, the observed pattern is very likely to reflect history rather than being a statistical quirk.

NATURE OF THE HOMININE ANCESTOR

Fossil evidence of the common ancestor of African apes and humans has yet to be found, not least because the hominoid fossil record in Africa between 4.5 and 8 million years is sparse (almost to the point of being non-existent). One question is, How would such a creature be recognized?

Ancestral anatomy can be inferred, based on comparisons among living and extinct hominoids. As a result, the common ancestor is now widely believed to have been intermediate in size between the gibbon and chimpanzee; it is imagined to have been principally (but not exclusively) arboreal and to have incorporated a significant amount of bipedalism in posture and locomotion, both in trees and on the ground. The ancestor is thought to have lacked the anatomical specialization of the African great apes (such as in the forelimbs and **axial skeleton**—that is, the vertebrae and ribs) that relate to knuckle-walking. The cranium would have been **prognathic** (protruding), as is seen in fossil and living apes. And because the cheek teeth in many fossil apes and (until recently) all known hominines are both large and covered with a thick enamel layer, the common ancestor has been assumed to fit this pattern. African apes, for example, have thin enamel, a presumed shared derived character.

David Pilbeam has recently proposed an alternative hypothesis, one influenced in part by the phylogeny suggested by the molecular data. If humans and chimpanzees are one another's closest relatives, and given that chimpanzees and gorillas share so many anatomical features, the common ancestor is likely to have been rather chimplike, says Pilbeam. (Such a pattern is more parsimonious than one involving parallel evolution of knuckle-walking in separate gorilla and chimpanzee lineages.) This proposed pattern would include a degree of knuckle-walking and thin-enameled teeth. The hominine lineage has lost many of these features, partly through its adaptation to bipedal locomotion and a change in diet. The recent discovery of a 4.5-million-year-old hominine, *Ardipithecus ramidus* (see unit 19), bolsters this view. This species is chimplike in some aspects of its dentition, including possessing thin enamel, and in its postcranial anatomy.

The suggestion of a chimplike ancestor has been resisted in the past and continues to inspire controversy because it would require "reversal" in the direction of evolution, particularly in the configuration of the vertebral column. For instance, African apes have four lumbar vertebrae, early hominines (as seen in two specimens of *Australopithecus africanus* and one *Homo erectus*) have six (presumably as an adaptation to bipedalism), and modern humans have five. An evolutionary progression along these lines would therefore involve an increase from four to six lumbar vertebrae, followed by a decrease to five. Anatomists consider such a progression as evolutionarily difficult, or at least unparsimonious. Pilbeam adduces new insights into the genetics of embryological development to argue that such transitions are actually achieved rather easily. For instance, experimental modification in the timing of expression of certain genes that control development (homeobox genes) in mice readily changes the number of lumbar vertebrae that develop.

If the common ancestor was actually chimplike, discerning the identity of a chimplike fossil from, for example, 6 million years ago would pose significant challenges. Such a specimen might be the common ancestor, but it might also be an early member of the modern chimp lineage.

CLASSIFICATION OF HOMINOIDS

The superfamily Hominoidea has traditionally been divided into three families: Hylobatidae (gibbons and siamangs), Pongidae (orangutan, gorilla, and chimpanzee), and Hominidae (humans). If, as Darwin believed and as modern systematists propose, classification should reflect genealogy, then classification of the Hominidea should be revised. Morris Goodman proposed such a revision in the early 1960s, based on his initial results. At the very least, he said, the Hominidae should include both humans and the African apes. The Hylobatidae family would remain intact, while orangutans would represent the sole occupants of Pongidae.

Goodman's proposal was vigorously resisted when it was introduced, and it continues to spur disagreement, although much less vehemently. One argument against it, promulgated successfully by the influential paleontologist George Gaylord Simpson, notes that placing humans and African apes in the same family obscures the evolutionary changes that have occurred in the past 5 million years or so, in which humans moved to a very different adaptation. A distinct ape **grade** exists, Simpson argued, which differs from the human grade. ("Grade" simply acknowledges similarities of adaption within a group of species.) To change the hominoid classification would not only discard this grade distinction, it is said, but also cause confusion.

The first point can be countered by pointing out that classification based on genealogy represents a more natural system. After all, a grade is an artificial construct of the human mind, having no fundamental biological basis. Classification based on clade, however, reflects evolutionary reality. The second point is correct: there *is* confusion over terminology. Until recently, everyone knew what was meant by a "hominid"—it included living and extinct human species. Now, some see it as including humans and African apes; to others, it means humans, African apes, and orangutans; and to still others, it signifies humans, African apes, orangutans, and Asian lesser apes. The differences depend on the classification preferred (see Table 15.1 for some examples).

These differences arise because, although the same philosophy of classification (that is, genealogy) is followed in all cases, the taxonomic levels chosen to reflect that reality may differ. For instance, in Goodman's classification, the family Hominidae includes humans and all

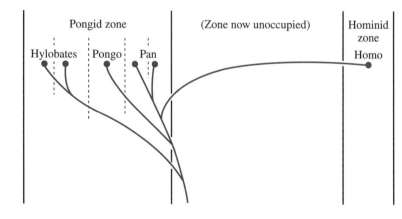

HOMINOID ADAPTATIONS: This diagram by George Gaylord Simpson expresses his rationale for supporting the traditional hominoid classification, in which all the great apes are members of a single family, the Pongidae. During evolution, hominines shifted their adaption to a very non-apelike pattern.

TABLE 15.1   Alternative classifications: *A* shows the traditional classification, now losing favor. *B* is an alternative proposed by Mann and Weiss. *C* is Goodman's most recent classification.

A	*B*	*C*
Superfamily Hominoidea	Superfamily Hominoidea	Superfamily Hominoidea
Family Hylobatidae	Family Hylobatidae	Family Hominidae
Hylobates	*Hylobates*	Subfamily Hylobatinae
Symphalangus	*Symphalangus*	*Hylobates*
Family Pongidae	Family Hominidae	Subfamily Homininae
Pongo	Subfamily Ponginae	Tribe Pongini
Pan	*Pongo*	*Pongo*
Gorilla	Subfamily Homininae	Tribe Hominini
Family Hominidae	Tribe Gorillini	Subtribe Gorillina
Homo	*Gorilla*	*Gorilla*
	Tribe Hominini	Subtribe Hominina
	Subtribe Panina	*Pan*
	Pan	*Homo*
	Subtribe Hominina	
	Homo	

apes; at the lowest level of the classification, the subtribe, gorillas belong to the Gorillina but human and chimpanzee species share the Hominina. Others suggest that this classification is too inclusive, although it does reflect the close evolutionary relationship of humans and chimpanzees. A less inclusive classification would allocate chimpanzees to the subtribe Panina and humans to Hominina.

A still less inclusive classification might give humans, chimpanzees, and gorillas their own subfamilies—Homininae, Paninae, and Gorillinae, respectively. The classification selected affects what we call living and extinct human species. For instance, humans would be called hominines under the latter scheme, which is the chosen route in this book. ✳

KEY QUESTIONS

- A belief that humans are special in the world of nature influenced earlier classifications of the Hominoidea. Does it continue to influence current thought?
- Why are the shared adaptations to knuckle-walking in African apes unlikely to be homoplasies?
- How can local molecular clocks for the Hominoidea be tested for accuracy?
- If, as some believe, the evolutionary tree for the African hominoids remains unresolved, what further data would clarify this uncertainty?

KEY REFERENCES

Andrews P. Evolution and environment in the Hominoidea. *Nature* 1992;360:641–646.

Andrews P, Martin L. Cladistic relationships of extant and fossil hominoid primates. *J Hum Evol* 1987; 16:101.

Bailey W. Hominoid trichotomy: a molecular overview. *Evol Anthropol* 1993;2:100–108.

Begun DR. Relations among the great apes and humans. *Yearbook Physical Anthropol* 1994;37:11–64.

Burke AC, *et al.* Hox genes and the evolution of vertebrate axial morphology. *Development* 1995;12: 333–346.

Goodman M. A personal account of the origins of a new paradigm. *Molec Phylogenet Evol* 1996;5: 269–285.

Groves CP, Patterson JD. Testing hominoid phylogeny with the PHYLIP programs. *J Hum Evol* 1991;20: 167–183.

Kim H-S, Takenaka O. A comparison of TSPY genes from Y-chromosomal DNA of the great apes and humans: sequence, evolution, and phylogeny. *Am J Physical Anthropol* 1996;100:301–309.

Li W-H, *et al.* Rates of nucleotide substitution in primates and rodents and the generation-time hypothesis. *Molec Phylogenet Evol* 1996;5:182–187.

Mann A, Weiss M. Hominoid phylogeny and taxonomy: a consideration of the molecular and fossil evidence in a historical perspective. *Molec Phylogenet Evol* 1996;5:169–181.

Marks J. Learning to live with a trichotomy. *Am J Physical Anthropol* 1995;98:212–213.

Moore WS. Inferring phylogenies from mtDNA variation: mitochondrial-gene trees versus nuclear-gene trees. *Evolution* 1995;49:718–726.

Pilbeam D. Genetic and morphological records of the Hominoidea and hominid origins: a synthesis. *Molec Phylogenet Evol* 1996;5:155–168.

Rogers J. The phylogenetic relationships among Homo, Pan, and Gorilla. *J Hum Evol* 1993;25: 201–215.

Ruvolo M. Molecular evolutionary processes and conflicting gene trees: the hominoid case. *Am J Physical Anthropol* 1994;94:89–114.

———. Seeing the forest and the trees. *Am J Physical Anthropol* 1995;98:218–232.

———. Molecular phylogeny of the hominoids. *Molec Biol Evol* 1997;14:248–265.

Shoshani J, *et al.* Primate phylogeny: morphological vs molecular results. *Molec Phylogenet Evol* 1996;5: 102–154.

Takahata N. A genetic perspective on the origin and history of humans. *Annu Rev Ecol Systematics* 1995;26:343–372.

ORIGIN OF THE HOMINOIDEA

16

The Hominoidea (apes and humans) is one of two superfamilies that constitute the infraorder Catarrhini; the second superfamily is the Cercopithecoidea (Old World monkeys). The infraorders Cararrhini and Platyrrhini (New World monkeys) together constitute the suborder Anthropoidea, or anthropoids, often called the higher primates. This unit will describe current thinking about the evolutionary history of anthropoids, and particularly the hominoids, including relationships between fossil and living species.

SOME GENERAL PATTERNS

Three key points stand out in any review of the evolution of the catarrhines.

First, the fossil record of the group generally does not overlap with the geographic areas where catarrhines are most abundant today. The early fossil record is concentrated in North Africa and Eurasia, with some specimens found in East and southern Africa. Modern Old World monkeys and apes are most abundant in the forests of sub-Saharan Africa and southeast Asia. This pattern may reflect real changes in the history of the group, or it may partly result from a biased fossil record: forest habitats are generally poor environments for fossil preservation.

Second, among living catarrhines, Old World monkeys are both more abundant and more diverse than apes. Some 15 genera and 65 species of Old World monkey exist, compared with five genera (*Pan, Gorilla, Pongo, Hylobates,* and *Homo*) and two dozen species of hominoid (a dozen of these are members of the Hylobates group, or gibbons). In earlier times, precisely the opposite

situation prevailed, with apes being more abundant and more diverse than monkeys.

Third, the early apes were not merely primitive versions of the species we know today. They combined various sorts of characters: some apelike, some monkeylike, and some that are unknown in modern large primates. In fact, most fossil apes are apelike only in their dentition, while much of the postcranial skeleton was monkeylike. Consequently, they are often referred to as "dental apes." Such anatomical novelties probably caused the early apes to be behaviorally distinct as well, as measured in terms of the way they moved and what they ate. This variation makes it much more difficult to predict the appearance and behavior of ancestral species, including the ancestor of the human lineage.

If the current fossil record is a reasonable reflection of catarrhine history, then a number of general trends—in body size, brain size, and locomotor and dietary adaptation—can be seen that are common to most groups undergoing adaptive radiation. First, an increase in body size occurs among the group as a whole and within certain lineages in the group, particularly the apes. Second, relative brain size is generally larger among the catarrhines than among the prosimian primates; and ape brains are larger than monkey brains. Third, the initial adaptive niche of quadrupedal, arboreal frugivory (fruit eating) broadens. Modes of locomotion come to include suspensory climbing in trees (apes only) and terrestriality (apes and monkeys); leaf eating (folivory) becomes steadily more important within the group as a whole (mainly monkeys).

Anthropoids are generally assumed to have originated in Africa approximately 50 million years ago, although some evidence points to Asia as their source. The most abundant early fossil evidence of anthropoids is found in North Africa, at the early Oligocene sites of the Fayum Depression, Egypt, where specimens range in age from 37 to 31 million years. The species found at these sites are thought to represent a time prior to the division between

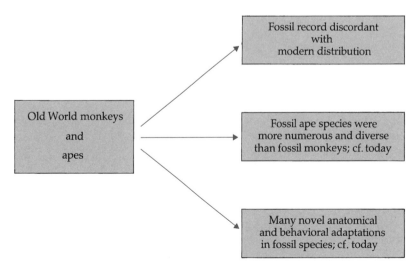

FEATURES OF THE CATARRHINE FOSSIL RECORD: A major lesson to be inferred from the main features of the catarrhine fossil record is that the present is not always a direct key to the past.

platyrrhines and catarrhines. The present fossil evidence strongly indicates an African origin of hominoids, occurring some 25 million years ago. Approximately 18 million years ago, hominoids migrated into Eurasia (following the joining of the continents through continental drift) and underwent a subsequent adaptive radiation there. The middle and later Miocene saw radiations of at least two groups of ape, one (the most common) in which the dentition resembles that of living hominoids and the postcranial skeleton is primitive, and the other in which the dentition is primitive and the postcranium more apelike.

No ancestors for living hominoids have been identified in the known fossil record, with the exception of the orangutan and possibly the gorilla.

EARLY ANTHROPOIDS

Algeripithecus minutus, a small primate that lived in North Africa perhaps as long as 50 million years ago (early Eocene), holds uncertain claim to be the earliest known anthropoid. It exhibits some anthropoid cranial characteristics but is otherwise rather primitive for primates. A little younger is the newly named Chinese genus *Eosimias* (dawn ape). A small creature (weighing between 2.5 and 3 ounces), *Eosimias* also possesses some dental characteristics of living hominoids but is prosimian in all other respects. Both species bear some resemblance to a possible basal anthropoid. The presence of other Eocene anthropoid species, *Amphipithecus* and *Pondaungia,* in Burma is taken by some to imply an Asian—not African—origin of anthropoids.

The anthropoid fossil record becomes relatively extensive only in the late Eocene/early Oligocene, at the Fayum Depression, where Elwyn Simons, of the Duke University Primate Center, has been working since the early 1960s. Currently one of the driest places on Earth, the region was covered with tropical forest bordering an inland sea 35 million years ago. The rich fauna and flora were typical of tropical forest and swamp ecosystems. Simons and his colleagues have recovered fossils of 11 anthropoid species, from beds dated at 37 to 31 million years ago. The species are assigned to two groups, the parapithecids and propliopithecids.

Parapithecids, which include *Qatrania, Serapia, Algeripithecus,* and *Apidium,* were small, marmoset-sized anthropoids that were mostly leaf-eaters. Like earlier putative anthropoids, the parapithecids exhibited a mix of anthropoid and prosimian features. They also possessed the New World monkey dental formula: two incisors, one canine, three premolars, and three molars on each side of the upper and lower jaws. (Cercopithecoids, by contrast, have only two premolars.) The New World dental structure may therefore have been primitive for all anthropoids. The Parapithecidae is not thought to have been ancestral to any later anthropoids.

The Propliopithecidae includes *Propliopithecus, Catopithecus,* and *Aegyptopithecus,* the largest of the Fayum anthropoids (males weigh as much as 13 pounds). Tooth structure indicates that members of this group were principally fruit-eaters. Males were significantly larger than females in this group, implying social systems in which males competed for females in some kind of polygynous structure (see unit 13). The 1995 announcement of a 37-million-year-old cranial and dental specimen of *Catopithecus browni* makes the species the earliest known undisputed anthropoid. The origin of this group (and the parapithecids) cannot be directly linked with known earlier Eocene primates, however. Some researchers consider the group to be ancestral to later cercopithecoids and hominoids.

Aegyptopithecus, or something like it, may therefore represent the form ancestral to Old World monkeys and apes. Some authorities consider it possible that a species akin to *Aegyptopithecus* and its contemporaries might represent the basic anthropoid condition prior to the split between Old World and New World anthropoids. According to Simons, *Aegyptopithecus* was "a generalized arboreal quadruped" with "no evidence whatever . . . of either arm swinging or upright walking tendencies."

THE EARLIEST HOMINOIDS

Hominoid fossils are known throughout much of the Miocene in Africa and Eurasia, with the earliest specimens of a species of *Proconsul* (dated at approximately 22 million years) coming from Africa, the likely region of origin for the clade. Although claims have been put forth for an even earlier *Proconsul* specimen, at 26 million years, their validity cannot be established because of the absence of reliably diagnostic parts. In any case, the clade apparently originated some time between 31 and 22 million years ago.

Hominoids underwent several adaptive radiations, producing a great abundance and variety of species that followed lifestyles not typical of modern apes. *Proconsul* itself produced several species, including one as small as a gibbon and another the size of a female gorilla. Miocene hominoids were creatures of tropical and subtropical forests. Climate change—the result of global cooling and local tectonic activity—greatly reduced hominoid habitat through the late Miocene in the Old World and was probably responsible for the drop in the diversity of hominoids. Cercopithecoid diversity increased in parallel with this change, and many monkey species came to occupy niches previously filled by hominoids.

It is worth repeating that the postcranial anatomy of extant hominoids—adaptations to a suspensory habit—evolved only recently; it is not seen in any fossil apes to any great degree. Most Miocene apes more closely resemble monkeys in terms of their posture and locomotion. We will begin with a description of *Proconsul,* which many consider to be the basal hominoid, before describing the archaic and modern hominoid radiations.

Proconsul fossils have been found at several sites in Kenya, and this species is probably the best-known Miocene ape. In its cranial and dental features, *Proconsul* is judged to be primitive; the thin enamel layer on its cheek teeth apparently reflects a nonhominoid origin. The brain was relatively large, and the increased surface area of the molars and broadening of the incisors imply a more

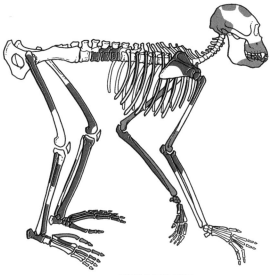

KNM-RU 2036

PROCONSUL AFRICANUS: This reconstruction is based on fossils found prior to 1959 (shaded) by Mary Leakey and in 1980, among the Nairobi Museum collections, by Alan Walker and Martin Pickford. This individual, a young female that lived approximately 18 million years ago, has characteristics of both modern monkeys (in its long trunk and arm and hand bones) and modern apes (in its shoulder, elbow, cranial, and dental characteristics). (Courtesy of Alan Walker.)

frugivorous diet. In its postcranial skeleton, *Proconsul* displays a mix of ape and monkey features. For instance, although it had no tail (like an ape), its thorax was narrow and deep, a characteristic seen in **pronograde** (body horizontal to the ground) monkeylike locomotion rather than **orthograde** (body more vertical to the ground) apelike locomotion. "In the forelimb skeleton, the shoulder and elbow region are remarkably apelike," notes Alan Walker of Pennsylvania State University, "but the arm and hand bones look more like those of some monkeys. In the hindlimb the reverse is true: the foot and lower leg bones are very apelike while the hip region looks less so." *Proconsul* would have moved more like a monkey than like the forelimb-dominated ape in terms of posture and locomotion. Interestingly, the hand had a large, opposable thumb, which makes *Proconsul* more like humans than either monkeys or apes. This feature suggests that *Proconsul* might have had considerable manipulative skills, perhaps including making and using simple tools, such as stripped twigs.

LATER HOMINOIDS

The earliest hominoid species have (so far) been recovered from African sites, indicating an African origin of the clade. The earliest known species outside of Africa is *Sivapithecus*, with specimens found in Pakistan, India, Nepal, and possibly Turkey, dating between 13 and 8 million years ago. In Europe, the earliest species is

Dryopithecus, which enjoyed a widespread radiation and dated from the same time period. *Dryopithecus* was one of the first fossil apes to be discovered, with a specimen located at a site in France in 1856. The presence of Miocene hominoids in Eurasia reflects faunal migrations (and subsequent adaptive radiations) from Africa after the continents joined through tectonic action, 18 million years ago. (Contact had been intermittent in earlier times.)

Hominoids later than *Proconsul* may be divided into archaic forms (hominoid dentition; primitive postcranium) and modern forms (primitive dentition; hominoid postcranium). *Proconsul* would be classified as primitive under this scheme. Some of the principal Miocene hominoids will now be described in these terms.

The archaic group is much larger and more geographically widespread than the modern group. *Afropithecus*, from northern Kenya, and *Heliopithecus*, from Saudi Arabia, are slightly younger than *Proconsul* but very similar to it in many ways. They differ, however, in having a long-faced, robust skull that resembles living great apes; they also possess thick enamel, unlike *Proconsul* and living apes. Thick enamel on cheek teeth probably represents an adaptation to a diet containing hard food items, such as tough fruits. The development of thick enameled teeth among hominoids might be interpreted in the context of the cooling Miocene climate, but no universal trend in this direction occurred through time—that is, thick and thin enamels are seen both early and late. For instance, *Kenyapithecus* (an archaic Kenyan species that lived from 15 to 12 million years ago) and *Dryopithecus* (a modern form that lived between 13 and 8 million years ago) have thick and thin enamel, respectively.

Other archaic hominoids include *Ouranopithecus* (Greece), *Lufengpithecus* (China), *Sivapithecus,* the recently discovered *Otapithecus* (a Namibian species from 15 million years ago), and *Ankarapithecus* (a Turkish species, dated at 9.8 million years). The first two lived approximately 8 million years ago. *Ouranopithecus (*also called *Graecopithecus*), had extremely thick enamel, whereas *Otapithecus* had thin enamel. *Ankarapithecus*, details of which were published late in 1996, exhibited a mix of gorilla-like and orangutan-like features in its cranial anatomy. A very large archaic hominoid, *Gigantopithecus*, lived in China, India, and Viet Nam from 8.6 to 0.2 million years ago, but for different periods of time in these parts of Asia. It had large, thickly enameled molar teeth, stood as high as 9 feet tall, and weighed as much as 600 pounds, making it the biggest hominoid ever.

Of all the Miocene hominoids, *Sivapithecus* holds the strongest claim to being ancestral to a living hominoid, the orangutan. This relationship is based on anatomical similarities in the structure of the face and palate.

Hominoids of modern aspect are rare in the fossil record. They include *Oreopithecus* (from Italy), *Morotopithecus bishopi* (from Uganda), and *Dryopithecus*.

Oreopithecus, the first specimens of which were found in the late nineteenth century, lived approximately 8 million years ago. Its dentition represents a mix of primitive

have recently been interpreted as implying a significant degree of habitual bipedal locomotion.

The Ugandan fossil, first found in the 1960s, recently dated to at least 20.6 million years, was similar to *Proconsul* and *Afropithecus* in terms of dentition and cranial anatomy. Parts of its postcranial anatomy, including shoulder and lumbar vertebrae, were derived in the direction of living apes and humans. Its evolutionary relationships are unclear.

Dryopithecus specimens have been found in Spain, Greece, Germany, and Hungary. They display a combination of primitive dentition and advanced postcranial anatomy that places them in the group of hominoids of modern aspect. *Dryopithecus* has been subject to many different phylogenetic interpretations since its discovery.

The January 1996 announcement of the discovery of an extraordinary partial skeleton of *Dryopithecus laietanus* from the site of Can Llobateres in Spain greatly increases our understanding of the species' postcranial anatomy and locomotor pattern, but it does not solve its phylogenetic affiliation. The newly discovered postcranial material is interpreted as reflecting more suspensory adaptation and orthograde posture (similar to living apes) than are seen in any Miocene ape. For instance, the lumbar vertebrae are proportionally shorter than in monkeys and most Miocene apes; the arms are powerful and capable of a wide range of movement; the hand is large and adapted for powerful grasping. The ratio of arm length to leg length (**intermembral index**) is larger than in living African apes and similar to that in the orangutan. The Spanish species is dated at 9.5 million years, indicating that the postcranial adaptations of living apes might have evolved by that date, depending on the still unsettled evolutionary relationship between *Dryopithecus* and the living apes.

This conservative discussion of the phylogenetic relationships of fossil hominoids leaves us with a tree with many branches; few, if any, of these branches appear to be joined to any other branches. Undoubtedly, the hominoid radiation was diverse and successful, and the later fossil species lived in drier, more open woodland habitats than either living hominoids or the early Miocene species. The African hominoid clade evolved at a time when climatic conditions were deteriorating in terms of preferred habi-

RUSINGA ISLAND IN THE EARLY MIOCENE: This community of apes, living 18 million years ago, illustrates something of the species diversity that would later become the characteristic of monkeys. Upper left, *Proconsul africanus;* upper right, *Dendropithecus macinnesi;* center, *Limnopithecus legetet;* lower, *Proconsul nyanzae.* (Courtesy of John Fleagle/Academic Press.)

and derived characters (but not like those of living hominoids); its trunk was short and the thorax broad, with long arms and short legs. Its elbow joints resembled those of modern apes. Its evolutionary relationships are unknown. Aspects of its lumbar, pelvic, and foot anatomy

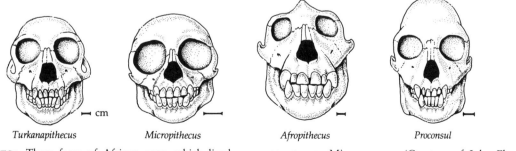

| Turkanapithecus | Micropithecus | Afropithecus | Proconsul |

MIOCENE APES: These faces of African apes, which lived some 18 million years ago, illustrate the diversity of morphology among Miocene apes. (Courtesy of John Fleagle/Academic Press.)

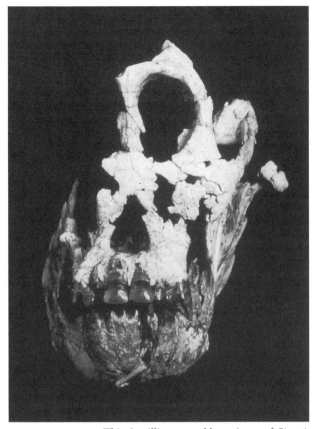

SIVAPITHECUS: This 8 million-year-old specimen of *Sivapithecus indicus* comes from the Potwar plateau in Pakistan. The animal was roughly the same size as a chimpanzee but had the facial morphology of an orangutan; it ate soft fruit (detected in the toothwear pattern) and was probably mainly arboreal. (Courtesy of David Pilbeam.)

tat for apes and when ape diversity was decreasing, perhaps as different adaptations or as reactions to that change. Today's African apes are complex woodland and forest creatures, while early hominines lived in more open environments. Open environments were once posited to be influential in the origin of the hominine clade, but recent fossil discoveries show the earliest known hominines lived in forest environments (see unit 19). ✳

KEY QUESTIONS

- Why have ancestral species of major hominoid radiations not been found or identified?
- Which set of anatomical characters of hominoids are likely to be more reliable for phylogenetic reconstruction: cranial and dental, or postcranial?
- How likely is it that hominoids arose in Africa? Explain your answer.
- Once abundant and diverse, hominoids were reduced to just a few genera. Should they be regarded as evolutionary failures compared with cercopithecoids?

KEY REFERENCES

Andrews P. Evolution and environment in the Hominoidea. *Nature* 1992;360:641–646.

Andrews P, Pilbeam D. The nature of the evidence. *Nature* 1996;379:123–124.

Beard KC, *et al.* Earliest complete dentition of an anthropoid primate from the late Middle Eocene of Shanxi Province, China. *Science* 1996;272:82–85.

Begun D. Relations among the great apes and humans. *Yearbook Physical Anthropol* 1994;37: 11–64.

Conroy GC, *et al.* Diamonds in the desert: the discovery of *Otapithecus namibiensis*. *Evol Anthropol* 1993;2:46–52.

de Bonis L, Koufos GD. Our ancestors' ancestor: *Ouranopithecus* is a Greek link in human ancestry. *Evol Anthropol* 1994;3:75–83.

Gebo DL, *et al.* A hominoid genus from the Early Miocene of Uganda. *Science* 1997;276:401–404.

Köhler M, Moyà-Solà S. Ape-like or hominid-like? The positional behavior of *Oreopithecus bambolii* reconsidered. *Proc Natl Acad Sci USA* 1997;94: 11747–11750.

Moyà-Solà S, Köhler M. A *Dryopithecus* skeleton and the origins of great-ape locomotion. *Nature* 1996;379:156–159.

Pilbeam D. Genetic and morphological records of the Hominoidea and hominid origins. *Molec Phylogenet Evol* 1996;5:155–168.

Rose MD. Functional and phylogenetic features of the forelimb in Miocene hominoids. In: Begun D, *et al.*, eds. Miocene hominoid fossils: functional and phylogenetic interpretations. New York: Plenum, 1996.

Simons EL, Rasmussen T. A whole new world of ancestors: Eocene anthropoideans from Africa. *Evol Anthropol* 1994;3:128–139.

ORIGIN OF BIPEDALISM

17

Although *Homo sapiens* is not the only primate to walk on two feet—for instance, chimpanzees, a small species of orangutan, and gibbons often use this form of posture in certain environmental circumstances—no other primate does so habitually or with a striding gait. The rarity of habitual bipedalism among primates—and among mammals as a whole—has given rise to the assumption that it is inefficient and therefore unlikely to evolve. As a result, anthropologists have often sought "special"—that is, essentially human—explanations for the origin of bipedalism. Strictly biological explanations are, however, more likely to be correct.

Human evolution is often cast in terms of four major novelties related to the basic hominoid adaptation: upright walking, reduction of anterior teeth and enlargement of cheek teeth, elaboration of material culture, and a significant increase in brain size. As the current fossil and archeological records indicate, however, these novelties arose at separate intervals throughout hominine evolution. In other words, hominines show a pattern of **mosaic evolution**.

Stone-tool making appears to have originated at roughly the same time as significant brain expansion, approximately 2.5 million years ago (see unit 23). The earliest hominine fossils discovered so far—from Ethiopia and Kenya—are dated 2 million years earlier (see unit 19); they show significant adaptation to bipedalism in combination with a hominine dental pattern that has distinct apelike overtones. It is therefore possible that the first hominine might have been apelike in all respects, apart from an adaptation to upright walking. If true, then bipedalism would represent the primary hominine adaptation.

In this unit we will examine some of the mechanics of bipedalism, the ecological context in which it might have

arisen, and the development of hypotheses that purport to account for its evolution.

BIOMECHANICS OF BIPEDALISM

The **striding gait** of human bipedalism involves the fluid flow of a series of actions—collectively, the **swing phase** and the **stance phase**—in which one leg alternates with the other. The leg in the swing phase pushes off using the power of the great toe, swings under the body in a slightly flexed position, and finally becomes extended as the foot again makes contact with the ground, first with the heel (the **heel-strike**). Once the heel-strike has occurred, the leg remains extended and provides support for the body—the stance phase—while the other leg goes through the swing phase, with the body continuing to move forward.

Two key features differentiate human and chimpanzee bipedalism. First, chimpanzees are unable to extend their knee joints—to produce a straight leg—in the stance phase. Thus, muscular power must be exerted in order to support the body. Try standing with your knees slightly bent, and you'll get the idea. The human knee can be "locked" into the extended position during the stance phase, thereby minimizing the amount of muscular power needed to support the body. The constantly flexed position of the chimpanzee leg also means that no toe-off and heel-strike occur in the swing phase.

Second, during each swing phase the center of gravity of the body must be shifted toward the supporting leg (otherwise one would fall over sideways). The tendency for the body to collapse toward the unsupported side is countered by contraction of the muscles (gluteal abductors) on the side of the hip that has entered the stance phase. In humans, because of the inward-sloping angle of the thigh to the knee (the **valgus angle**), the two feet at rest are normally placed very close to the midline of the body. Therefore, the body's center of gravity need not be shifted very far laterally back and forth during each phase of walking.

Modern human anatomy is a fully terrestrial adaptation, although the earliest hominines also demonstrated some arboreal adaptation. As we shall see later, these differences have implications for energetic efficiency.

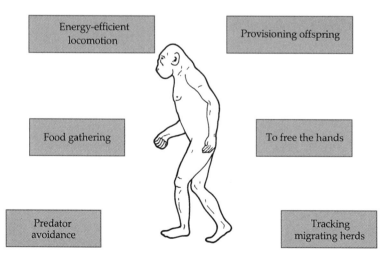

HYPOTHESIZED CAUSES OF BIPEDALISM:
Perhaps the defining characteristic of hominines, bipedalism has inevitably long been the focus of speculation as to its evolutionary cause. Some of the main ideas are shown here.

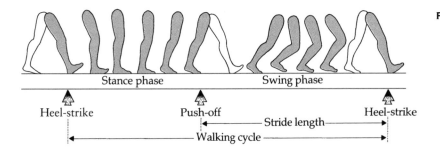

Stance phase   Swing phase

Heel-strike   Push-off   Heel-strike

Stride length

Walking cycle

PHASES OF BIPEDALISM: Upright walking in humans requires a fluid alternation between stance phase and swing phase activity for each leg. Key features are the push-off, using the great toe, at the beginning of the swing phase, and the heel-strike, at the beginning of the stance phase.

The suite of anatomical adaptations that underlie human bipedalism includes the following characters:

- Curved lower spine;
- Shorter, broader pelvis and an angled femur, which are served by reorganized musculature;
- Lower limbs and enlarged joint surface areas;
- Extensible knee joint;
- Platform foot in which the enlarged great toe is brought in line with the other toes; and
- Movement of the foramen magnum (through which the spinal cord enters the cranium) toward the center of the basicranium.

ECOLOGICAL CONTEXT OF THE ORIGIN OF BIPEDALISM

The nature of the evolution of bipedalism in hominines depended, of course, on the nature of the locomotor adaptation of the immediate ancestor. The ancestor might have been a knuckle-walker, like the chimpanzee, or a

species much more arboreally adapted. In any case, the quadrupedal to bipedal transformation is not as dramatic a shift as it might at first appear, because primates are not

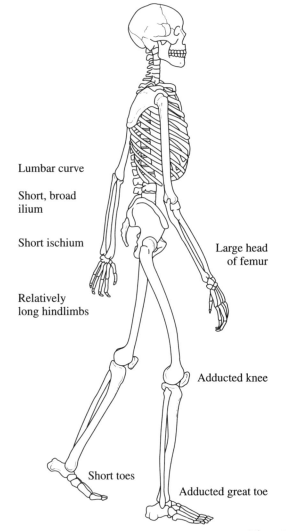

Lumbar curve

Short, broad ilium

Short ischium

Relatively long hindlimbs

Large head of femur

Adducted knee

Short toes

Adducted great toe

ANATOMICAL ADAPTATIONS TO BIPEDALISM: The principal adaptations involve a lumbar curve of the spine; a short, broad pelvis; and long hindlimbs. These characters bring the knees closer to the center of the body (adduction) to form the valgus angle of the femur, and bring the great toe in line with the other toes (adduction).

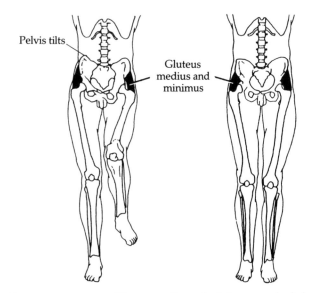

Pelvis tilts

Gluteus medius and minimus

THE PELVIC TILT: Gluteus medius and minimus muscles link the femur (thigh bone) with the pelvis. They contract on the side in stance phase, preventing a collapse toward the side of the unsupported limb. Nevertheless, the pelvis tilts during walking. (Courtesy of David Pilbeam.)

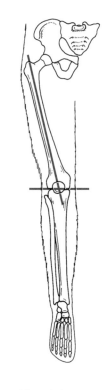

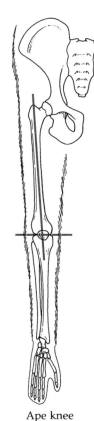

THE VALGUS ANGLES IN HUMANS, APES, AND AN EARLY HOMININE: The angle subtended by the femur at the knee, the valgus angle, is critical to bipedal locomotion. With the femur angled as in humans, the foot can be placed underneath the center of gravity while striding. An ape's femur is not angled in this way, causing the animal to "waddle" during bipedal locomotion. The valgus angle of *Australopithecus afarensis*, a 3 million-year-old (or older) hominine, is humanlike, indicating its commitment to bipedality. Also note the humanlike shape of the *afarensis* pelvis. (Courtesy of Luba Gudz.)

Human knee   *Afarensis* knee   Ape knee

true quadrupeds (like a horse), and body posture is often relatively upright, such as in tree-climbing.

The earliest hominines appear to have evolved under ecological circumstances (that is, heavily wooded) similar to those typical for living and extinct apes (see unit 19), not in relatively open savannah, as has long been assumed.

DEVELOPMENT OF IDEAS ON THE ORIGIN OF BIPEDALISM

As we saw in unit 3, Darwin essentially equated hominine origins with human origins, proposing an evolutionary package that included upright walking, material culture, modified dentition, and expanded intelligence. In

the 1960s, this incipient "Man the hunter" scenario found an added advantage in bipedalism: although humans are slower and less energy-efficient than quadrupeds when running at top speed, at a slow pace bipedalism allows for great stamina such as might be effective in tracking and killing a prey animal. Recently, with the replacement of the "Man the hunter" image by "Man the scavenger" (see unit 26), it has been suggested the endurance locomotion provided by bipedalism enabled the earliest hominines to follow in the wake of migrating herds, opportunistically scavenging the carcasses of the unfortunate young and the infirm old.

One problem arises with both these explanations: not only do stone tools that are required for cutting meat

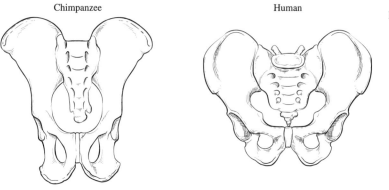

Chimpanzee   Human

PELVIC ANATOMY: In apes, the pelvis is long and narrow; in humans, it is short and broad.

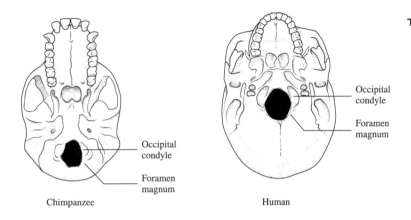

Chimpanzee

Occipital condyle

Foramen magnum

Human

Occipital condyle

Foramen magnum

THE BASICRANIUM: Because the skull is perched atop a vertical spine in a biped, the foramen magnum (through which the spinal cord enters the cranium) is located toward the center of the cranium; it is found toward the back in apes. The occipital condyles articulate with the first vertebra (atlas vertebra) of the axial spine.

from carcasses apparently postdate hominine origins by as much as 3.5 million years, but also no indication of regular meat eating has been found in the dentition of the earliest known hominines. In fact, evidence from microwear patterns on the surface of teeth (see unit 18) shows that hominine diets remained predominantly vegetarian until approximately 1.8 million years ago—that is, until the origin of *Homo erectus/ergaster.*

Other explanations offered for the origin of bipedalism have included the following:

- Improved predator avoidance, as the biped would be able to see further across the "open plain" than the quadruped;
- Display or warning;
- A shift in diet, such as seed eating; and
- Carrying things.

The last explanation has been featured in two hypotheses in recent years: the "Woman the gatherer" hypothesis, and the "Man the provisioner" model.

The "Woman the gatherer" hypothesis, advanced initially in the early 1970s, shifted putative evolutionary novelty from hunting meat (a male activity) to gathering plant foods (a female activity), which might have required technological innovations such as digging sticks and means of carrying many small items. As often happens in modern chimpanzees, females are envisaged as having foraged together and with their offspring, with who they shared food. Males were peripheral, socially (see unit 13). The "Woman the gatherer" hypothesis is more conservative than the "Man the hunter" model, in that the first hominines are viewed as being basically apelike rather than already essentially human. Nevertheless, it focuses on the need to carry things: specifically, food for sharing within infants.

Another hypothesis that focuses on the need to carry things is "Man the provisioner," in which males gathered food and returned it to some kind of home base; there, the food was shared with females and offspring, specifically "his" female and offspring. Proposed in 1981 by Owen Lovejoy of Kent State University, this model envisages pair bonding and sexual fidelity between male/female couples, with the male providing an important part of the dietary resources. Such a provisioning pattern would enable females to reproduce at shorter intervals, thus giving them a selective advantage over other large hominoids, which, says Lovejoy, were reproducing at a dangerously slow rate. The system would work only if a male could be reasonably certain that the infants he was helping to raise were his—hence the need for pair bonding and sexual fidelity. Although it received widespread attention, Lovejoy's hypothesis has been widely criticized, not least because the very large degree of sexual dimorphism in body size seen in these creatures is very difficult to explain, given the putative monogamous social structure proposed (see units 12 and 13).

ENERGETICS OF BIPEDALISM: POSSIBLE IMPLICATION IN ITS ORIGIN

A more parsimonious—and therefore more scientifically attractive—explanation of bipedalism was proposed by Peter Rodman and Henry McHenry of the University of California at Davis in a 1980 publication. Very simply, they suggest that bipedalism might have evolved, not as part of a change in the *nature* of the diet or social structure, but merely as a result of a change in the *distribution* of existing dietary resources. Specifically, in the more open habitats of the late Miocene, hominoid dietary resources became more thinly dispersed in some areas; the continued exploitation of these resources demanded a more energy-efficient mode of travel—hence the evolution of bipedalism. In this scenario, the evolution of bipedalism reflects the improved locomotor efficiency associated with foraging, and nothing else. (Interestingly, efficiency of foraging is invoked in a recent description of bipedal features in the Miocene ape *Oreopithecus bambolii*.)

Rodman and McHenry's proposal is based on a few simple points. First, although human bipedalism is less energy-efficient than conventional quadrupedalism at high speeds, it is just as efficient—or more so—at walking speeds. Second, chimpanzees are roughly 50 percent less energy-efficient than conventional quadrupeds when walking on the ground, whether they employ knuckle walking or move bipedally. Therefore, noted Rodman and McHenry, "there was no energetic Rubicon separating

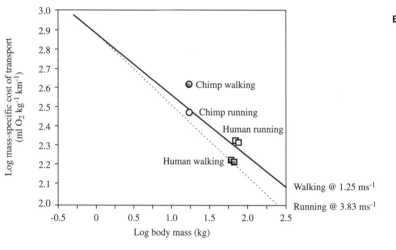

ENERGETICS OF LOCOMOTION: The solid line represents the energy cost of running (at 3.83 meters per second) in mammals of different body size; the dotted line shows the cost of walking (at 1.25 meters per second). Note that chimpanzees are less efficient than other mammalian quadrupeds at both running and walking, while humans are less efficient at running but more efficient at walking.

hominoid quadrupedal adaptation from hominine bipedalism."

For bipedalism to evolve among hominoids, only a selective advantage favoring improved energetic efficiency of locomotion was necessary. A more dispersed food resource could provide such a selection pressure.

Rodman and McHenry's hypothesis has recently been challenged on several counts, particularly by Karen Steudel of the University of Wisconsin. She points out that this scenario implicitly assumes that the common ancestor of humans and African apes was a knuckle-walker, which may not be correct (see unit 15). In addition, the postcranial skeleton of the early hominines differed from that of modern humans, specifically in including a significant degree of arboreal adaptation. The energy efficiency of bipedal locomotion in these creatures is therefore likely to have been lower than in modern humans, upon which the above energetic calculations were based. There is, concludes Steudel, "no reason to suppose that our quadrupedally-adapted ancestors would have reaped energetic advantages when they shifted to an upright stance." Rod-

man and McHenry maintain that, although their hypothesis may have oversimplified the situation, it remains valid.

Lynne Isbell, of Rutgers University, and Truman Young, of Fordham University, recently extended the evolutionary context of the energy-efficiency hypothesis to other African hominoids. If, as the hypothesis argues, Miocene climate change made hominoid dietary resources less densely distributed, then hominoids would have been forced to become more efficient in exploiting them. Isbell and Young accept that bipedalism represents one potential adaptation to this situation, which inevitably requires an increase in the daily travel distance while foraging for dispersed resources. A second strategy is to reduce the required daily travel distance, which is achieved by diminishing group size. (A large group requires more total food resources each day than a small group, and must therefore travel further to harvest it.) This strategy, argue Isbell and Young, was adopted by chimpanzees, which exhibit a fission-fusion group structure. As part of their argument, they cite field observations of gorillas and chimpanzees in

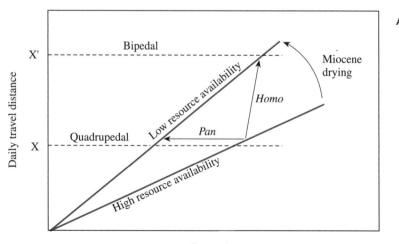

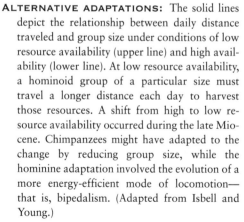

ALTERNATIVE ADAPTATIONS: The solid lines depict the relationship between daily distance traveled and group size under conditions of low resource availability (upper line) and high availability (lower line). At low resource availability, a hominoid group of a particular size must travel a longer distance each day to harvest those resources. A shift from high to low resource availability occurred during the late Miocene. Chimpanzees might have adapted to the change by reducing group size, while the hominine adaptation involved the evolution of a more energy-efficient mode of locomotion—that is, bipedalism. (Adapted from Isbell and Young.)

Gabon, where the apes feed heavily on fruits. When these resources become scarce, gorillas maintain their group size, but switch their dietary emphasis to leaves. In contrast, chimpanzees continue to eat fruits, but forage in smaller groups or even alone.

Isbell and Young's analysis is important because it puts hominine bipedalism within a general evolutionary ecology context of different behavioral adaptations by African hominoids to the same environmental circumstances, that is food resources becoming more widely distributed as a result of climate change. A key issue, of course, is what exactly were the environmental conditions when bipedalism originated, not what it was when the new mode of locomotion was well developed.

An hypothesis, developed by Kevin Hunt, of Indiana University, shifts the focus away from foraging efficiency to feeding efficiency. From more than 600 hours of field observations of chimpanzees and their bipedal behavior—which included stationary feeding of fruits from bushes and low branches in small trees, and locomotion from one spot to another—Hunt made the following observations: 80 percent of bipedal behavior was related to stationary feeding; only 4 percent was observed during direct locomotion. Hunt suggests, therefore, that the hominine bipedal adaptation was primarily a feeding adaptation; only later in hominine history did it become a specifically locomotor adaptation.

The plethora of hypotheses offered to explain the evolution of bipedalism reflects both a fertility of ideas among anthropologists and the difficulty of using available evidence to discriminate between them. Any attempt to test hypotheses must encompass the possibility that hominine bipedalism arose in a heavily wooded or forested environment, rather than in open woodland or grassland savannah that was once thought to be the case. ✵

KEY QUESTIONS

- What does the rarity of primate bipedalism imply, other than that it is "difficult" to evolve?
- Given the energetic differences between hominoid quadrupedalism and human bipedalism, would an evolutionary transformation be *necessarily* fast or slow?
- Which hypotheses would suffer adversely if bipedalism evolved in a wooded or even forested context?
- Could a hominoid that was completely apelike apart from being bipedal be classified as a hominine?

KEY REFERENCES

Hunt KD. The evolution of human bipedality: ecology and functional morphology. *J Hum Evol* 1994;26:183–202.

Isbell LA, Young TP. The evolution of bipedalism in hominids and reduced group size in chimpanzees. *J Hum Evol* 1996;30:389–397.

Lovejoy CO. Evolution of human walking. *Sci Am* Nov 1988:118–125.

Cahiers de Paleoanthropologie, Editions du CNRS, 1991:133–141.

Köheler M and Moyà-Solà S. Ape-like or hominid-like? The positional behavior of *Oreopithecus bambolii* reconsidered. *Proc Natl Acad Sci USA* 1997;94:11747–11750.

Rodman PS, McHenry HM. Bioenergetics of hominid bipedalism. *Am J Physical Anthropol* 1980; 52:103–106.

Shreeve J. Sunset on the savannah. *Discover* July 1996:116 –125.

Steudel K. Limb morphology, bipedal gait, and the energetics of hominid locomotion. *Am J Physical Anthropol* 1996;99:345–355.

JAWS AND TEETH

18

Jaws—particularly lower jaws—and teeth are by far the most common elements recovered from the fossil record. The reason is that, compared with much of the rest of the skeleton, jaws and teeth are very dense (and teeth very tough), which increases the likelihood that they will survive long enough to become fossilized.

Because jaws usually serve as an animal's principal food-processing machine, the nature of a species' dentition can yield important clues about its mode of subsistence and behavior. Overall, however, the dental apparatus is evolutionarily rather conservative, with dramatic changes rarely appearing. For instance, human and ape dentition retains roughly the basic hominoid pattern established more than 20 million years ago. Moreover, different species facing similar selection pressures related to their feeding habits may evolve superficially similar dental characteristics, as we shall see, for example, in the matter of enamel thickness. Similar sets of jaws and teeth may therefore arise in species with very different biological repertoires.

In this unit we will examine four facets of hominoid dentition: the overall structure of jaws and teeth; the pattern of eruption; the characteristics of tooth enamel; and the indications of diet that are to be found in microwear patterns on tooth surfaces.

BASIC ANATOMY

Perhaps the most obvious trend in the structure of the primate jaw (and face) throughout evolution is its shortening from front to back and its deepening from top to bottom, going from the pointed snout of the tarsier to the flat face of *Homo sapiens*. Structurally, this change involved the progressive tucking of the jaws under the brain case, which steadily reduced the angle of the lower jaw

bone (mandible) until it reached the virtual "L" shape seen in humans. Functionally, the change involved a shift from an "insect trap" in prosimians to a "grinding machine" in hominoids. Grinding efficiency increases as the distance between the pivot of the jaw and the tooth row decreases, with hominines being closest to this position.

The primitive dental pattern for anthropoids includes (in a half-jaw) two incisors, one canine, three premolars, and three molars, giving a total of 36 teeth. This pattern is seen in modern-day New World anthropoids, while Old World anthropoids possess two premolars (not three), giving them a total of 32 teeth. Overall, the modern ape jaw is rather rectangular in shape, while the human jaw more closely resembles an arc. One of the most striking differences, however, is that apes' conical and somewhat blade-shaped canine teeth are very large and project far beyond the level of the tooth row; in these animals, males' canines are substantially larger than those found in females, an aspect of sexual dimorphism with significant behavioral consequences (see unit 13).

When an ape closes its jaws, the large canines are accommodated in gaps (diastemata) in the tooth rows: between the incisor and canine in the upper jaw, and between the canine and first premolar in the lower jaw. As a result of the canines' large size, an ape's jaw is effectively "locked" when closed, with side-to-side movement being limited. By contrast, human canines—in both males and females—are small and barely extend beyond the level of the tooth row. As a result, the tooth rows have no diastemata, and a side-to-side "milling" motion is possible, which further increases grinding efficiency. The upper incisors of apes are large and spatula-like, which is a frugivore adaptation. In contrast, human upper incisors are smaller and more vertical, and, with the small, flat canines, they form a slicing row with the lower teeth.

The single-cusped first premolar of apes is highly characteristic, particularly the lower premolar against which the huge upper canine slides. Ape molar teeth are larger than the premolars and include high, conical cusps. In humans, the two premolars assume the same shape and have become somewhat "molarized." The molars them-

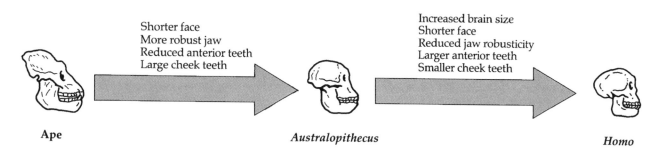

Ape **Australopithecus** **Homo**

Shorter face
More robust jaw
Reduced anterior teeth
Large cheek teeth

Increased brain size
Shorter face
Reduced jaw robusticity
Larger anterior teeth
Smaller cheek teeth

EVOLUTIONARY TRENDS IN DENTITION: The transition from ape to *Australopithecus* and from *Australopithecus* to *Homo* involved some changes that were continuous and others that were not. For instance, the face became increasingly shorter throughout hominine evolution, while robusticity of the jaw first increased and then decreased. The combined increase in cheek tooth size and decrease in anterior tooth size that occurred between apes and *Australopithecus* was also reversed with the advent of *Homo*.

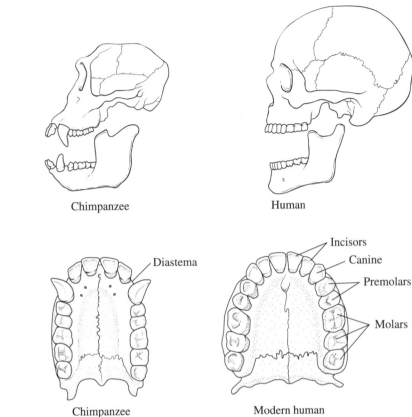

JAWS AND TEETH: Note the longer jaw and more projecting face in the chimpanzee, the protruding incisors, and large canines.

Chimpanzee   Human

Diastema

Incisors
Canine
Premolars
Molars

Chimpanzee   Modern human

selves are large and relatively flat, with low, rounded cusps—characteristics that are extremely exaggerated in some of the earlier hominines (see unit 20).

The hominine dental package as a whole can therefore be regarded as an extension of a trend toward a more effective grinding adaptation. In the earliest known hominines—*Ardipithecus ramidus* and *Australopithecus anamensis* from approximately 4.5 million years ago (see unit 19)—the dentition remains strikingly apelike, with a significant degree of sexual dimorphism. Within 2 million years, however, the canines in several hominine species have become smaller and flattened, looking very much like incisors (see unit 20).

ERUPTION PATTERNS

The pattern of eruption of permanent teeth in modern apes and humans is distinctive, as is its overall timing. Recently anthropologists have debated this aspect of hominoid dentition, specifically asking how early hominines fit into this picture. Were they more like humans or more like apes? Although the issue remains to be fully resolved, indications are that until rather late in hominine history, dental development was in many ways rather apelike, particularly in its overall timing.

The ape tooth eruption pattern is M1 I1 I2 M2 P3 P4 C M3; the corresponding human pattern is M1 I1 I2 P3 C P4 M2 M3. The principal difference, therefore, is that in apes the canine erupts after the second molar, while in

humans it precedes the second molar. Associated with the prolonged period of infancy in humans is an elongation of the time over which the teeth erupt. The three molars appear at approximately 3.3, 6.6, and 10.5 years in apes, whereas the ages are 6, 12, and 18 years in humans.

Thus, a human jaw in which the first molar has recently erupted indicates that the individual was roughly 6 years old. An ape's jaw with the first molar just erupted would indicate an individual a little more than 3 years old. The question is, How old is an early hominine jaw in this state? Is it 3 years old or 6 years old? As it happened, the first australopithecine to be discovered—the Taung child, *Australopithecus africanus* (see unit 20)—had just reached this state of development.

University of Michigan anthropologist Holly Smith recently analyzed tooth eruption patterns in a series of fossil hominines and concluded that most of the early species were distinctly apelike. For *Homo ergaster,* which lived from 1.9 million until approximately 400,000 years ago, her results implied that early members of this species showed a pattern that was intermediate between humanlike and apelike. For instance, in 1985 a remarkably complete skeleton of *Homo ergaster* (denoted KNW-WT 15,000) was discovered on the west side of Lake Turkana, Kenya. The individual was a youth whose second molar was in the process of erupting. A human pattern of development would imply an age of 11 or 12 years when he died, while an ape pattern would give 7 years. In fact,

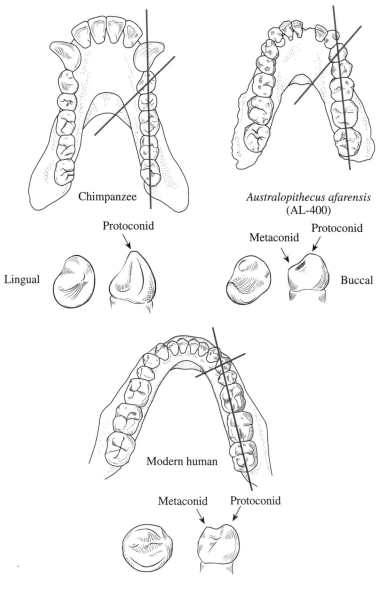

Chimpanzee

Protoconid

Lingual

Australopithecus afarensis
(AL-400)

Metaconid   Protoconid

Buccal

Modern human

Metaconid   Protoconid

EARLY HOMININE DENTITION: The first pre-molar in apes is characteristic in having one cusp (protoconid); in humans, the tooth has two cusps (the protoconid and metaconid). In apes, the axis of the premolar in relation to the tooth row is more acute than in modern humans. In *Australopithecus afarensis*, an early hominine, the tooth is intermediate in shape between humanlike and apelike, but its axis resembles that seen in apes.

Smith's analysis suggests that he was probably 9 years old. The fully human pattern of dental development did not evolve until in later descendants of *Homo ergaster*, that is *Homo erectus*.

Smith's conclusion has been challenged by University of Pennsylvania anthropologist Alan Mann, who a decade earlier had proposed that all hominines followed the human pattern of development. Nevertheless, Smith's position received support in late 1987, when Glenn Conroy and Michael Vannier of Washington University School of Medicine published results of their computed tomography (CT) analysis of the Taung child's skull. The two were able to "see" the unerupted teeth within the jaw bone, and consequently concluded that the teeth would have emerged in an apelike pattern.

The debate has been extended further by two researchers at University College London, who claim to be able to determine the exact age of a tooth by counting the number of lines—striae of Retzius—within the enamel. Although this technique is not universally accepted, the two researchers, Timothy Bromage and Christopher Dean, believe that the lines represent weekly increments—thus giving them an anthropological equivalent of tree rings, which measure yearly increments.

When Bromage and Dean applied their technique to a series of australopithecine and early *Homo* fossils, they obtained ages that were between one-half and two-thirds of what would be inferred if a human standard of dental development had been applied. If they and Smith are correct, then hominines followed a distinctly apelike pattern of dental development until relatively recently in evolutionary history. This concept has important implications for the period of infant care. Once infant care becomes prolonged, which becomes necessary when significant postnatal brain growth takes place (see unit 31), then social life becomes greatly intensified. The dental evidence indicates that this prolongation may have begun with *Homo erectus*, which is in accord with data on increased brain size.

ENAMEL THICKNESS

The relative thickness of enamel on cheek teeth has played an important role in anthropology, not least because Elwyn Simons interpreted *Ramapithecus* as being an early hominine through identification of this character. Modern humans carry a thick enamel coat on their teeth, whereas the African apes exhibit thin enamel (in the orangutan, the enamel layer is of intermediate thickness). Until the 1994 discovery of *Ardipithecus ramidus* changed the picture, all known fossil hominines also possessed thick enamel. Thick enamel was therefore assumed to be a shared character for the African hominoid clade. Thin enamel was seen as an adaptation to fruit eating, while thick enamel was envisioned as an adaptive response to processing tougher plant foods.

As we saw in unit 16, the evolution of thin and thick enamel followed a complex path throughout hominoid history. Thin enamel appears to be a primitive character for the hominoid clade as a whole, but thick enamel has arisen several times independently during the history of the group. What about the African hominoid clade? As already indicated, thick enamel was traditionally considered to be a characteristic of this clade, with the chimpanzee and gorilla having reverted to a primitive state of thin

enamel. The most recent analysis of enamel formation in hominoids and a reevaluation of late Miocene hominoids in Africa have turned this view around, however. It now seems likely that the common ancestor of modern African hominoids had thin enamel, that the earliest hominines also possessed thin enamel (with thick enamel developing only later in the clade's history), and that chimpanzees and gorillas represent the primitive state of the group, not a reversal. The thick enamel of later hominines and, for instance, the late Miocene ape *Sivapithecus* reflects independent evolution, not homology.

TOOTHWEAR PATTERNS

The surface of tooth enamel bears an animal's primary contact with its food, and to some extent at least a signature of that contact is left behind. Using a scanning electron microscope, Alan Walker of Johns Hopkins Medical School has produced images of a range of characteristic toothwear patterns: for grazers, browsers, frugivores, bone-crunching carnivores, and so on. The teeth of grazers, for instance, are etched with fine lines that are produced by contact with tough silica inclusions (phytolyths) in grasses; browsers' teeth are smoothly worn, as are those of fruit-eaters; scavengers' teeth are often deeply marked as a result of bone crushing. In a series of comparisons, all early hominines appear to fit into the frugivore category, along with modern chimpanzees and

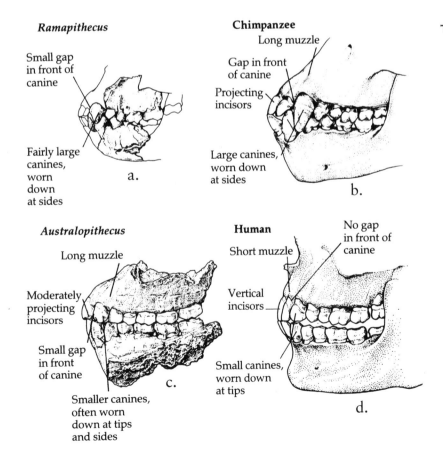

TOOTH CHARACTERISTICS: This diagram shows some of the major characteristics in (a) a Miocene ape, (b) a chimpanzee, (c) *Australopithecus afarensis,* and (d) *Homo sapiens.* (From *Our Fossils Ourselves,* courtesy of the British Museum [Natural History].)

orangutans. This pattern entails a rather smooth enamel surface into which are etched a few pits and scratches.

A major shift occurs, however, with *Homo ergaster/erectus,* whose enamel is heavily pitted and scratched. Such a pattern resembles a cross between a hyena (a bone-crunching carnivore) and a pig (a rooting omnivore). Although it is not yet possible to interpret precisely the implications for the *Homo erectus* diet, it is significant that toothwear patterns indicate some sort of abrupt change in hominine activities at this point in history—perhaps significant brain expansion, reduction in body size dimorphism, systematic tool making, use of fire, or migration out of Africa.

In spite of their limitations, then, teeth clearly have the ability to yield information about hominine history that goes far beyond what simply went down our ancestors' throats. ✳

KEY QUESTIONS

- How reliable are teeth as indicators of a species' diet?
- What other information would one need to assess the significance of the reduction of overall size and loss of sexual dimorphism in hominine canines?
- How would you recognize the jaws and teeth of the first hominines?
- How reliable a phylogenetic indicator is enamel thickness?

KEY REFERENCES

Beynon AD, *et al.* On thick and thin enamel in hominoids. *Am J Physical Anthropol* 1991;86:295–309.

Bromage TG, Dean MC. Re-evaluation of the age at death of immature fossil hominids. *Nature* 1985; 317:525–527.

Conroy GC, Vannier MW. Dental development of the Taung skull from computerized tomography. *Nature* 1987;329:625–627.

Macho GA, Wood BA. Role of time and timing in hominid dental evolution. *Evol Anthropol* 1995;4:17–31.

Mann AE, *et al.* Maturational patterns in early hominids. *Nature* 1987;328:673–675.

Smith BH. Dental development in *Australopithecus* and early *Homo. Nature* 1986;323:327–330.

———. The physiological age of KNM-WT 15,000. In: Walker A, Leakey R, eds. The Nariokotome *Homo erectus* skeleton. Cambridge, MA: Harvard University Press, 1993:195–220.

Teaford M. Dental microwear and dental function. *Evol Anthropol* 1994;3:17–30.

Ungar PS, Grine FE. Incisor size and wear in *Australopithecus africanus* and *Paranthropus robustus. J Hum Evol* 1991;20:313–340.

Walker A, Teaford M. Inferences from quantitative analysis of dental microwear. *Folia Primatologia* 1989;53:177–189.

Genetic evidence implies that the hominine clade arose between 5 and 6 million years ago (see unit 15). Fossil evidence of the clade is plentiful late in its history, but becomes progressively sparser toward its origin, particularly earlier than 4 million years ago. The earliest putative hominine fossils were located at Lothagam in northern Kenya and Tabarin in central Kenya and are dated at 5.6 and 5 million years, respectively. Both fossils consist of jaw fragments, neither of which displays much anatomical detail. Until recently, the earliest known hominine for which sufficient diagnostic anatomical evidence was available was *Australopithecus afarensis*, fossils of which have been found in Ethiopia, Tanzania, and Kenya, and most of which date between 2.9 and 3.9 million years. (Meave Leakey and colleagues found a 4-million-year-old piece of mandible and several isolated teeth at a site on the east side of Lake Turkana, northern Kenya, which they say closely resemble those associated with *afarensis*.)

New finds of fossils as old or older than *A. afarensis* have been made in Ethiopia, Kenya, and Chad (central Africa). These specimens, which are sufficiently different from *A. afarensis* to have been named new species, include the following: *Ardipithecus ramidus* from Ethiopia, dated at 4.4 million years; *Australopithecus anamensis* from Kenya, with an age range of 4.2 to 3.9 million years; and *Australopithecus bahrelghazali* from Chad, with an age estimate of 3 to 3.5 million years. The known early history of the hominine group has therefore become more complex, as was predicted in earlier editions of this book.

ANATOMY OF *AUSTRALOPITHECUS AFARENSIS*

The first *afarensis* fossils were found in the mid-1970s in the Hadar region of Ethiopia, by an international team led by Donald Johanson, of the Institute of Human Origins (IHO), and Maurice Taieb, a French paleontologist. The extensive fossil collection includes the famous partial skeleton known as Lucy. A dozen jaw fragments recovered from Laetoli, Tanzania, soon after the Ethiopian finds are also included in *A. afarensis*. The initial interpretation of the fossils was controversial and remains so today, albeit to a lesser degree. While many anthropologists accept that the multitude of fossil specimens that have been attributed to *afarensis* do indeed represent a single, sexually dimorphic species, others believe that the fossils belong to two, and perhaps more, species.

Superficially, *A. afarensis* is essentially apelike above the neck and essentially humanlike below the neck.

The cranial capacity of *A. afarensis* ranges between 380 and 450 cm³, or not much bigger than the 300 to 400 cm³ range found in chimpanzees. The cranium itself is long, low, and distinctly similar to that of an ape, having a pronounced ridge (the nuchal crest) at the back to which were attached powerful neck muscles that balanced the head; the larger individuals (males?) have a sagittal crest. As in apes, the upper part of the *A. afarensis* face is small, while the lower part is large and protruding. The projecting (prognathous) lower face partly explains why powerful neck muscles are required to balance the head atop the vertebral column: in physical terms, this structure is a matter of moments.

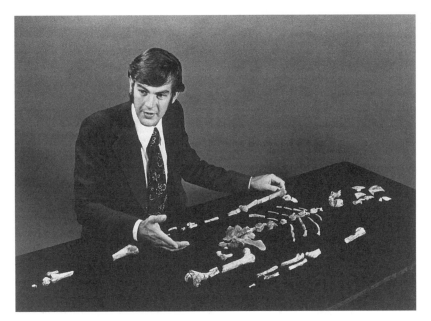

SKELETON OF "LUCY": This 40 percent complete skeleton, shown with her discoverer, Donald Johanson in 1975, is one of the smallest specimens of *Australopithecus afarensis*. Her anatomy combines ape and human characteristics. Obviously adapted for considerable bipedalism, Lucy nevertheless had somewhat apelike limb proportions (short legs and long arms), and an apelike cranium and dentition. (Courtesy of the Cleveland Museum of Natural History.)

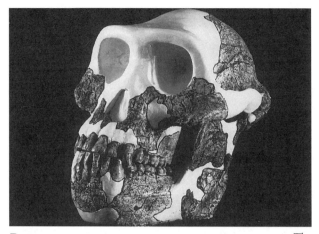

RECONSTRUCTION OF AN *AFARENSIS* CRANIUM: The apelike features of *Australopithecus afarensis* are particularly evident in this cranium, which was constructed from fragments of several different crania. A relatively complete cranium was discovered in 1993, showing anatomy very much like this one. The increased robusticity in the jaws, the slightly enlarged cheek teeth, and reduced canines provide major clues to its hominine status. (Courtesy of the Cleveland Museum of Natural History.)

Many details of the underside of the *A. afarensis* cranium (the basicranium) signify its hominine status, including the central positioning of the foramen magnum (see unit 17), through which the spinal cord passes. The hominine status of *A. afarensis* is even more clearly seen in the jaws and teeth, however.

A comparison of a modern ape's dentition (the dentition of a chimpanzee, for example) with that of modern humans reveals some striking differences (see unit 18). In most respects, *A. afarensis* is somewhat intermediate between these two patterns. Although reduced, the canines are still large for the typical hominine and significant sexual dimorphism is present; a diastema is required to accommodate each canine in the opposite jaw. In many individuals, the first premolar is distinctly apelike in having a single cusp, but the development of a second cusp can sometimes be discerned. Although the molars are characteristically hominine in overall pattern, they do not resemble the grinding millstones that are apparent later in the hominine lineage (see unit 20).

BEHAVIOR OF *AUSTRALOPITHECUS AFARENSIS*

As we saw in unit 17, the bipedal adaptation imprints itself in many different ways on the postcranial skeleton. The question is, How well does *A. afarensis* measure up as a biped? Functional analyses of various parts of the postcranial skeleton have been carried out by a large number of researchers, working in the United States, England, and France.

Owen Lovejoy collaborated with Johanson and his colleagues to concentrate on the pelvis and lower limbs. The pelvis of *A. afarensis* is undoubtedly more like that

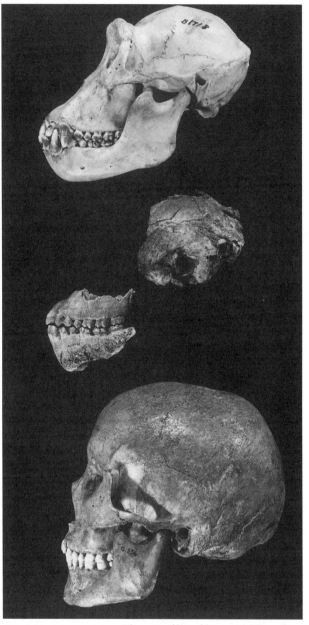

CRANIA COMPARED: These profiles of human, chimpanzee, and *afarensis* crania show how very apelike the first known hominine was. (Courtesy of the Cleveland Museum of Natural History.)

of a hominine than that of an ape, being squatter and broader, but significant differences exist as well, such as the angle of the iliac blades (hip bones). These differences were not functionally significant in terms of achieving the balance required for bipedal locomotion, concluded Lovejoy. Combined with the architecture of the femoral neck and the pronounced valgus angle of the knee, this character would permit a full, striding gait, essentially like modern humans in overall pattern if not in every detail. In other words, *A. afarensis* was said to be a fully committed terrestrial biped, with any apelike anatomy being genetic baggage and not functionally significant.

Meanwhile, other researchers began to see indications of arboreal adaptation in the *A. afarensis* anatomy. French researchers Christine Tardieu and Brigitte Senut studied the lower limb and upper limb, respectively, and inferred a degree of mobility that would be consistent with arboreality. Russell Tuttle, of the University of Chicago, pointed out that the bones of the hands and feet were curved like those of an ape, which could be taken as indicative of climbing activity. William Jungers reported that although *A. afarensis* arms are hominine in terms of length, its legs remain short, like those of an ape, which favors a climbing adaptation. Examining certain *A. afarensis* wrist bones, Henry McHenry concluded that the joint would have been much more mobile than in modern humans, a character consistent with an arboreality.

Following a more wide-ranging survey, Jungers, Jack Stern, and Randall Susman (all of SUNY, Stony Brook) argued that the full suite of postcranial anatomical adaptations indicated that, although *A. afarensis* was bipedal while on the ground, it spent a significant amount of time climbing trees, for sleeping, escaping predators, and foraging. Moreover, they concluded, while the animal was moving on the ground it could not achieve a full striding gait, as Lovejoy had argued, but instead adopted a bent-hip, bent-knee posture. Such a posture would clearly have important biomechanical and energetic implications for *A. afarensis*.

The differences of opinion in the *A. afarensis* locomotor debate stem partly from a lack of agreement over how to define the anatomy in certain instances and partly from differences in functional interpretation of other aspects of the anatomy. The opposing views were aired on an equal footing at a scientific symposium organized by the Institute of Human Origins in Berkeley in 1983. Since then, most publications have favored the partially arboreal, bent-hip, bent-knee bipedal locomotor posture.

The key anatomical features cited in support of a partially arboreal adaptation include the following:

- Curved hand and foot bones,
- Great mobility in the wrist and ankle,
- A shoulder joint (the glenoid fossa) that is oriented toward the head more than in humans, and
- Short hind limbs.

Opponents of arboreal adaptation dispute the degree of mobility in the *A. afarensis* ankle, and cite the loss of the opposable great toe, which has become aligned with the other toes, a clear adaptation to bipedality (but see the discussion below).

Anatomical features that might imply a less than human style of bipedality are found in several parts of the body. For instance, although the forelimbs have assumed hominine proportions, thus improving weight distribution and balance required for bipedalism, the legs are short, as in an ape. Short legs mean short stride length. In addition, the foot is long relative to the leg, meaning that clearance could be achieved only by increasing knee flexion during walking (like trying to walk in oversized shoes).

The SUNY researchers and Maurice Abitol, of the Jamaica Hospital, New York, independently interpret the angle of the iliac blade of the pelvis in *A. afarensis* to imply a method of balance during bipedalism more like that of a chimpanzee than a human—that is, involving a bent hip. The SUNY group also claims that the lunate articular surface of the socket (the acetabulum) into which the head of the femur fits in the pelvis is less complete in *A. afarensis* than in modern humans. This incompleteness arises in a region that takes stress in humans when the fully extended hindlimb passes beneath the hip joint. Ergo, this kind of stride does not occur in *A. afarensis*.

Completing the case for a bent-hip, bent-knee walking posture is the suggestion by the SUNY researchers that the *A. afarensis* knee joint cannot lock in a fully flexed position, as it does in modern humans. The Kent State researchers dispute three points of this description of the anatomy, ultimately rejecting the functional interpretation. The shape of the joint surfaces of certain bones in the foot (the metatarsals) can be taken to imply a greater ability for flexion, which would be useful for climbing, and a poorly developed stability when in a toe-off position. If *A. afarensis* did employ a bent-hip, bent-knee posture, then it would not have used the toe-off step to the degree that occurs in the modern human striding gait (see unit 17).

Finally, Jungers has examined the size of hindlimb joints—particularly the femoral head—in modern apes, humans, and *A. afarensis*. The rationale was that distributing body weight on four limbs for most of the time—as chimpanzees and gorillas do, for instance—would not require the joint surfaces of the lower limbs to be as extensive, relatively speaking, as they must be if full weight was permanently balanced on the hindlimbs, as occurs in humans. Sure enough, humans have much larger femoral head surfaces than would an African ape of the same size. Although the femoral head surface in *A. afarensis* is larger than that of an ape of the same size, it does not even approach the human range. This finding leads Jungers to conclude that "the adaptation to terrestrial bipedalism in early hominines was far from complete and not functionally equivalent to the modern human condition." Such an anatomically and functionally intermediate stage in *A. afarensis* should not be too surprising, especially since the postcranial anatomy of its predecessor, *Ardipithecus ramidus*, is reported to be distinctly apelike.

Another aspect of the postcranial anatomy worth noting in relation to the biology of *A. afarensis* is the structure of the hands. Although they have often been characterized as "surprisingly modern," they are actually rather apelike in manipulative capacity and overall curvature. For instance, the thumb is shorter than in the human hand, and the fingertips are much narrower. Human fingertips are broad, a trait related to the high degree of innervation required to perform fine manipulative tasks. It should be noted that the earliest stone tools recognized from the fossil record date to approximately 2.5 million years, which is post-*afarensis* (see unit 23).

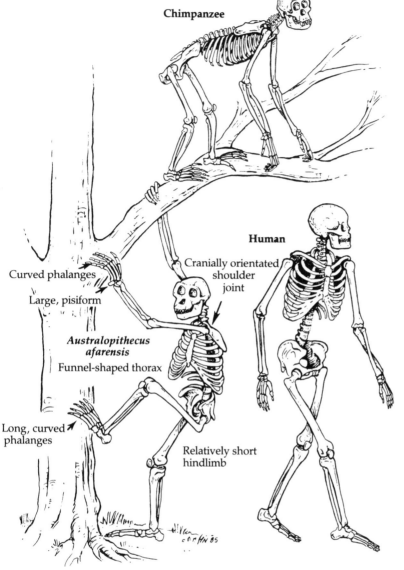

Chimpanzee

Curved phalanges

Large, pisiform

Australopithecus afarensis

Funnel-shaped thorax

Long, curved phalanges

Cranially orientated shoulder joint

Human

Relatively short hindlimb

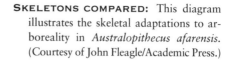

SKELETONS COMPARED: This diagram illustrates the skeletal adaptations to arboreality in *Australopithecus afarensis*. (Courtesy of John Fleagle/Academic Press.)

Although bipedal in posture, *A. afarensis* retained several apelike aspects, particularly in body proportions. As can be seen in the accompanying diagram, its legs are relatively shorter and its arms relatively longer than in modern humans. In addition, as Peter Schmid and Leslie Aiello have demonstrated independently, the shape of the trunk is apelike in being bulky relative to stature.

Overall, then, *A. afarensis* anatomy—and presumably behavior—is somewhat intermediate between that of an ape and a human, a pattern that does not exist today.

NEW FOSSIL DISCOVERIES

The discoveries of *Ardipithecus ramidus* and *Australopithecus anamensis* simultaneously dislodged *afarensis* as the earliest known hominine species and threw doubt on its status as the ancestor of all later hominines. The emerging picture of early hominine evolution is therefore one involving an early bushy adaptive radiation, with considerable uncertainty about how some of the species might be linked together phylogenetically.

The *A. ramidus* specimens include part of a child's mandible, some isolated teeth, a fragment of basicranium, and three bones of a left arm of a single individual. The dentition is more primitive (that is, more apelike) than in *afarensis*, with narrower molar teeth capped with thin enamel, unlike the condition in all other known hominines; the canines are larger, but not as large as in living apes. The arm is both apelike and non-apelike, from which the species' discoverers, Tom White and his colleagues, conclude that the mode of locomotion cannot confidently be determined. Nevertheless, the position of the foramen magnum, through which the spinal cord passes in the basicranium, indicates that the creature employed some sort of bipedal posture.

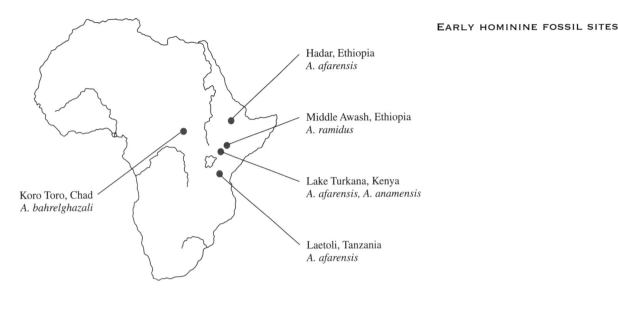

Hadar, Ethiopia
A. afarensis

Middle Awash, Ethiopia
A. ramidus

Lake Turkana, Kenya
A. afarensis, A. anamensis

Laetoli, Tanzania
A. afarensis

Koro Toro, Chad
A. bahrelghazali

In August 1995, Meave Leakey, Alan Walker, and two colleagues published details of hominine fossils from two sites in northern Kenya, Kanapoi and Allia Bay, which they named *Australopithecus anamensis* ("anam" means "lake" in the local Turkana language). The fossils (9 from Kanapoi and 12 from Allia Bay) include upper and lower jaws, cranial fragments, and the upper and lower parts of a leg bone (tibia). The dentition is less apelike than in *ramidus*, having thick enamel on the molar teeth but relatively large canines. The tibia implies that *anamensis* was larger than *ramidus* and *afarensis*, with an estimated weight of 46 to 55 kilograms; its humanlike anatomy implies that *anamensis* was bipedal in posture and locomotion. The Kanapoi fossils have been dated at 4.2 million years and those at Allia Bay at 3.9 million years.

Because the history of australopithecine discoveries was, until recently, located exclusively in eastern or southern Africa, many anthropologists assumed that it reflected a real difference in the distribution of hominines and apes. That is, hominines were seen as being restricted to east of the Great Rift Valley, with apes remaining mainly in the west. The relatively continuous forest cover of central and western Africa was thought to provide an unsuitable habitat for hominines. At the end of 1995, however, this picture changed, with the announcement of the discovery of a hominine mandible in Chad, central Africa, which is 2500 kilometers west of the Rift Valley. The mandible, which has thick-enameled teeth, has been dated by faunal correlation to between 3 and 3.5 million years old. Michel Brunet (of the University of Poitier, France), David Pilbeam (of Harvard University), and several colleagues initially described the jaw as being similar to that of *Australopithecus afarensis*. On further study, however, they identified differences that signaled a different species, which they named *Australopithecus bahrelghazali*.

The interpretation of the evolutionary relationships among these early hominines remains uncertain, but is focused principally on *ramidus* and *anamensis*. Some schol-

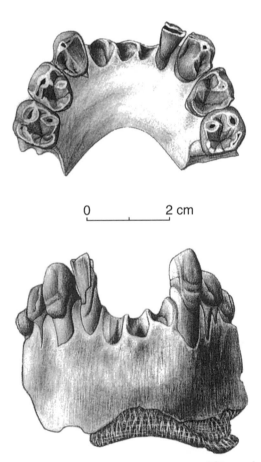

AUSTRALOPITHECUS BAHRELGHAZALI: This newly discovered partial mandible from Chad, central Africa, is the first australopithecine to be found west of the Rift Valley, overturning the assumption that hominine habitat was restricted to areas east of the Rift Valley. The drawings show the top and front view of the mandible. (Courtesy of M. Brunet.)

0   2 cm

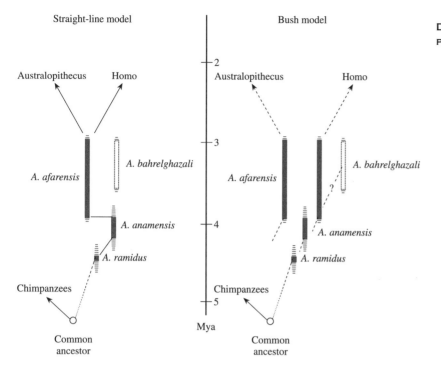

Straight-line model

Bush model

Mya

DIFFERENT PHYLOGENETIC INTER-PRETATIONS: Some scholars feel confident in making phylogenetic links between the recently discovered early hominines, shown in a straight-line model (*left*); others feel a looser interpretation is more realistic at present, which gives a bush model (*right*).

ars, such as White, have suggested an ancestor-descendant relationship, with *ramidus* being ancestral to *anamensis*, and *anamensis* being ancestral to *afarensis*. In their paper announcing the discovery of *anamensis*, Leakey and Walker stated that the species might be ancestral to *afarensis*, but conceded the possibility of several species coexisting at this early period in hominine history, making firm phylogenetic reconstruction premature at this stage.

The discovery of *bahrelghazali* further complicates the picture. In their 1996 publication, Brunet, Pilbeam, and their colleagues note that, because of differences between the newly named species and the recently discovered *Australopithecus anamensis* and *Ardipithecus ramidus*, *A. bahrelghazali* probably belongs to a clade that was separate since at least 4 million years ago and possibly longer. Because it is more gracile than other hominines of the time, the authors say, this species may be related to the ancestry of *Homo*. If correct, a phylogeny of hominines that entails *afarensis* being ancestral to all later hominines is likely to oversimplify hominine evolution.

PALEOENVIRONMENTS OF THE EARLY HOMININES

Analysis of the geology of the Aramis site and the fossils of other creatures found there indicates that this area was a closed woodland or forest setting at the time that these hominines lived there. For instance, 30 percent of the vertebrates were colobine monkeys, which are forest animals. The Allia Bay hominines apparently lived in or near gallery forest associated with a large river; at Kanapoi, the environment was more open, but probably close to gallery forest. The Hadar *afarensis* population lived in a woodland habitat, while Laetoli was much more open, possibly even grassland savannah. The *bahrelghazali* species apparently lived in a lakeside environment, incorporating rivers and streams and associated woodland. It is therefore apparent that the earliest hominines occupied a diversity of habitats, including closed forest and open terrain. This consideration is important in assessing competing hypotheses for the origin of bipedalism (see unit 17). 🌼

KEY QUESTIONS

- Why might the identification of a very early hominine species be difficult based on limited fossil evidence?
- Why was the discovery of fossils older than than *Australopithecus afarensis* to be expected?
- How does the knowledge of extant species constrain the interpretation of behavioral scenarios for paleospecies in general and in *A. afarensis* in particular?
- What are the implications of the discovery that the earliest known hominines lived in heavily wooded or forest environments?

KEY REFERENCES

Abitol MM. Lateral view of *Australopithecus afarensis*. *J Hum Evol* 1995;24:211–229.

Aiello LC. Variable but singular. *Nature* 1994;368:399–400.

Andrews P. Ecological apes and ancestors. *Nature* 1995;376:555–556.

Brunet M, *et al. Australopithecus bahrelghazali*, a new species of early hominid from Koro Toro region, Chad. *Comptes Rendues*, series II 1996;322:907–913.

———. The first australopithecine 2,500 kilometers west of the Rift Valley (Chad). *Nature* 1995;378:273–274.

Crompton RH, *et al.* The mechanical effectiveness of erect "bent-hip, bent knee" bipedal walking in *Australopithecus afarensis. J Hum Evol* 1998;35:55–74.

Johanson DC, White TD. A systematic assessment of early African hominids. *Science* 1979;203:321–330.

Jungers WL. Relative joint size and hominid locomotor adaptations with implications for the evolution of hominid bipedalism. *J Hum Evol* 1988;17:247.

Kimbel WH, *et al.* The first skull and other discoveries of *Australopithecus afarensis* at Hadar, Ethiopia. *Nature* 1994;368:449–451.

Leakey MG, *et al.* New four-million-year-old hominid species from Kanapoi and Allia Bay, Kenya. *Nature* 1995;376:565–571.

Lewin R. Bones of contention. *New Scientist* Nov 4, 1995:14–15.

Richmond BG, Jungers WL. Size variation and sexual dimorphism in *Australopithecus afarensis* and living hominoids. *J Hum Evol* 1995;29:229–245.

Susman RL, Stern JT, Jungers WL. Arboreality and bipedality in the Hadar hominids. *Folia Primatologica* 1984;43:113–156.

White TD, *et al.* New discoveries of *Australopithecus* at Maka in Ethiopia. *Nature* 1993;366:261–265.

———. *Australopithecus ramidus*, a new species of early hominid from Aramis, Ethiopia. *Nature* 1994;371:306–312.

Wood B. The oldest hominid yet. *Nature* 1994;371:280–281.

THE HOMININE ADAPTATION

PART 5

The Australopithecines
Early Homo
Hominine Relations
Early Tool Technologies

THE AUSTRALOPITHE-CINES

20

If one were able to go back to Africa at a time between 3 and 2 million years ago, one would find several hominine species, perhaps sharing the same habitat, much as some species of Old World monkeys do today, or perhaps occupying different habitats, as do modern chimpanzees and gorillas. How many hominine species existed on the continent during that period remains a matter of debate and uncertainty—no less than six, and maybe more.

However many hominine species existed 2 million years ago, they could be classified into two groups: one composed of animals with relatively large brains and small cheek teeth, and a second comprising species with relatively small brains and large cheek teeth. The large-brained species were members of the genus *Homo*, of which several species may have coexisted. The second group are the australopithecines (members of the genus *Australopithecus*); they all became extinct.

In this unit, we will discuss the discovery of some of the major australopithecine fossils, and the anatomy and biology of these creatures; unit 21 will address the earliest members of the *Homo* group; and unit 22 will describe current hypotheses explaining how these various hominine species were related to one another—their evolutionary tree, or phylogeny.

MAJOR SITES OF AUSTRALOPITHECINE FOSSILS: SOUTH AFRICA

Raymond Dart, an Australian anatomist at the University of the Witwatersrand, Johannesburg, South Africa, described the first australopithecine in November 1924 and published his interpretation of it in the journal *Nature* in February 1925. He named the specimen *Australopithecus africanus*, or southern ape from Africa. The fossil, which had been collected by workers at a lime quarry at Taung, southwest of Johannesburg, was that of an immature apelike individual, who, based on an apelike pattern of tooth development (see unit 18) died at the age of three years. The specimen consists of the face, part of the cranium, the complete lower jaw, and a **brain endocast**, formed when sand inside the skull hardened to rock, recording the shape of the brain. An expert in neuroanatomy, Dart considered the brain to have a humanlike rather than apelike, configuration; he also noted that the formamen magnum was placed centrally in the basicranium, as it is in humans, and not toward the rear, as is the case in apes. Moreover, the canine teeth were small—a humanlike character. He concluded that the creature was a biped and was therefore a primitive form of human.

A decade passed before further hominine discoveries were made, when Robert Broom, a Scottish paleontologist, joined Dart in Johannesburg and initiated further exploration. During the next several decades, a rich collection of hominine specimens—cranial and postcranial—was recovered from three cave sites near to Johannesburg (Sterkfontein, Swartkrans, and Kromdrai) and another hundred miles to the northeast (Makapansgat). Sterkfontein and Makapansgat yielded further *A. africanus specimens,* while remains of a more heavily built species, *A. robustus,* were recovered from Swartkrans and Kromdrai. (Broom actually gave the more robust species by a different genus name, *Paranthropus*; a generic distinction between the two species has recently become supported.)

Dating the South African hominines has proved difficult because their cave context is not appropriate for radiometric dating. A combination of paleomagnetic dating and faunal correlation (see unit 7) has yielded ranges of 3.5 to 2.5 million years for the gracile australopithecines and 2.0 to 1 million years for the robust species. A recent reassessment of the ecology of the australopithecines indicates, for instance, that their habitat at Makapansgat consisted of a mixture of forest and thick bush, rather than open savannah once assumed to have prevailed in the area. At the caves near to Johannesburg, the habitat was more open.

Anthropologists at first balked at the suggestion that australopithecines were part of human evolution, and instead viewed them as a form of ape. Acceptance finally

TAUNG CHILD: Initially thought to have died at the age of seven years (based on a human pattern of development), the Taung child actually lived to be only three years old (based on an ape pattern of development). (Courtesy of Peter Kain and Richard Leakey.)

MAJOR AUSTRALOPITHECINE DISCOVERIES: EAST AFRICA

The first hominine discovery in East Africa was made in mid-1959 at Olduvai Gorge in Tanzania, when Mary Leakey found a skull (but no lower jaw) that was similar to the robust australopithecines of South Africa, but even more heavily built. Because of the differences between the Olduvai hominine and those discovered in South Africa, Louis Leakey gave it the name of a new genus species, *Zinjanthropus boisei*. This was later changed to *Australopithecus boisei*. The age of the Olduvai fossil was soon established as 1.75 million years via the first application of radiometric dating (potassium/argon) in anthropology.

Although specimens of other hominines have been found at Olduvai (*Homo habilis*, unit 21, and *H. erectus*, unit 24), no unequivocal remains of *A. africanus* have been found there. The discovery of *A. afarensis* at the Laetoli site, near Olduvai Gorge, was described in unit 19.

The Leakeys' work at Olduvai Gorge helped establish East Africa as an important source of early hominines, but it was their son, Richard, who built on that foundation and made the region preeminent in paleoanthropology. In his first full season of prospecting on the east side of Lake Turkana, in northern Kenya in 1969, Richard Leakey found a complete, intact skull of *A. boisei*. This find initiated an almost uninterrupted period of discovery,

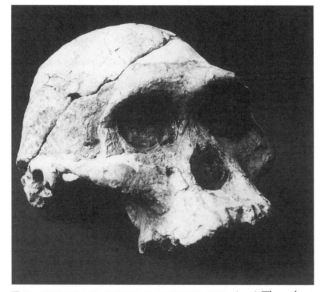

TWO FORMS OF AUSTRALOPITHECINE: (*top*) The robust form of australopithecine, from Swartkrans; (*bottom*) the gracile form, from Sterkfontein. (Courtesy of Peter Kain and Richard Leakey.)

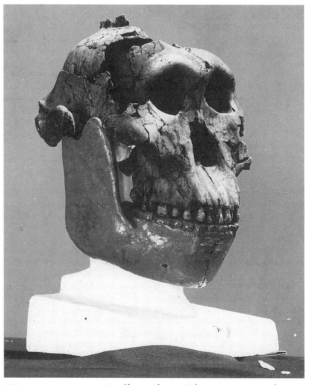

"ZINJANTHROPUS": Shown here with a reconstructed mandible, the cranium was reconstructed from a jigsaw of hundreds of fragments. (Courtesy of Bob Campbell.)

came at the end of the 1940s. Modern paleoanthropology was therefore established in South Africa, where at least two species of australopithecine thrived early on, in coexistence with early species of *Homo*, as was learned in later decades (see units 21 and 22).

Robust australopithecine at Lake Turkana:
Richard Leakey found this intact cranium of *Australopithecus boisei* (KNM-ER 406) on the first major season of work on the east side of Lake Turkana. (Courtesy of Peter Kain and Richard Leakey.)

which continues today under the direction of Leakey's wife, Meave.

Since the early 1980s, collections have also been made on the west side of Lake Turkana. These finds include a complete cranium of a 2.6-million-year-old robust australopithecine, which some term *Australopithecus aethiopicus*. The type specimen of this species had been found earlier by French researchers, in the Omo Valley, Ethiopia. Because the sediments around Lake Turkana are interleaved with volcanic tuffs, the fossils of the region can

now be securely dated. The collection shows the coexistence of several hominine species (*Australopithecus* and *Homo*) between 3 and 2 million years ago, but no unequivocal *A. africanus*. Many consider the latter to be an exclusively South African species, with *A. robustus* and *A. boisei* being geographical variants of the robust form. The discovery of *A. anamensis* around Lake Turkana was described in unit 19.

The important discoveries of *A. afarensis* in the Hadar region of Ethiopia, was noted in unit 19. The discovery of a 1.4-million-year-old specimen of *A. boisei* from the Konso region of Ethiopia was reported at the end of 1997, which in most of its characters is *boisei*-like, but the cheek bones are *robustus*-like, and the back of the cranium is *humans*-like. The newly announced *Australopithecus bahrelghazali* from Chad, which is a contemporary of *A. afarensis*, was also described in unit 19.

No undisputed australopithecine fossil has been found outside the African continent. Most scholars agree that hominines did not leave Africa until approximately 2 million years ago, when *Homo ergaster/erectus* expanded its range to include Eurasia (see unit 24).

Australopithecine biology

Like all early hominines, the australopithecines were essentially bipedal apes with modified dentition. The hominine mode of locomotion and dental apparatus are likely to have been adaptations to a habitat—and therefore diet—that increasingly differed from the environments associated with apes (see unit 17). The later australopithecines appear to have lived in a more open environmental setting—not the open plains of traditional stories, but bushland and woodland savannah. Food was probably located in widely scattered patches and, judging from the structure of these species' teeth and jaws, appears to have required more grinding than an ape's diet.

Australopithecine sites mentioned in the text

Koro Toro, Chad
A. bahrelghazali

Hadar, Ethiopia
A. afarensis

Audipithecus ramidus

Omo, Ethiopia
A. aethiopicus

Turkana, Kenya
A. boisei, A. aethiopicus, A. anamensis

Olduvai, Tanzania
A. boisei

Laetoli,
A. afarensis

Taung, South Africa
A. africanus

Makapansgaat, South Africa
A. africanus

Sterkfontein
A. africanus

Swartkrans
A. robustus

Kromdrai
A. robustus

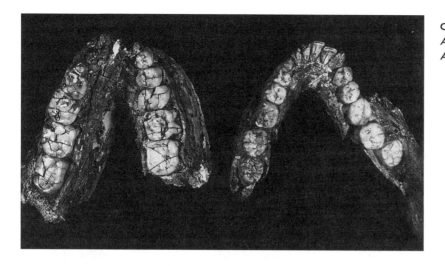

COMPARISON OF LOWER JAWS OF *AUSTRALOPITHECUS ROBUSTUS* AND *A. AFRICANUS*: Note the massive molar teeth in the *A. robustus* mandible from Swartkrans (*left*) compared with that of *A. africanus* from Sterkfontein (*right*). (Courtesy of Milford Wolpoff.)

Analysis of microwear patterns of australopithecine teeth gives some insight into diet. For instance, using scanning electron microscopy, Alan Walker found that the microwear pattern in robust australopithecines resembled that of chimpanzees and orangutans, both of which eat various forms of fruit. More recently Frederick Grine, of the State University of New York at Stony Brook, and Richard Kay, of Duke University, concluded that the robust species consumed foods that were tougher than those eaten by the gracile species. The difference, they suggested, matches that found between the modern-day spider monkey, which eats fleshy fruits, and the bearded saki, which lives on seeds encased in a tough covering.

AUSTRALOPITHECINE ANATOMY

The terms "gracile" and "robust" appear to imply substantial anatomical differences between the two forms, with one being small and delicately built and the other exhibiting a larger and generally more massive form. In recent years, however, scholars have come to realize that the difference between the two forms lies mainly in the dental and facial adaptations to chewing: the robust forms have larger grinding teeth, more robust jaws, and more bulky chewing muscles and muscle attachments.

Recent body weight and stature estimates for australopithecines are as follows:

- *A. africanus*: 41 kilograms for males and 30 kilograms for females, with statures of 138 and 115 centimeters, respectively
- *A. robustus*: 40 kilograms for males and 32 kilograms for females, with statures of 132 and 110 centimeters, respectively
- *A. boisei*: 49 kilograms for males and 34 kilograms for females, with statures of 137 and 124 centimeters, respectively.

Estimates of brain size, which are based on a small number of specimens, typically give the robust species an edge over their gracile cousins. In fact, both were considered to be very close to 500 cm³ (see unit 31). However, recent analysis by Glenn Conroy, of Washington University School of Medicine, and others, using computerized tomography (or CT scanning), indicates that brain capacities of australopithecines have been consistently overestimated by as much as 10 percent.

The teeth, jaw, and cranial anatomy is really one functional complex. As we saw in unit 18, the hominine dental adaptation can be described in general as moving in the direction of producing a grinding machine. The two forms of australopithecine differ in that the robust species have taken this adaptation to an extreme, having enormous, flat molars and relatively small, bladelike incisors and canines.

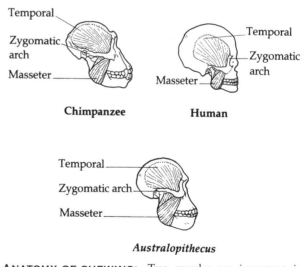

ANATOMY OF CHEWING: Two muscles are important in moving the lower jaw during chewing: the masseter, which is attached to the zygomatic arch (cheek bone), and the temporal, which passes through the arch. The larger the masseter and temporal muscles, the larger the arch. Chimpanzees have approximately three times as much chewing-muscle bulk than modern humans, and the australopithecines even more.

This exaggeration of the hominine dental adaptation is most extreme in the robust australopithecine group. For instance, all hominines have a tooth row that is tucked under the face more than in apes, giving them a less projecting facial profile and increasing chewing efficiency. In the robust australopithecines, this is particularly marked. The extra muscle power necessary for this chewing action in the robust species has two anatomical consequences. First, one of the muscles that powers the lower jaw—the temporal muscle—is anchored to a raised bony crest that runs along the top of the cranium, front to back. This sagittal crest, which is also found in gorillas, is absent in gracile australopithecines. Second, the great size of the temporal muscle in robust australopithecines and the existence of a second chewing muscle, the masseter, cause the cheek bones (the zygomatic arch) to become exaggerated and flared forward. This feature and the strengthening of the central part of the face by pillars of bone give the robust australopithecine face a characteristic "dished" appearance.

In terms of function and overall size, the postcranial skeletons (that is, from the neck down) of gracile and robust australopithecines are very similar to one another, as far as can be deduced from the limited amount of fossil material available. The australopithecine pelvis of 2 mil-

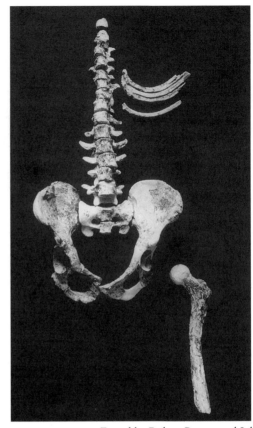

PARTIAL SKELETON: Found by Robert Broom and John Robinson in the late 1940s (and partially reconstructed by Robinson), these bones clearly show the bipedal anatomy of *A. africanus* (museum number, Sts 14). (Courtesy of Peter Kain and Richard Leakey.)

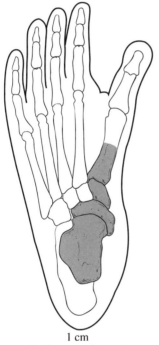

1 cm

ANCIENT FOOT: The drawing shows the recently discovered four articulating foot bones (dark areas) of *Australopithecus africanus;* dated at 3.5 million years, this species is the oldest known hominine in South Africa. The angle of articulation of the bones implies that the great toe diverges from the other toes, as in apes, but to a lesser degree. This feature might have been an adaptation to a degree of arboreality. (Courtesy of R. J. Clarke.)

lion years ago was very much like that of Lucy, who lived a million years earlier. The thigh bone of australopithecines diverges from the typical *Homo* pattern: the head of the femur is smaller than in *Homo* and is attached to a longer, more slender neck.

A recent analysis of an *A. africanus* partial skeleton, discovered in 1987 and published in 1997, revealed that the joints of the arm bones of this specimen were more robust than in modern humans. This implies that this species probably climbed trees as a significant part of its daily routine. The recent discovery of four articulating foot bones from Sterkfontein also implies some arboreality in early australopithecines.

Recent evidence from an unusual anatomical source—the inner ear—also implies that australopithecine locomotion was not identical to that of a fully committed biped. Three bony tubes arranged as arches at right angles to one another form an important organ of balance, known as the semicircular canals or vestibular system. Fred Spoor, an anatomist at the University of Liverpool, England, measured the dimensions of these three arches (the anterior, posterior, and lateral semicircular canals) in living primates, including humans, and found an important dif-

THE BLACK SKULL: Found by Alan Walker in 1984, the skull shows extreme features of australopithecine robusticity, but is dated at 2.6 million years. It is considered by some to be a member of *Australopithecus aethiopicus*. (Courtesy of Alan Walker.)

ference between humans and apes. In humans, the anterior and posterior canals are larger than in apes, while the lateral canal is smaller. Spoor interprets the difference in humans as an adaptation to the demands of bipedal locomotion. Spoor used computerized tomography to measure the dimensions of semicircular canals in a series of hominine fossils. In all australopithecines, the pattern was apelike; in contrast, it was humanlike in early *Homo*. He concluded that australopithecines did not move bipedally in the same way as modern humans or even early *Homo*.

An analysis of the trunk of *Australopithecus* (as seen in Lucy) implies that, however well adapted this species was at bipedal walking, bipedal running was not part of its repertoire. Peter Schmid, of the Anthropological Institute, Zurich, concluded that Lucy's chest was funnel-shaped, not barrel-shaped as in modern humans. The shoulders, trunk, and waist are important elements in

human running: the shoulders enable arm swinging and balance, and in Lucy these features were more apelike than humanlike. In other words, Lucy and other australopithecines may have been bipeds, but active, prolonged running was an adaptation that came only with *Homo*.

RELATIONSHIP BETWEEN ROBUST AND GRACILE AUSTRALOPITHECINES

The gracile and robust australopithecines have often been viewed as basically the same animal, but built on different scales. Functionally speaking, this notion is accurate in many respects. The relationship may also be viewed in terms of evolutionary progression, however, with the gracile species being seen as ancestral to the robust species, in whom the australopithecine traits had become extremely exaggerated: specifically, the chewing apparatus became increasingly robust. If true, then the fossil record should have revealed a steady increase through time in dental, facial, and jaw robusticity.

The 1985 discovery of the *Australopithecus aethiopicus* cranium (KNM-WT 17,000) from the west side of Lake Turkana finally put to rest this simple relationship. The cranium was as robust as any yet known, but was 2.5 million years old. Clearly, the huge molars, flared cheek bones, and dished face could not be the end-product of an evolutionary line if it were present at the origin of that supposed line. How this discovery affects the shape of the hominine family tree remains under discussion (see unit 22).

This cranium, known colloquially as the "black skull," was surprising not only because of its great age but also because it contained an unexpected combination of anatomical characteristics. Although the face was distinctly like that of that most robust of robust australopithecines, *Australopithecus boisei*, the cranium—particularly the top and back—were not: they were similar to those of *Australopithecus afarensis*. Such anatomical combinations in these species surprised many people, and reminds us that hominine biology of 3 to 2 million years ago was more complicated than current hypotheses have allowed. Similarly, the mosaic set of features seen in the Konso *A. boisei* cautions against simple categorizations. ✳

KEY QUESTIONS

- What is the likely locomotor pattern in australopithecines?
- Why do evolutionary biologists not favor reversals, such as would be the case in a progression from *afarensis* to *africanus* to *Homo*, with respect to the robusticity of the joints of the arm?
- What kind of evidence might settle the issue of whether australopithecines made and used stone tools?

- What is the likely relationship between robust and gracile australopithecines?

KEY REFERENCES

Clarke RJ, Tobias PV. Sterkfontein member 2 foot bones of the oldest South African hominid. *Science* 1995;269:521–524.

Conroy GC. Endocranial capacity in an early hominid cranium from Sterkfontein, South Africa. *Science* 1998;280:1730–1731.

Dart R. Adventures with the missing link. New York: Viking Press, 1959.

Grine FE, ed. Evolutionary history of the robust australopithecines. New York: Aldine, 1989.

Grine FE, Kay RF. Early hominid diets from quantitative image analysis of dental microwear. *Nature* 1988;333:765–768.

Lewin R. Bones of contention. New York: Simon and Schuster, 1987.

McHenry HM. How big were early hominids? *Evol Anthropol* 1992;1:15–20.

———. Behavioral implications of early hominid body size. *J Hum Evol* 1994;27:77–87.

McHenry HM, Berger L. Apelike body proportions in *Australopithecus africanus* and their implications for the origin of *Homo*. *Am J Physical Anthropol* 1996;22 (suppl):163–164.

———. Body proportions in *Australopithecus afarensis* and *A. africanus* and the origin of the genus *Homo*. *J Hum Evol* 1998;35:1–22.

Spoor F, *et al*. Evidence for a link between human semicircular canal size and bipedal behavior. *Evol Anthropol* 1996;30:183–187.

Walker A, *et al*. 2.5-Myr *Australopithecus boisei* from west of Lake Turkana, Kenya. *Nature* 1986;322:517–522.

Wood B. Early hominid species and speciation. *J Hum Evol* 1992;22:351–365.

EARLY *HOMO*

21

In the earliest known specimens of *Homo*, the brain size is significantly larger than in australopithecines: 640 cm³ as compared with approximately 500 cm³. Body size was slightly larger, too—albeit not enough to account for the larger brain size. For the first time, simple stone tools are found in the record (see unit 22), and diet may have shifted to include more meat, procured either by scavenging, simple hunting, or a combination of both (see unit 26). The archeological evidence of this shift in subsistence patterns is often assumed to be associated with behaviors unique to *Homo*, although this point remains to be definitively demonstrated. The taxonomic interpretation of early *Homo* fossils was considered contentious when they were first found, and in many ways it remains so today.

THE FIRST DISCOVERIES

The first discoveries of early *Homo* fossils were made at Olduvai Gorge, not long after Mary Leakey had found Zinjanthropus (later known as *Australopithecus boisei*) and Louis Leakey pronounced it to be the maker of the gorge's stone tools. Between 1960 and 1963, a series of fossils were uncovered close to the Zinj site, including hand and foot bones, a lower jaw, and parts of the top of a cranium. The fossils, which were judged to be slightly older than Zinj (therefore older than 1.75 million years), were less robust than Zinj; in addition, the teeth were smaller and the brain was calculated to be significantly larger, with a volume estimated at 640 cm³.

Much of the analysis of these fossils was carried out by John Napier, of London University, and Phillip Tobias, of the University of the Witwatersrand, Johannesburg. In April 1964 Leakey, Napier, and Tobias published a paper in *Nature* announcing *Homo habilis* (handy man), a name that had been suggested to them by Raymond Dart. The publication provoked near outrage among anthropologists, for two reasons: (1) the naming of *habilis* as *Homo* required a redefinition of the genus, including reducing the brain size required to qualify as *Homo*; and (2) many argued that insufficient "morphological space"

divided *Australopithecus africanus* (the presumed ancestor of *habilis*) and *Homo erectus* (the presumed descendant).

The second objection flowed from the prevailing ethos of "lumping" rather than "splitting." In the early days of paleoanthropology, the discovery of hominine specimens was often accompanied by the proposal of a new species. The tendency to name new species on the basis of small anatomical differences between specimens is known as splitting. By the 1960s, anthropologists recognized what they should already have known—namely, that considerable anatomical variation appears within populations. The tendency to designate significant anatomical variation between specimens as intraspecific rather than interspecific variation is known as lumping. Splitters see many species in the record; lumpers see few.

For *Homo habilis* to be a valid species, it must be intermediate between the *A. africanus* and *Homo erectus*, because it was of intermediate age. Lumpers expected considerable anatomical variation in both *africanus* and *erectus*, which left little or no room for an equally variable intermediate. The putative *Homo habilis* fossils therefore had to be either *Australopithecus africanus* or *Homo erectus*. Unfortunately, the critics of *habilis* could not decide to which species it belonged; some said that it

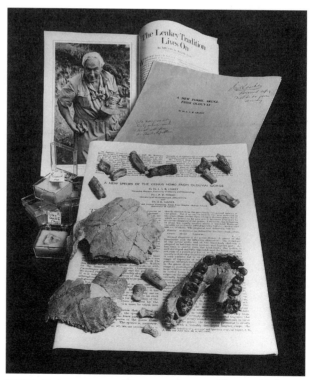

TYPE SPECIMEN OF *HOMO HABILIS*: The establishment of the species *Homo habilis* in 1964 involved a redefinition of the genus *Homo*. This development, among other things, provoked a strong reaction to its validity. (Courtesy of John Reader.)

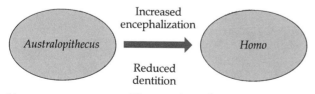

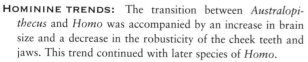

HOMININE TRENDS: The transition between *Australopithecus* and *Homo* was accompanied by an increase in brain size and a decrease in the robusticity of the cheek teeth and jaws. This trend continued with later species of *Homo*.

was a large *africanus*, while others argued that it was a small *erectus*.

Eventually, *Homo habilis* was accepted by most anthropologists as a valid species, partly through the discovery of other, similar specimens, and partly because of a recognition of the excessive lumping tendency. Nevertheless, the species' history in the science has been rocky, principally because of the large degree of anatomical variation found among specimens that are intermediate between *africanus* and *erectus*, which are therefore putative members of *habilis*. Ironically, a current resolution of this dilemma that is gaining much favor involves a recognition of *two* species of *Homo* at this early time (close to 2 million years ago)—not just one, the point to which earlier workers objected so stridently.

FURTHER FINDS, MORE PUZZLES

In 1972 Richard Leakey announced the discovery of a fossil that was to make him world-famous. That fossil, KNM-ER 1470, was the larger part of a cranium pieced together from hundreds of fragments, and has been dated at 1.9 million years old. The face was large and flat, the palate was blunt and wide, and, judging by their roots, the absent teeth would have been large. These features are reminiscent of australopithecines. Nevertheless, the cranium was large, estimated at 750 cm³, which betokened *Homo*. Eventually, the fossil was described in a *Nature* publication as *Homo*, but with its species undetermined.

A year after the announcement of 1470's discovery, a second cranium was found at Lake Turkana, which was to play an important role in the resolution of early *Homo*. Known as KNM-ER 1813, its face and palate are similar to those of *Homo habilis* from Olduvai and different from those of 1470; the brain is small, however—not much more than 500 cm³. Despite this disparity, 1813 has been described by some as a female *Homo habilis*, though Leakey himself has not made this claim.

In 1986, Donald Johanson, Tim White, and a large team of colleagues discovered an extremely fragmented hominine skeleton at Olduvai Gorge, comprising part of the upper jaw, some cranial fragments, most of the right arm, and parts of both legs. The following year they published details of the fossils, code-named OH 62, which they attributed to *Homo habilis*, and dated at between 1.85 and 1.75 million years old. An influential reason why they designated the specimen as *Homo habilis* was the resemblance of the palate to that of a skull found at Sterkfontein a decade earlier, code-named Stw 53, which was assigned to *habilis*. Cranial remains were insufficient to estimate a brain size. The limb proportions, however, were both interesting and surprising.

OH 62 was a small, mature female, comparable to Lucy in being approximately 1 meter tall. As with Lucy, the arms were long and the legs short, compared with later *Homo*. The unexpected aspect, as shown by Robert Martin and Sigrid Hartwig-Scherer, of the Anthropological Institute in Zurich, was that OH 62's arms were even longer than those possessed by Lucy, and its legs shorter.

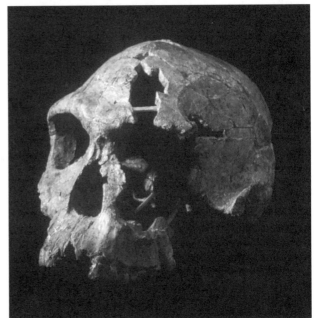

TWO SKULLS FROM KOOBI FORA, KENYA: The cranium KNM-ER 1470 (*top*) was found in 1972 and recognized as belonging to the genus *Homo*, although no species attribution was made initially. A second smaller cranium, KNM-ER 1813 (*bottom*), was found a year later and was thought by some to be *Homo* and others *Australopithecus*. It is now attributed to *Homo* by most observers. (Courtesy of Richard Leakey and Peter Kain.)

Thus, the specimen was even more apelike than *afarensis*, its presumed ancestor.

The year before OH 62 was found, the *Homo ergaster* youth had been unearthed on the west side of Lake

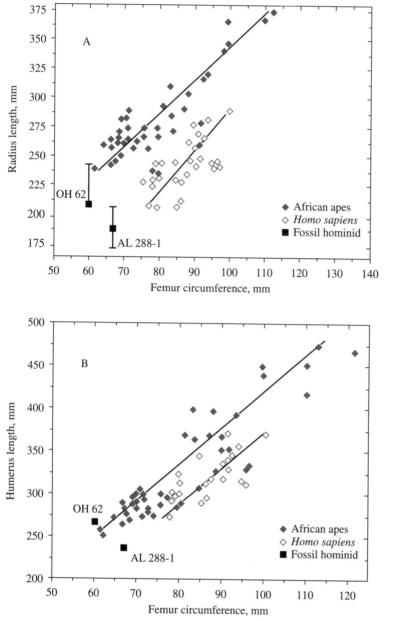

BODY PROPORTIONS OF LUCY AND "LUCY'S CHILD": Comparisons of radius length against femur circumference (A) and humerus length against femur circumference (B) indicate that "Lucy's child" (OH 62) is more apelike than its presumed ancestor, Lucy (Al 288–1). (Courtesy of S. Hartwig-Scherer and R. D. Martin.)

Turkana (see unit 24). This specimen was tall (almost 2 meters) and had very humanlike limb proportions, but lived only 200,000 years later than OH 62. If *Homo habilis* is ancestral to *Homo ergaster*, then evolution from an apelike to a humanlike condition must have occurred very rapidly, which is conceivable under a punctuated equilibrium mode of evolution (see unit 4). It would also require an evolutionary reversal, from moderately apelike limb proportions in *afarensis*, to more apelike proportions in *Homo habilis*, to humanlike proportions in *Homo ergaster*—that is, if OH 62 was indeed a member of *Homo habilis*.

By this time (the mid-1980s), *Homo habilis* had become something of a grab bag of specimens different from its presumed ancestor and its presumed descendant. The question, "Do you accept *Homo habilis* as a valid species?," would likely draw the response, "Well, it depends on which specimens you want to include." As a result, some scholars began to contemplate splitting "*Homo habilis*" into more than one species.

THE EARLIEST KNOWN *HOMO*

The strongest claims for evidence of *Homo* earlier than 2 million years come from the recent reassessment of a cranial fragment from Kenya and a recently discovered mandible from the site of Uraha in Malawi, which lies between east Africa and south Africa.

In 1967, the temporal bone (side of the head) of a hominine was discovered in the Chemeron formation near Lake Baringo, in central Kenya. The structure around the ear—specifically the mandibular fossa, or jaw joint—is diagnostic of *Homo*. The fossil has recently been dated at 2.4 million years old, making it close to that of the oldest-known stone tools, from Kenya and Ethiopia.

In October 1993, an international team, led by Friedemann Schrenk, a German paleontologist, and Timothy Bromage, of Hunter College, New York, published a description of a partial hominine mandible that they had discovered near Lake Malawi. The mandible is less robust than that in australopithecines and the cheek teeth smaller, indicating its association with *Homo;* the specimen has been dated by faunal correlation at between 2.5 and 2.3 million years old, an age comparable to that of the Chemeron hominine. The authors assigned the Malawi specimen to *Homo rudolfensis*, a contemporary of *Homo habilis* that is also found at Lake Turkana (as described later in this unit).

The evolution of *Homo* has been associated with the climatic cooling that occurred approximately 2.5 million years ago, and these two early specimens are consistent with that hypothesis (see unit 5). Because single specimens provide only a loose guide to a species' first appearance, one can say only that *Homo* appeared *at least* 2.4 million years ago; how much earlier it arose is a matter of speculation. It is also impossible to say whether the origin took place in east Africa, further south, or at some other unknown location.

ANATOMY AND BIOLOGY OF EARLY *HOMO*

As previously noted, the brain capacity of early *Homo* is larger than that of the australopithecines, a change that produces several associated anatomical characteristics. For instance, the temple areas in australopithecines narrow markedly (best seen from top view), forming what is known as the **postorbital constriction**. In early *Homo*, this constriction is much reduced because of the expanded

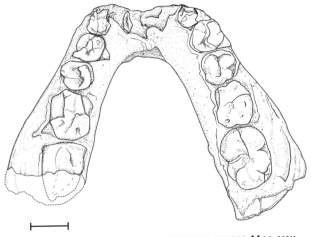

MANDIBLE OF *HOMO RUDOLFENSIS* FROM MALAWI
(Courtesy of F. Schrenk and T. G. Bromage.)

brain. In addition, the face of an australopithecine is large relative to the size of its cranial vault, a ratio that is reduced in the larger-brained *Homo* species. The cranial bone itself is thicker in *Homo* than in *Australopithecus*.

The tooth rows in early *Homo* are tucked under the face as in other early hominines, a feature that becomes even more exaggerated in later species of *Homo*. The jaw and dentition of *Homo*, however, are less massive than in the australopithecines. Although the teeth are capped with a thick layer of enamel, their overall appearance gives less of an impression of a grinding machine than appears in the small-brained hominines: the cheek teeth are smaller and the front teeth larger than in australopithecines, and the premolars are narrower. The patterns of wear on early *Homo* teeth are, however, indistinguishable from those of the australopithecines: the pattern is that of a generalized fruit-eater. Only with the evolution

SITES OF EARLY *HOMO* FOSSIL FINDS:
The species attributions are those suggested by B. Wood (1992).

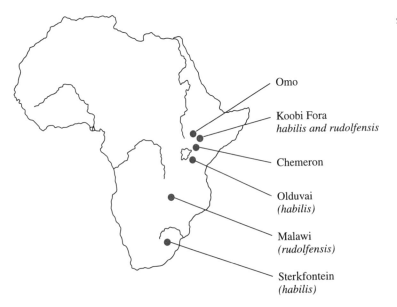

Omo

Koobi Fora
habilis and rudolfensis

Chemeron

Olduvai
(habilis)

Malawi
(rudolfensis)

Sterkfontein
(habilis)

of *Homo ergaster* 1.9 million years ago does the tooth-wear pattern make a dramatic shift, perhaps indicating the inclusion of a significant amount of meat in the diet.

The original set of *Homo habilis* fossils from Olduvai Gorge included a relatively complete hand; its structure was compatible with an ability to make and use tools. The evolution of technological skills associated with stone-tool making has always appeared to be a satisfactory explanation for the expansion of brain capacity in the *Homo* lineage. If australopithecines were equally skillful, then this explanation fails. Presumably, some selection pressure on mental skills must have separated the *Homo* and australopithecine lineages. Whether this separation was associated with the development of more complex subsistence activities or lay in the realm of more complex social interaction (see unit 31) is difficult to determine.

In an analysis of the body proportions of the early hominines, Leslie Aiello found a distinctly human form—that of small body bulk for stature—as well as an apelike form—that of high body bulk for stature. All australopithecines are characterized by the apelike form; *Homo erectus/ergaster* is humanlike, as are certain specimens attributed to *Homo habilis*. OH 62, however, would fit best in the apelike group. The shift from apelike body proportions to humanlike proportions is seen only in *Homo*, and is assumed to be associated with an adaptive shift that includes greater routine activity.

TAXONOMIC TURMOIL

As noted earlier, the OH 62 partial skeleton, with its primitive postcranium, was influential in spurring a revision of the *Homo habilis* taxon. At the time of its discovery, the taxon included dozens of specimens (from Olduvai, Lake Turkana, and Sterkfontein) that displayed an uncomfortably wide range of anatomical variation. Several workers had already expressed the opinion that the fossils belonged to two species, not one. In a major cladistic analysis published in *Nature* in February 1992, Bernard Wood formally proposed two species, a proposal that is widely accepted at present.

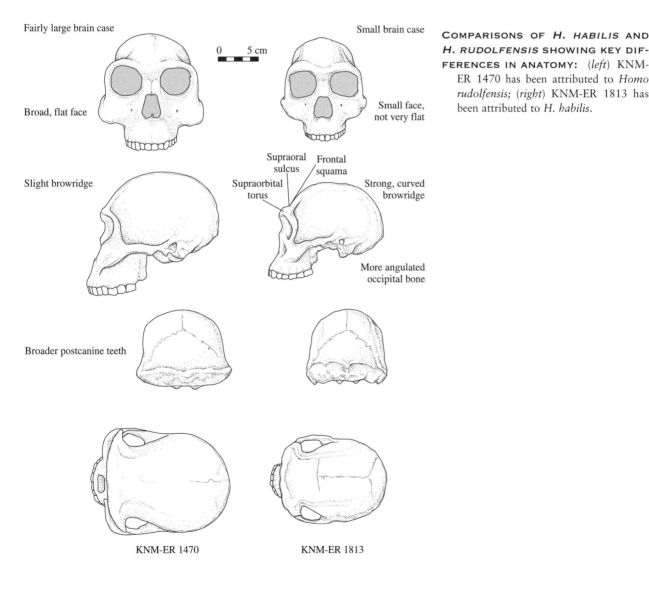

Fairly large brain case

Broad, flat face

Slight browridge

Broader postcanine teeth

0   5 cm

Small brain case

Small face, not very flat

Supraoral sulcus

Frontal squama

Supraorbital torus

Strong, curved browridge

More angulated occipital bone

KNM-ER 1470   KNM-ER 1813

COMPARISONS OF *H. HABILIS* AND *H. RUDOLFENSIS* SHOWING KEY DIFFERENCES IN ANATOMY: (*left*) KNM-ER 1470 has been attributed to *Homo rudolfensis*; (*right*) KNM-ER 1813 has been attributed to *H. habilis*.

The two species proposed by Wood are *Homo habilis* and *Homo rudolfensis*. They are distinguished as follows: *Homo rudolfensis* has a "flatter, broader face and broader postcanine teeth with more complex crowns and roots and thicker enamel." *H. rudolfensis* also has a larger cranium. Wood includes all non-australopithecine specimens at Olduvai in *Homo habilis*, whereas the Lake Turkana fossils are divided between *H. habilis* and *H. rudolfensis*. The small, enigmatic cranium 1813 is included in *H. habilis*, as is a partial skeleton, KNM-ER 3735, which has primitive limb proportions like those of OH 62. The famous 1470 skull is designated as *Homo rudolfensis*, together with a collection of other specimens that includes examples of modern-looking leg bones. The Malawi hominine is designated as *H. rudolfensis*. The Chemeron hominine does not possess characters that are diagnostic of either species.

Other workers, such as Christopher Stringer and Richard Leakey, agree that two species existed. They suggest, however, that the Olduvai specimens should be split into two species: *Homo habilis*, as originally designated by the type material, and a smaller, more archaic form represented by OH 13 and OH 62. Other suggestions have been put forth as well.

Whatever form a consensus might eventually take, there is now general agreement that two species of *Homo* coexisted 2 million years ago. Although Wood's taxonomic distinction is based principally on certain cranial and dental characters, it is useful to think of *Homo habilis* as a smaller-brained creature with archaic postcranium, and *H. rudolfensis* as larger-brained with a more modern postcranium. Which of the two (if either) gave rise to later *Homo* is still debated. *H. rudolfensis* appears to have a good claim based on brain size and modern postcranium, but some insist that its facial and dental anatomy disqualify it from this role; *H. habilis* has a better claim in this latter respect, but its smaller brain and archaic postcranium militate against it.

Progress has definitely be made with the *Homo habilis* muddle, but a consensus on the details remains to be reached. ✳

KEY QUESTIONS

- How strong is the evidence that two species of early *Homo* coexisted?
- How strong is the evidence as to which of these species might have been ancestral to later *Homo*?
- How could one resolve the question of who made the tools?
- What shift in subsistence strategies might be consistent with the change in body proportions between *Australopithecus* and *Homo*?

KEY REFERENCES

Aiello LC. Allometry and the analysis of size and shape in human evolution. *J Hum Evol* 1992;22: 127–147.

Hartwig-Scherer S, Martin RD. Was "Lucy" more human than her "child"? *J Hum Evol* 1991;21: 439–449.

Hill A, *et al*. Earliest *Homo*. *Nature* 1992;355: 719–722.

Johanson DC, *et al*. New partial skeleton of *Homo habilis* from Olduvai Gorge, Tanzania. *Nature* 1987;327:205–209.

Lieberman DE, *et al*. Homoplasy and early *Homo*: an analysis of the evolutionary relationships of *H. habilis* sensu stricto and *H. rudolfensis*. *Am J Physical Anthropol* 1996;51:15–34.

McHenry HM, Berger LR. Body proportions in *Australopithecus afarensis* and *A. africanus* and the origin of the genus *Homo*. *J Hum Evol* 1998; 35:1–22.

Schrenk F, *et al*. Oldest *Homo* and Pliocene biogeography of the Malawi Rift. *Nature* 1993;365: 833–835.

Stanley SM. An ecological theory for the origin of *Homo*. *Paleobiology* 1992;18:237–257.

HOMININE RELATIONS

22

This unit will explore recent developments and current thinking about how early hominines were evolutionarily related to one another. This subject—phylogeny—has always attracted the attention of anthropologists, often overshadowing the more basic questions of hominine biology, such as subsistence strategies and behavior.

THE PROBLEMS OF "LUMPING"

During the first half of this century, scholars commonly assigned a new species name to virtually each new fossil unearthed. In this "splitting" paradigm, each variant in anatomical structure was taken as indicating a separate species. The result was a plethora of names in the hominoid record. In 1965, Elwyn Simons and David Pilbeam, both then at Yale University, rationalized this paleontological mess and reduced the number of genera and species to a mere handful (the "lumping" paradigm).

Lumping became the guiding ethic of anthropology. Taken to its extreme, it led to the "single-species hypothesis," which became popular during the 1960s and early 1970s (see unit 3). Although the single-species hypothesis is no longer considered valid, there is a persisting tendency to interpret anatomical differences as within-species variation rather than among-species variation. One reason for this trend is that, because of the nature of the system, no practical guide has been developed to explain how much anatomical difference between two fossils signals the existence of separate species. "The reason for this is, of course, that there is no direct relationship, indeed no consistent relationship at all, between speciation and morphological change," says Ian Tattersall, an anthropologist at the American Museum of Natural History (see unit 4).

In other words, a daughter species might sometimes diverge from the parental species but develop very little obvious anatomical difference, while considerable differences might arise in other cases. Unless the living animals are available so that you can observe their behavior, it is often impossible to know whether the individuals belong to one species or two. As a result, it is obviously easier to subsume anatomical differences under within-species variation rather than to argue for separate species. This tendency has certainly become a tradition in anthropology. The result, argues Tattersall, "is simply to blind oneself to the complex realities of phylogeny." In other words, the true hominine family tree—the one that actually happened in evolutionary history—almost certainly is more bushy than the ones currently drawn by anthropologists.

Although most anthropologists would regard Tattersall's position as somewhat extreme, many are coming to accept that hominine phylogeny is more complex than it is usually portrayed. This view was emphasized by the rethinking provoked by the 1985 discovery of the "black skull," a robust australopithecine that did not immediately fit into the prevailing phylogenetic picture (see unit 20), and by other recent discoveries (see unit 19). Cladistic methodology appears to offer the most promising approach for overcoming the problem of lumping (see unit 8).

WHICH DATA ARE THE MOST RELIABLE PHYLOGENETIC INDICATORS?

Paleontologists reconstruct phylogenies from comparisons of anatomical similarities present in fossil specimens. As discussed in unit 8, only those similarities that result from a shared evolutionary history (homologies) can reliably lead to accurate phylogenies. Similarities that result from independent, parallel evolution (homoplasies) may lead to erroneous phylogenies. Most anthropologists now accept that homoplasy has been common in hominine evolution but, as we will see later in this unit, less agreement has been reached regarding which traits are homoplasies between certain lineages and which are not. Again, cladistic analysis should, in principle, help resolve this issue.

A further obstacle to accurate phylogenetic reconstruction arises from the way in which different traits are treated. In anthropology, phylogenetic reconstruction is based almost exclusively on cranial traits, for the very

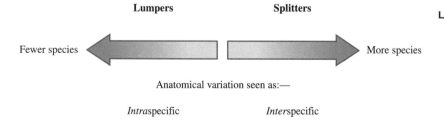

Lumpers   Splitters

Fewer species ⬅   ➡ More species

Anatomical variation seen as:—

*Intra*specific   *Inter*specific

LUMPERS AND SPLITTERS: Different philosophical and methodological approaches yield different views of the species richness of the fossil record. In its early years, anthropology was dominated by splitters, which yielded a plethora of species. Sentiment then switched to lumping, which underestimated species richness. Recently, a swing away from lumping has occurred, but not a return to the previous excesses.

SPLITTERS AND LUMPERS: Louis Leakey (seated) was a keen splitter, reflecting the philosophy of his time; his son Richard Leakey was more cautious, reflecting changing times. (Courtesy of the L. S. B. Leakey Archives.)

good reason that postcranial fossils are much rarer. In one of the more complete cladistic analyses of hominine phylogenetics, Randall Skelton, of the University of Montana, and Henry McHenry, of the University of California, Davis, employed 77 such traits. In their 1992 paper, Skelton and McHenry addressed the issue of the values assigned to these traits, identifying two problems: the independence of the traits, and sample bias.

If all 77 traits were independent, then they would provide information on 77 evolutionary transformations, forming a powerful body of evidence. Anatomical traits are not independent, however, but form parts of trait complexes. For instance, an important trend in early hominine evolution was toward heavy chewing in order to process tough plant foods. This development is seen, for instance, in an increase in the size of molar teeth and in the thickness and depth of the mandible. Bigger teeth and more powerful chewing also require a more robust mandible, changes in face structure, and possibly alterations in the mechanics of muscles that move the jaws. Changes in the size of molar teeth and the robusticity of the mandible are therefore linked as part of an evolutionary package and are not independent of one another.

Thus, phylogenetic analyses should logically group traits into functional packages, rather than treat them as independent. In their analysis of hominine phylogeny, Skelton and McHenry identified five such functional complexes among the 77 traits: heavy chewing (34 traits), anterior dentition (11 traits), basicranium flexion (11 traits), prognathism/orthognathism (8 traits), and encephalization (3 traits). Grouped in this way, the 77 traits give phylogenetic information on just five evolutionary transformations—not 77. Even these five functional complexes are not completely independent, however, because the masticatory system involves many parts of the cranium. For instance, the evolution of traits associated with anterior dentition is linked in part to the evolution of heavy chewing, as is the shape of the

face and certain cranial traits, such as the possession of a sagittal crest.

The second problem of bias sampling is evident from the traits listed above—namely, some aspects of anatomy are more widely represented than others in the fossil record. Traits associated with heavy chewing are obviously the most common, because teeth and jaws are the most resilient parts of the cranium and consequently become part of the fossil record much more frequently. For this reason, anthropologists have concentrated much of their work on teeth and jaws, including basing phylogenetic reconstruction on them. Teeth and jaws, however, are particularly susceptible to homoplasy: species with

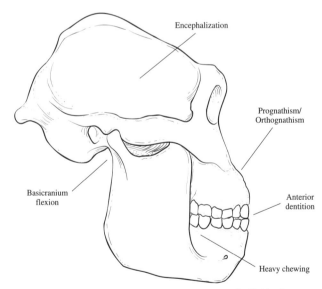

INTERDEPENDENCE OF CHARACTERS: Individual anatomical traits are typically parts of functional complexes and are not evolutionarily independent. These five functional complexes are associated with the hominine cranium.

similar diets will develop similar dentition through natural selection. Teeth and jaws, and their *interpretation*, may therefore receive more attention than their phylogenetic reliability justifies.

KEY QUESTIONS IN HOMININE PHYLOGENY

Three key questions arise in a phylogenetic reconstruction of early hominines:

- The relationship of *Australopithecus afarensis* to earlier and later hominines.
- The relationships among the robust australopithecines (*A. aethiopicus, robustus,* and *boisei*).
- The origin of the genus *Homo*.

As we saw in unit 19, two decades after the first specimens of *A. afarensis* were discovered no consensus had been reached on whether they represent one extremely sexually dimorphic species or two less variable species (one large and one small). Until recently, the majority view held that just one species was present between 3.9 and 2.9 million years ago, and that this species was ancestral to all later hominines. The recent discovery of more *A. afarensis* fossils from Ethiopia did not resolve this difference of opinion.

The 1994 and 1995 reports of the discovery of *Ardipithecus ramidus* and *Australopithecus anamensis* prove false the often implicit assumption that *A. afarensis* was the founding species of the hominine clade. The likelihood that *ramidus* and *anamensis* were part of a bushy phylogeny prior to *afarensis*, rather than being stages in a single, transforming lineage, impacts the status of *afarensis*. (The notion of a single, transforming lineage does have its supporters, however.) It is unlikely that a phylogenetically bushy clade would be reduced to a single species, which then gives rise to further bushiness. Unlikely—but not impossible. Further fossil finds in the period 5 to 3 million years ago will be necessary to resolve this issue.

The question of robust australopithecine relationships affects the placement of *A. aethiopicus* in the evolutionary tree: Is it ancestral to the two later robust australopithecines, or is it separate from them? The issue of the origin of the genus *Homo* concerns the identity of its direct ancestor: Is it *A. afarensis, A. africanus*, or some as yet unknown third species? These two questions will be considered through Skelton and McHenry's cladistic analysis, not because it is universally accepted (it is not, but it is widely respected), but because it offers a strategy for addressing some key problems, particularly that of homoplasy.

THE SKELTON/MCHENRY ANALYSIS

Skelton and McHenry performed a cladistic analysis of the 77 cranial traits in several ways: they treated the traits as if they were independent; they compared the five functional complexes discerned; and they grouped the traits by anatomical region (face, anterior dentition, posterior dentition, mandible, palate, basicranium, and cranial vault), which is another way of overcoming linkage between traits. They then compared the results from these various analyses. Their study was performed prior to the discovery of *Ardipithecus ramidus* and *Australopithecus anamensis*, and it took the conservative position that *Australopithecus afarensis* is indeed a single species. The analysis of the later hominines is unaffected by these recent discoveries. One of the most important, and controversial, conclusions of their work was that traits associated with heavy chewing in hominines are subject to homoplasy.

Mentioned earlier was the trend in early hominine evolution toward ever-heavier chewing. Traits associated with heavy chewing are least developed in *A. afarensis* and most strongly developed in *A. boisei*. The black skull, *A. aethiopicus*, also possesses large cheek teeth and a robust mandible, which many anthropologists interpret as indicating an ancestral relationship to *A. boisei* and the South African robust australopithecine, *A. robustus*. The anterior dentition of *A. aethiopicus*, however, is more similar to that of *A. afarensis* than to that of the other robust australopithecines. The degree of prognathism in *A. aethiopicus* resembles that in *A. afarensis*, while the other robust australopithecines are much less prognathic and more similar to *Homo*. The most parsimonious tree from a phylogenetic analysis using only traits related to the functional complex of heavy chewing gives a cladogram that links all three robust australopithecines as a clade. Analyses using posterior dentition, an anatomical region associated with heavy chewing, produce the same phylogenies.

By contrast, most other types of analysis (taking the 77 traits independently, and assessing the other func-

Key questions in early hominine evolution:

Relationship of *A. afarensis*
to early and later hominines

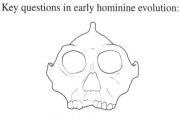

Relationships among
the robust australopithecines

Origin of the genus *Homo*

KEY QUESTIONS IN EARLY HOMININE EVOLUTION

tional and regional complexes, either independently or grouped) yield a different series of possible cladograms, with one being most common. In this tree, *A. aethiopicus* is not ancestral to the other australopithecines, but rather a large-toothed form of *A. afarensis* that became extinct with no descendants. The persistence of this particular cladogram is evidence of its strength, say Skelton and McHenry, which implies that the traits associated with heavy chewing shared by *A. aethiopicus* and the other two robust australopithecines are homoplasies—not the result of common ancestry. A second aspect of Skelton and McHenry's phylogeny that differs from phylogenies constructed by other workers is its proposal of a close link between the other robust australopithecines (*A. boisei* and *robustus*) and earliest *Homo* (discussed below). The proposed phylogeny requires three hypothetical ancestors—species that are as yet unknown, but are implied by the evolutionary transitions in the phylogeny.

Skelton and McHenry's phylogeny is as follows. *A. afarensis* is the most primitive early hominine after *Ardipithecus ramidus* and *Australopithecus anamensis*, from which it probably derived (see unit 19). They propose that *afarensis* gave rise to an as yet unknown species that was *aethiopicus*-like in some ways (in traits not related to heavy chewing); this species was the common ancestor of *aethiopicus* on one hand, and gave rise to *A. africanus*, early *Homo*, and the later robust australopithecines on the other. *A. aethiopicus* is therefore viewed as a side branch that became extinct, while *A. afarensis* was ancestral to all later hominines (but was not their common ancestor). *A. africanus* is derived from the *aethiopicus*-like ancestor, and in its turn gave rise to another proposed *africanus*-like species; this species was the common ancestor of earliest *Homo* on one hand and the robust australopithecines (via a proposed *robustus*-like common ancestor) on the other. Many anthropologists agree that *robustus*-like anatomy is likely to be ancestral to *boisei*. The close relationship between *Homo* and *A. robustus* and *A. boisei* (they share a common ancestor to the exclusion of other hominines) is reflected in a more flexed cranial base, a deeper jaw joint, less prognathism, and greater encephalization compared with *A. africanus*.

This phylogenetic scheme, like other proposed alternatives, implies considerable homoplasy in hominine evolution, particularly in the heavy chewing complex. In contrast to Skelton and McHenry's proposal, other schemes have proposed that *A. aethiopicus* was ancestral to the other robust australopithecines, and that heavy chewing traits are homologous (not homoplasic). A recent cladistic analysis by David Strait and Frederick Grine, at the State University of New York, Stony Brook, strongly supports this view (the monophyly of the three robust australopithecines). This phylogeny shifts the requirement for homoplasy to other traits—namely, anterior dentition, basicranial flexion, encephalization, and prognathism/orthognathism—that *A. aethiopicus* shares with other species.

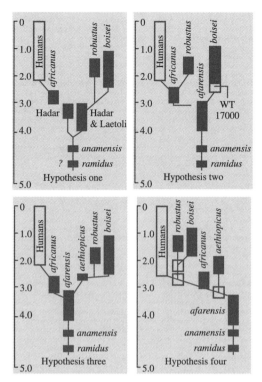

A FOREST OF HOMININE EVOLUTIONARY TREES: Numerous phylogenetic interpretations of hominine history have been proposed. Hypothesis 4 is based on Skelton and McHenry's analysis, and shows the hypothetical ancestors as open boxes. Hypothesis 3 shows the three robust australopithecines as being monophyletic.

A second area of homoplasy appears in the evolution of *Homo*. The shape of the face and small cheek teeth superficially resemble those of *A. afarensis*. Thus, these traits in *Homo* must have resulted from the retention of primitive traits present in *afarensis*, in which case *afarensis* would the be the direct ancestor of *Homo*, or via a reversal of the hominine trend, in which case *africanus* would be the ancestor. A study of the **ontogeny** of facial development reveals that the formation of facial anatomy in *Homo* is unique, not a primitive retention. The well-documented reduction in the size of cheek teeth later in the *Homo* lineage also leads to the conclusion that this trend began with early *Homo*, and thus was not a primitive retention at this stage. If, as Skelton and McHenry point out in their analysis, the face and dentition of *Homo* are indeed uniquely derived, then these traits provide no useful information about the large-toothed australopithecine (known or yet to be discovered) from which it evolved; other, shared traits, such as basicranial flexion and orthognathism, are necessary to link *Homo* to *A. africanus*.

Skelton and McHenry's preferred phylogeny is one of several that can be seen in the anthropological literature; its strength, however, lies in its cladistic methodology and thoughtful treatment of potential biases. Many other schemes derive more than one lineage from *A. afarensis*, for instance, and designate *A. aethiopicus* as the ancestor of the other robust australopithecines. The most contro-versial aspect of the Skelton/McHenry phylogeny is its suggestion that the robust australopithecines are not monophyletic. ✺

KEY QUESTIONS

- Why is homoplasy so pervasive in hominine evolution?
- How would one test, for instance, the hypothesis that *Homo habilis* derived from *A. africanus* rather than from *A. afarensis*?
- Why is the proposed affinity of *Homo*, *Australopithecus robustus*, and *A. boisei* considered controversial?
- What kind of fossil discovery would most upset current views of hominine phylogeny?

KEY REFERENCES

McHenry HM. Homoplasy, clades, and hominid phylogeny. In: Meikle WE, *et al.*, eds. Contemporary issues in human evolution. San Francisco: California Academy of Sciences, Memoir 21, 1996.

Simons E. Human origins. *Science* 1989;245: 1343–1350.

Skelton RR, McHenry HM. Evolutionary relationships among early hominids. *J Hum Evol* 1992;23: 309–349.

Strait DS, *et al.* A reappraisal of early hominid phylogeny. *J Hum Evol* 1997;32:17–82.

Tattersall I. Species concepts and species recognition in human evolution. *J Hum Evol* 1992;22:341–349.

Wood B. Origin and evolution of the genus *Homo*. In: Meikle WE, *et al.*, eds. Contemporary issues in human evolution. San Francisco: California Academy of Sciences, Memoir 21, 1996.

EARLY TOOL TECHNOLOGIES

23

Stone artifacts have been collected by amateurs and professionals alike for centuries and studied as evidence of earlier societies. The mode of study, however, often focused on the implements as phenomena in themselves, with a great emphasis on classification of types. Today a strong interest has developed in studying artifacts within the subsistence context of early hominines. In addition to attempting to understand the functions of individual artifact types, archeologists use these relics to answer the following kinds of questions: How broad was the diet? Specifically, to what extent was hunting an important subsistence activity? Did the social context of subsistence activity include a "home base," such as occurs in modern foraging people (see unit 26)? How did the hominines exploit their range, and how large was it? Thus, experimental archeology, once practiced by a small group of experts in a limited way, has emerged as an important research technique, allowing researchers to use stone implements with the aim of understanding early tool technologies.

Stone-tool assemblages have been classified into five categories, or modes, that are defined by characteristic artifacts in them. These categories appear sequentially through time, but may overlap when earlier modes persist after the appearance of later modes. Mode I technology, the earliest, is based on simple chopping tools that are made by knocking a few flakes off a small cobble. Mode II is characterized by tools that require more extensive conceptualization and preparation, such as bifacial handaxes. In mode III, large cores are preshaped by the removal of large flakes and then used as a source of more standardized flakes that are retouched to produce a large range of artifacts. Mode IV technology is characterized by narrow stone blades struck from a prepared core. Mode V consists of microlith technology, which, as implied, constitutes the production of small, delicate artifacts.

This classification system, which was developed by J. Desmond Clarke, of the University of California, Berkeley, permits a description of the characteristics of archeological assemblages, not of archeological time period. For instance, mode I technology appeared in Africa some 2.6 million years ago and persisted (as an opportunistic practice) until historical times. Moreover, the first appearance of a particular mode often differs in Africa and Eurasia. For instance, mode IV (blade tools) were produced in Africa nearly 250,000 years ago, but did not enter the European record until 40,000 years ago. Such differences almost certainly reflect the dynamics of the origin and migration of anatomically modern humans (see units 27 through 29).

For reasons related to the development of the science of archeology, a different terminology is used to describe archeological time periods in sub-Saharan Africa and those in Eurasia. In Africa, the time before 10,000 years ago (the time of the agricultural revolution, or Neolithic; see unit 36) is known as the **Stone Age**. It is divided into three parts: the Earlier Stone Age (ESA), the Middle Stone Age (MSA), and the Later Stone Age (LSA). In northern Africa and Eurasia, stone-tool cultures prior to the Neolithic are termed the **Paleolithic** and are divided into three stages that are roughly equivalent to those in the Stone Age: the Lower Paleolithic, the Middle Paleolithic, and the Upper Paleolithic. These stages have been defined according to cultural evolution—a somewhat confusing system given that, while the boundaries between the stages are relatively clear in Eurasia, Africa has been associated with a more continuous flow of development. This difference in character may reflect local cultural change in Africa and evidence of population incursions in Eurasia. As mentioned earlier, the timing of first appearance of characteristic cultural artifacts (such as blades) often varies between the two geographic regions.

Bearing in mind the elasticity of stage boundaries, technology development unfolded as follows. The beginning of the ESA corresponds with the first appearance of mode I tools, 2.6 million years ago; the entire ESA includes the first appearance of mode II, approximately 1.5 million years ago, and ends with the first appearance of prepared cores (mode III), which also marks the beginning of the MSA, 300,000 years ago). Traditionally, the LSA was characterized by the first appearance of blade tools and artifacts of personal adornment (mode IV), such as beads, some 60,000 years ago. The recent discovery of blade industries as old as 250,000 years ago, however, has produced the paradox of LSA artifacts within the MSA (see unit 29).

In Eurasia, the Lower Paleolithic begins when humans moved beyond Africa, perhaps close to 2 million years ago (see unit 27), and ends with the first appearance of prepared cores (mode III), some 200,000 years ago. The Middle Paleolithic begins with the first appearance of mode III and ends with the first appearance of mode IV tools, 40,000 years ago. The Upper Paleolithic begins with the first appearance of mode IV, encompasses the appearance of mode V, and ends with the agricultural revolution.

This unit will focus on the first part of the African Earlier Stone Age. Unit 25 will describe the technologies in the remainder of the ESA and the MSA, and the Lower and Middle Paleolithic of Eurasia. The archeology associated with the origin of modern humans (the MSA and LSA of Africa and the Upper Paleolithic of Eurasia) is the subject of unit 29.

THE EARLIEST KNOWN TOOLS

The oldest stone tools in the archeological record are dated to approximately 2.6 million years ago, and are known from sites in the Lower Omo Valley, the Hadar

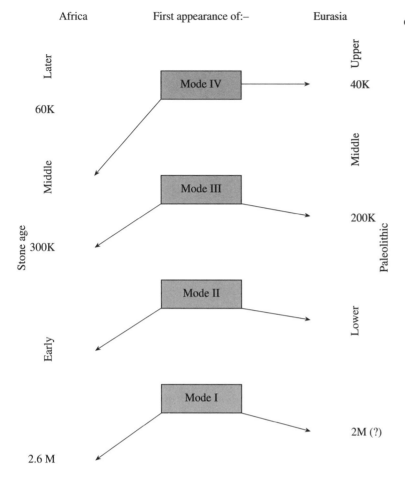

CULTURAL PERIODS: For reasons related to the history of the science of archeology and the impact of new discoveries, the classification of the different periods and stages of cultural development in sub-Saharan Africa and Eurasia represents an uneven mixture of cultural innovation and chronology. (See text for details.)

region, and the Gona region of Ethiopia, and the western shore of Lake Turkana, Kenya. The artifacts from the Lower Omo Valley are atypical, in that they are small quartz pebbles that were shattered to yield sharp-edged implements. Most tools dating from the period 2.6 million to 1.5 million years ago were made from lava cobbles, and constitute a range of so-called core tools and small, sharp flakes. Generically, the technology is known as Oldowan, after Olduvai Gorge, Tanzania. (The gorge was once called Oldoway Gorge; hence the derivation of the tool technology's name.)

The technology, which is mode I, was defined based on the artifact assemblages found in bed I and lower bed II at Olduvai Gorge (1.9 to 1.6 million years old) through the long and meticulous work of Mary Leakey. The artifacts fall into four categories:

- Tools, which include types such as scrapers, choppers, discoids, and polyhedrons
- Utilized pieces, such as large flakes produced in the manufacture of tools, having sharp edges useful for cutting
- Waste, or small pieces produced in the manufacture or retouching of tools and utilized pieces that had no use

- Manuports, which are pieces of rock carried to a site but not modified

The half-dozen or so tool types named in the typical Oldowan assemblage were not tightly restricted categories such as would be produced by a stone knapper with distinct mental templates for specific implements. The different forms tended to flow into one another typologically, and they carry an air of opportunistic production. This process contrasts with later finds in the archeological record, which exhibit evidence of tighter control over the production of specific tool types.

Frequently the labels applied to the various core forms implied function, such as scrapers and choppers. The small flakes removed from the cores were initially assumed to be waste, but may sometimes have proved useful as cutting tools. In the early 1980s, however, a series of experimental studies by Indiana University archeologist Nicholas Toth led to the conclusion that the real tools in the Oldowan assemblages were the flakes, and that the core forms represented the by-products of flake production. Toth discovered that undirected flaking of cobbles of different shapes led automatically to specific core forms, depending on the shape of the cobble used.

EXPERIMENTAL ARCHEOLOGY: These artifacts were made by Nicholas Toth as a way of understanding the principles of manufacturing the Oldowan assemblage. (*top row*) Hammerstone, unifacial chopper, bifacial chopper, polyhedron, core scraper, bifacial discoid. (*bottom row*) Flake scraper, six flakes. An actual tool kit would comprise mainly flakes. (Courtesy of Nicholas Toth.)

Toth did not suggest that the core forms were never used as tools; rather he concluded that they were not manufactured specifically for use as scrapers, choppers, or similar tools. In experimental butchering, Toth found that the most effective implement for slicing through hide was a small flake; a similar finding applied to dismembering and defleshing. For chopping residual dried meat from a scavenged carcass, however, a heavier implement was best, such as a large flake or a sharp-edged core (for example, a chopper). A heavy core or unmodified cobble was effective for breaking bone to gain access to marrow or brain. The manufacture of digging sticks was achieved with a range of implements: a sharp-edged chopper was

useful for cutting a suitable limb from a tree, a flake or a flake scraper for fashioning the point, and a rough stone surface for honing the point. Flakes and scrapers offered an effective method for removing fat from hide. Nuts could be cracked easily with an unmodified stone hammer and anvil.

Direct evidence of the application of an ancient tool is difficult to obtain, not least because the coarse nature of lava flakes does not sustain clear signals of the material with which it has been in contact. Nevertheless, Toth and Lawrence Keeley, of the University of Illinois, examined 54 flakes from a 1.5 million-year-old site from Koobi Fora, on the eastern side of Lake Turkana, and found evidence of use-wear on nine of them. Four had been used in butchering, three were applied to wood, and two were associated with soft vegetation.

These and other studies give a sense of the variety of subsistence activities that became possible with the adoption of simple stone-tool technology. The small, sharp flake is, however, probably the most important implement and represents a technological and economic revolution. It allowed hominines to slice through hide and gain access to meat, with the stone flake literally opening up a new world of resources: potentially significant quantities of meat. The use of digging sticks permitted more efficient access to underground food sources, such as tubers. By broadening the diet in this way, hominines enriched and introduced a potential stability into their source of energy, which was important in the further expansion of the brain (see unit 31).

SKILLFUL OLDOWAN TOOL MAKERS

The hominines' skill at producing flakes represented a technological revolution. Although the Oldowan industry is technically rather crude, the regular production of flakes is not a matter of chance. Three conditions must be met by a stone knapper who wishes to produce flakes routinely by **percussion**. First, the core must have an acute

CORES COMPARED: A simple chopper from an archeological site (light color) compared with the same tool made recently. (Courtesy of Nicholas Toth.)

TOOL PROFILE: A chopper and the flakes produced during its manufacture. (Courtesy of Nicholas Toth.)

edge, one less than 90 degrees, near which the hammer can strike. Second, the core must be struck with a glancing blow about 1 centimeter from the acute edge. Third, the blow must be directed through an area of high mass, such as a ridge or a bulge. By examining the composition of cores and flakes at archeological sites, Toth could infer that the tool makers of 2.6 million to 1.5 million years ago had indeed mastered the percussion stone-knapping skill.

Similar comparative studies have shown that the ancient tool makers used the percussion technique exclusively to produce flakes. Toth demonstrated that of the three possible techniques for producing flakes—percussion, anvil (striking the core on a stationary anvil), and

bipolar (striking the core with a hammerstone while it rests on an anvil)—percussion was the most efficient. Again, the ancient tool makers showed their skill, as they also did in avoiding flawed cobbles, which flake in unpredictable ways.

A debate over how much skill is required to carry out this simplest of stone knapping has recently been addressed in a most interesting fashion: by asking a bonobo (pygmy chimpanzee) to make Oldowan tools. This debate was initiated by Thomas Wynn, an archeologist, and William McGrew, a primatologist. In 1989, the two researchers published a paper called "An ape's view of the Oldowan," in which they asked the following question: "When in human evolution did our ancestors cease behaving like apes?" In other words, given the opportunity and motivation, could an ape make Oldowan tools?

Toth had an opportunity to test this experimentally, when he collaborated with Sue Savage-Rumbaugh, of Georgia State University. Savage-Rumbaugh had spent 10 years working with a male bonobo, Kanzi, who had learned to use a large vocabulary of words displayed on a computerized keyboard and understood complex spoken English sentences. Toth encouraged Kanzi to make sharp stone flakes in order to gain access to favored food items enclosed in a box that was secured with string. Kanzi was an enthusiastic participant in the experiment over a period of several years. Despite being shown the percussion knapping technique, however, he never used it. Sometimes Kanzi produced flakes by knocking cobbles together, but without the precission inherent in the Oldowan technique; often he would simply smash the cobble by throwing it at another hard object, including the floor. Kanzi knew what he needed

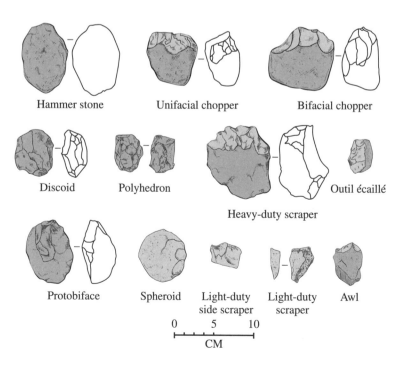

Hammer stone   Unifacial chopper   Bifacial chopper

Discoid   Polyhedron   Outil écaillé

Heavy-duty scraper

Protobiface   Spheroid   Light-duty side scraper   Light-duty scraper   Awl

0   5   10
CM

OLDOWAN ARTIFACTS: The manufacture of these simple pebble tools requires considerable skill.

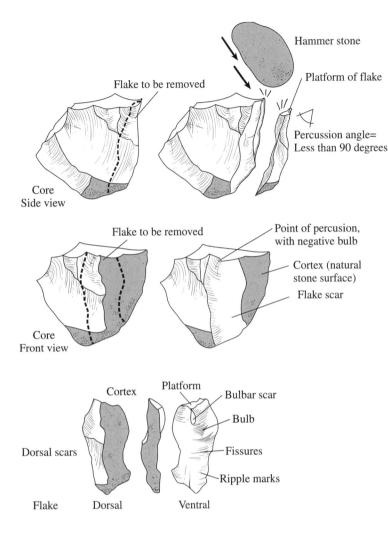

Hammer stone

Flake to be removed

Platform of flake

Percussion angle=
Less than 90 degrees

Core
Side view

Flake to be removed

Point of percusion,
with negative bulb

Cortex (natural
stone surface)

Flake scar

Core
Front view

Cortex   Platform
Bulbar scar

Bulb

Dorsal scars

Fissures

Ripple marks

Flake   Dorsal   Ventral

DIAGNOSTIC FEATURES OF FLAKING BY PERCUSSION: Producing sharp, usable flakes requires the delivery of forceful blows at the correct angle and at the correct location on the core. Flakes produced in this manner have certain features produced by concoidal fracture. (Courtesy of Nicholas Toth.)

(sharp flakes) and figured out ways to obtain them (banging or throwing rocks), but he was not an Oldowan tool maker.

Thus, a clear difference separates the stone-knapping skills of Kanzi and the Oldowan tool makers, which appears to imply that these early humans had indeed ceased to be apes.

WHO MADE THE TOOLS?

In the period 2.6 to 1.5 million years ago, several hominine species (*Homo* and *Australopithecus*) lived as contemporaries (see unit 22). How, then, is the identity of the tool maker to be discerned? Was it *Homo, Australopithecus,* or both? After some 1 million years ago, when only *Homo* existed, tool-making technology certainly continued—some of it very Oldowan-like (see unit 26). The argument from parsimony, therefore, would be that the earliest technology was also the product of *Homo*. In addition, the earliest evidence of stone-tool making coincides with the first appearance of *Homo*, approximately 2.5 million years ago (see unit 21).

Randall Susman, of the State University of New York, Stony Brook, argues that robust australopithecines also

had the manipulative potential to make tools. He bases his contention on the anatomy of the hand bones, and particularly the thumb, gathered from deposits in the cave of Swartkrans, South Africa. The deposits, which are thought to date to roughly 1.8 million years ago, also contain stone tools and putative digging sticks. The breadth of the thumb and the fingertips in the Swartkrans fossils indicates a degree of vascularity and innervation consistent with increased manipulative skill. Recent detailed studies of the thumb have that it was capable of forming a power grip, which is important in percussion stone knapping. The fingertips of modern apes and of *Australopithecus afarensis* are narrow; those of modern humans are broad. Susman concludes that, although early australopithecines were unable to make tools, later species, including early *Homo,* may have possessed this capacity.

Complicating the putative attribution of the finger bones to the *Australopithecus* species at Swartkrans is the fact that the same sedimentary layers have yielded fragments of *Homo*. As Susman points out, 95 percent of the hominine cranial bones found are those of *Australopithecus,* suggesting "an overwhelming probability" that the hand bones are indeed remnants of this species. He also notes apparent differences in the morphology of the thumb in the Swartkrans material and thumb morphol-

ogy in a known *Homo erectus* specimen. Some observers contend that this evidence is too tenuous for definitive conclusions to be drawn. ✵

KEY QUESTIONS

- What kind of evidence could settle the identity of the earliest stone-tool makers?
- What new questions might be tackled by experimental archeology that are not available to traditional approaches?
- What are the implications of the stasis and lack of innovation in early tool technologies?
- What ecological circumstances might have encouraged the initial evolution of stone-tool technology?

KEY REFERENCES

Gibbons A. Tracing the identity of the first tool makers. *Science* 1997;276:32.

Gibson KR, Ingold T, eds. Tools, language and cognition in human evolution. Cambridge: Cambridge University Press, 1993.

Keeley LH, Toth N. Microwear polishes on early tools from Koobi Fora, Kenya. *Nature* 1981;293: 464–465.

Schick KD, Toth N. Making silent stones speak. New York: Simon and Schuster, 1993.

Semaw S, *et al.* 2.5-million-year-old stone tools from Gona, Ethiopia. *Nature* 1997;385:333–336.

Susman RL. Fossil evidence for early hominid tool use. *Science* 1994;265:1570– 1573.

———. Hand function and tool behavior in early hominids. *J Hum Evol* 1998;35:23–46.

Toth N. The Oldowan reassessed. *J Archeol Sci* 1985;12:101–120.

Toth N, *et al.* Pan the tool-maker; investigations into the stone tool-making and tool using capabilities of a bonobo (*Pan paniscus*). *J Archeol Sci* 1993;20: 81–91.

Wood B. The oldest whodunnit in the world. *Nature* 1997;385:292–293.

Wynn T, McGrew WC. An ape's view of the Oldowan. *Man NS* 1989;24:383–398.

PART 6

OUT OF AFRICA

The Changing Position of Homo erectus
New Technologies
Hunter or Scavenger?

THE CHANGING POSITION OF *HOMO ERECTUS*

24

This unit deals with the species of *Homo* that is assumed to be intermediate between early *Homo* (*habilis/rudolfensis*) and modern-day humans, *Homo sapiens*. Until recently, the story would have been portrayed as relatively straightforward: Early *Homo* gave rise to a larger-bodied, larger-brained species, *Homo erectus*, approximately 2 million years ago, in Africa. Roughly 1 million years ago, *Homo erectus* expanded its range beyond Africa, first into Asia and then into Europe, developing geographically variable populations. *Homo erectus* then became the direct ancestor of *Homo sapiens*, either by a speciation event in a single population in Africa, which then spread throughout the Old World and replaced established populations of *Homo erectus* (the "out of Africa" or single-origin model), or by a gradual, worldwide (excluding the Americas and Australia) evolutionary transformation of all populations of *Homo erectus* (the multiregional evolution model). (See units 27 through 30.)

Much that was assumed to have been settled about the earlier events in this scenario has been overturned in recent years, through the discovery of new fossils and the redating and reinterpretation of known fossils. It would be helpful to give a snapshot of evolutionary events as currently viewed by most anthropologists.

Early *Homo* gave rise to a large-bodied, large-brained species in Africa approximately 2 million years ago, but this species is now called *Homo ergaster* by some anthropologists. *Homo ergaster* expanded its range beyond Africa and into Asia soon after its origin and at least by 1.8 million years ago; it then gave rise to *Homo erectus* in those areas. *Homo erectus* expanded its range throughout Asia, back into Africa, and presumably into Europe, although few unequivocal fossils have been found (most evidence takes the form of the stone-tool technology often associated with the species). Approximately 150,000 years ago, a speciation event in Africa gave rise to *Homo sapiens* (probably from *Homo ergaster* but possibly from *Homo erectus*), which then spread into the rest of the Old World, and subsequently into Australia and the Americas.

A BRIEF HISTORY OF DISCOVERY

The first discoveries of *Homo erectus* were made in 1891 and 1892 in Java, Indonesia, by Eugene Dubois, a Dutch medical doctor, who had gone there specifically to search for "the missing link." The specimens were of a skull cap and a complete thigh bone, or femur, which indicated that the creature had walked upright. Although he was initially ambivalent over the human nature of his

fossil find, Dubois eventually came to name the species *Pithecanthropus erectus*, or upright ape man, inspired in part by Ernst Haeckel's speculations on human ancestry (see unit 3). Great controversy greeted Dubois's announcement, and no agreement could be reached as to whether *Pithecanthropus* was human, ape, or something in between.

The rehabilitation of *Pithecanthropus erectus* as an important discovery in human evolution coincided with discoveries in China, at the Choukoutien (now Zoukoudian) site near Peking (now Beijing). In 1927, Davidson Black, the Canadian-born director of the Peking Medical College, recognized the human affinities of a tooth that had been found at the site. He named it *Sinanthropus pekinensis*, or Chinese man from Peking. An immense effort was mounted toward uncovering more fossils. Within a decade a rich haul had accumulated, including 14 partial or fragmentary crania, 14 mandibles, more than 100 teeth, and many other fragments. Black concluded that *Sinanthropus* and *Pithecanthropus* were similar creatures, having a long, low, thick-boned skull, with a brain size intermediate between that of a human and an ape. Black died prematurely of a heart attack in 1934, and his work was continued by the German anatomist Franz Weidenreich.

Meanwhile, fossil prospecting was continuing in Java, under the eye of the German anatomist G. H. Ralph von Koenigswald. Many *Pithecanthropus* teeth, jaw, and cranial fragments were recovered, including the almost complete cranium of a child from the Modjokerto site. One problem with fossil collecting in Java was that it was often performed by local farmers, who came across specimens in their work or developed a talent for finding them. The issue of **provenance** of the fossil, or its exact location in the sediments from which it was recovered, was therefore

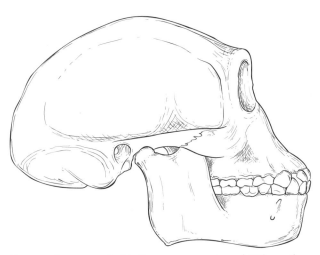

DUBOIS'S VIEW: In his first reconstruction of *Pithecanthropus* (1896), Dubois reflected his ambivalence over the human nature of the fossil, and chose to emphasize an apelike nature, seen in the prognathism and large canines.

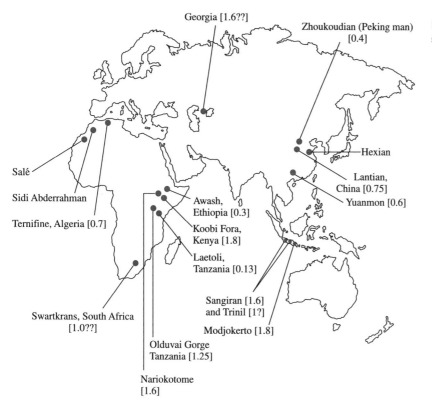

Georgia [1.6??]
Zhoukoudian (Peking man) [0.4]
Salé
Sidi Abderrahman
Ternifine, Algeria [0.7]
Hexian
Lantian, China [0.75]
Yuanmon [0.6]
Awash, Ethiopia [0.3]
Koobi Fora, Kenya [1.8]
Laetoli, Tanzania [0.13]
Sangiran [1.6] and Trinil [1?]
Modjokerto [1.8]
Swartkrans, South Africa [1.0??]
Olduvai Gorge Tanzania [1.25]
Nariokotome [1.6]

MAJOR *HOMO ERECTUS/ERGASTER* SITES: Figures in parentheses indicate the estimated age (where known). Until recently, no fossil specimen outside of Africa was dated as being older than approximately 1 million years. The recent redating of fossils from Java, if correct, suggests that this species expanded its range beyond Africa soon after it evolved.

often a serious problem. Accurate provenance is essential if the fossil is to be reliably dated. This caveat applies particularly to the Modjokerto skull, found in 1936, as we will see below.

In 1951, *Sinanthropus* and *Pithecanthropus* were subsumed under a single nomen, *Homo erectus*, which was recognized as a widespread species that exhibited significant geographical variation.

Since the 1950s, discoveries of *Homo erectus* fossils have been made sporadically, principally in Africa, but also in Asia (see map for sites of the principal finds). The first of these discoveries took place at Ternifine, in Algeria, where three jaws, a cranial bone, and some teeth of *Homo erectus*, dated at between 600,000 and 700,000 years old, were discovered in the mid-1950s. Later finds in northern Africa were made at Sidi Abderrahman (a jaw), in Morocco soon after the first Ternifine find, and at Salé (cranial fragments), also in Morocco, in 1971. Meanwhile, several specimens attributed to *Homo erectus* were made at Olduvai Gorge, in East Africa, including a rather robustly built, large-brained cranium, OH 9, initially dated at 1.2 million years (although it is probably younger). The South African cave site of Swartkrans also yielded *Homo erectus* fossils, which were originally classified as *Telanthropus capensis*. Fossil prospecting in Java contributed an important cranium (Sangiran 17) in 1969 and a face and cranium (Sangiran 27 and 31) in the late 1970s in the Sangiran dome region of the island.

The richest source of fossils, however, has been the Lake Turkana region of northern Kenya, both on the east side (Koobi Fora) and the west side. These sites have yielded both the oldest known and the most complete specimens. In 1975, an almost complete cranium was recovered from Koobi Fora (KNM-ER 3733), with an age of 1.8 million years, and a brain size of 850 cm³. A decade later the virtually complete skeleton of a nine-year-old *Homo ergaster* boy was unearthed at Nariokotome, on the west side of the lake (KNM-WT 15,000). The boy stood more than 5 feet tall when he died, and would have exceeded 6 feet had he lived to maturity. His cranial capacity was 880 cm³. And his body stature and proportions—tall, thin, long arms and legs—are typical of humans adapted to open, tropical environments (see unit 11).

CHANGING VIEWS: DATES AND EVOLUTIONARY PATTERN

As the finds of putative *Homo erectus* fossils accumulated, two conclusions seemed to emerge. First, anatomical variations, which were seen initially in Asia, appeared to have proliferated elsewhere. Second, the species appeared to have originated in Africa close to 2 million years ago, and first set foot outside of Africa not much earlier than 1 million years ago. In recent years, both of these assumptions have been challenged.

Few of the Asian *Homo erectus* fossils have secure radiometric dates, with faunal correlation and paleomag-

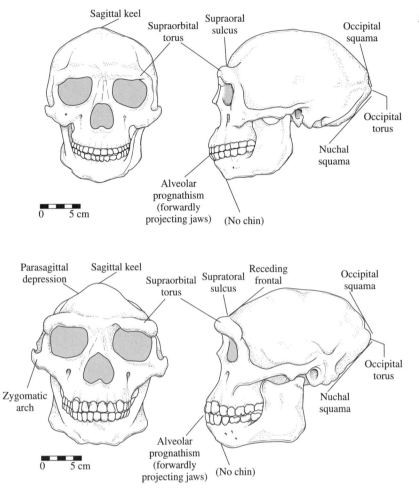

HOMO ERECTUS: These two reconstructions by Weidenreich of Zoukoutian *Homo erectus* (*top*) and Indonesian *Homo erectus* (*bottom*) show some of the anatomical variations present in Asian *Homo erectus*.

netic dating often used to approximate their age instead. Even where the presence of volcanic tuffs makes radiometric dating possible, as in Java, uncertainty has arisen over the reliability of such dates because of questions about provenance, as explained earlier. Consequently, the consensus was that no *Homo erectus* specimen outside of Africa was older than approximately 1 million years. The Beijing fossils were estimated to be roughly 300,000 years old (but have recently been shown to be at least 400,000 years old), but another Chinese site, Lantian, may be more than 700,000 years old. The oldest non-African *Homo erectus* sites were held to be in Java, with estimates of a little more than 1 million years for the Modjokerto child and something close to 750,000 years for Sangiran 27/31.

Until recently, the oldest specimen attributed to *Homo erectus* in Africa was KNM-ER 3733, from Koobi Fora, which was radiometrically dated to 1.8 million years. It was therefore assumed that *Homo erectus* originated in Africa and then, after a delay of almost 1 million years, spread into Asia. This apparent delay constituted a major puzzle to be explained in the overall history of *Homo erectus*. Some suggested that early *erectus* populations

lacked a sufficiently sophisticated technology for moving beyond the traditional hominine geographic range. This technology, the Acheulean industry (see unit 25), is first seen in the archeological record some 1.4 million years ago (a date that still left an apparent, albeit smaller, delay). A new fossil find in 1992 and the redating of certain Javan fossils in 1994 implied one of two things: either no delay occurred, and *Homo erectus* expanded its range beyond Africa as soon as it evolved there, or *Homo erectus* evolved in Asia, not Africa, close to 2 million years ago.

In 1992, two German researchers announced the discovery of a *Homo erectus* mandible at Dmanisi, in Georgia, eastern Asia. Its age—inferred from faunal correlation—was said to be 1.6 to 1.8 million years. (More recent dating work suggests, however, that this figure might be an overestimate.) Then, in early 1994, Carl Swisher and Garniss Curtis, of the Geochronology Center, Berkeley, announced new dates (based on single-crystal laser fusion; see unit 7) for the Modjokerto and Sangiran fossils: 1.8 and 1.6 million years, respectively. If correct, the Modjokerto skull would be equivalent in age to KNM-ER 3733, from Koobi Fora.

Many anthropologists are reluctant to accept the new dates, however, because of the lingering uncertainties about the provenance of the Modjokerto find. If correct, however, the new work changes the question anthropologists must answer about *Homo erectus*: there is now no delay to be explained, but the pattern of the species' origin is less clear. Although KNM-ER 3733 and the Modjokerto skull are of equivalent age, a sufficient margin of error exists in the dates to permit a gap in age of at least 100,000 years. A quick calculation shows that, even at the glacial pace of population expansion of 10 miles per generation, *Homo erectus* could move from East Africa to East Asia in a mere 25,000 years. An African origin followed by population expansion into Asia is therefore consistent with the dates as currently known. Some anthropologists argue that an alternative pattern is equally plausible—with *Homo erectus* originating in Asia and then moving into Africa.

CHANGING VIEWS: ANATOMY AND EVOLUTIONARY PATTERN

As mentioned earlier, many anthropologists have recently concluded that the anatomical variations seen between different geographical populations of *Homo erectus* reflect the existence of more than one species, a view that is supported by cladistic analysis. The early African specimens, such as KNM-ER 3733, the slightly younger 3883, and WT-15000 (the Turkana boy) have been assigned to a new species, *Homo ergaster*, while the Asian specimens remain as classic *Homo erectus*. The two species are viewed as having an ancestor/descendant relationship, with *ergaster* originating in Africa close to 2 million years ago and then quickly expanding its range into Asia, where it probably gave rise to *erectus*. In this hypothesis, the later presence of *erectus* in Africa (such as the robust OH 9 from Olduvai Gorge) is interpreted as an Asia-to-Africa population expansion. Alternatively, *ergaster* might have given rise to *erectus* in Africa.

Many aspects of *ergaster* and *erectus* anatomy are, of course, similar, with the principal differences being a higher cranial vault, thinner cranial bone, absence of a sagittal keel, and certain cranial base characteristics in *ergaster*. Other distinguishing features include a long, low cranium (particular in *erectus*), thick cranial bone (particular in *erectus*), the presence of brow ridges, a shortened face, and a projecting nasal aperture, suggesting the first appearance of the typical human external nose with the nostrils facing downward.

The body size of *ergaster/erectus* also represents an increase relative to that of early *Homo*, and reached nearly 1.8 meters and 63 kilograms in males and about 1.55 meters and 52 kilograms in females; this size compares with 52 and 32 kilograms, respectively, for male and female *habilis*. The larger body size is consistent with a more wide-ranging subsistence strategy. Equally significant is the fact that the difference in body size between males and females is far less than that observed in all earlier hominines. This change probably reflects a change in social structure and dynamics. For instance,

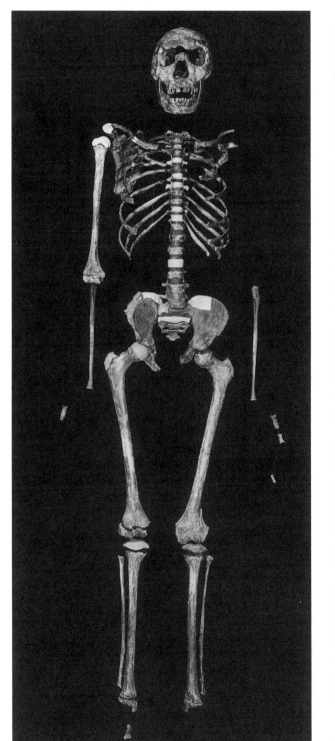

THE TURKANA BOY: Discovered in 1984 on the west side of Lake Turkana, Kenya, this virtually complete specimen includes many skeletal elements not previously known. (Courtesy of Alan Walker/National Museums of Kenya.)

Homo sapiens

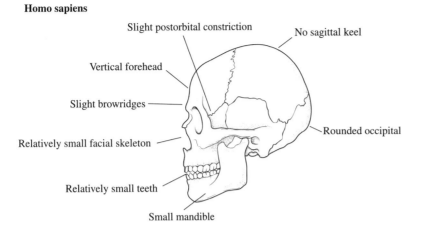

Slight postorbital constriction

No sagittal keel

Vertical forehead

Slight browridges

Relatively small facial skeleton

Rounded occipital

Relatively small teeth

Small mandible

Homo erectus

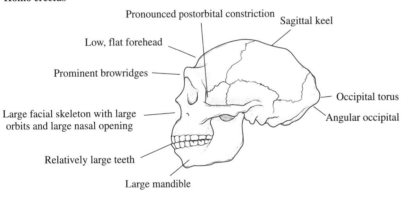

Pronounced postorbital constriction

Sagittal keel

Low, flat forehead

Prominent browridges

Occipital torus

Large facial skeleton with large
orbits and large nasal opening

Angular occipital

Relatively large teeth

Large mandible

perhaps the greater complexity of *ergaster/erectus* lifeways included a degree of male-male cooperation (see unit 13).

Until the discovery of the Turkana boy skeleton, which is dated at 1.6 million years old, the postcranial anatomy was known from only a few elements, such as the femur and pelvis. The wealth of information provided by the boy's skeleton indicates that the postcranium of *H. ergaster* is similar to that of modern humans, but more robust and heavily muscled; this structure implies routine heavy physical exertion. The thigh bone is unusual, in that the femoral neck is relatively long but the femoral head—part of the ball-and-socket joint with the pelvis—is large. This combination represents something of a mix between modern human and australopithecine anatomy: modern humans have a short femoral neck attacked to a large head, while australopithecines possessed a long neck and a small head.

In the cervical and thoracic vertebrae, the hole through which the spinal cords runs is significantly smaller than in

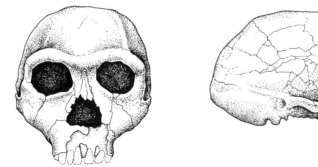

HOMO ERGASTER: This cranium, KNM-ER 3733 from Koobi Fora, Kenya, is 1.8 million years old. It shares many similarities with Asian *Homo erectus* (more particularly with the Chinese specimens), but is judged by some anthropologists to be a different species, *Homo ergaster*.

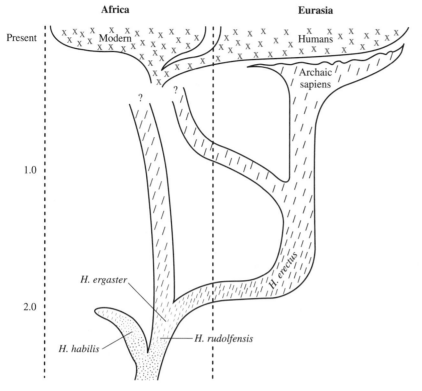

Africa   **Eurasia**

A POSTULATED PHYLOGENY: *Homo ergaster* is seen here as being the descendant of *H. rudolfensis* in Africa approximately 2 million years ago, and immediately expanding its range into Asia, where it gives rise to classic *H. erectus*. *H. erectus* persists in Eurasia, where it gives rise to archaic *sapiens* species, including Neanderthals; it also enters Africa, where it or *H. ergaster* gives rise to modern humans, which replace established archaic populations.

modern humans, which presumably indicates a smaller demand for nerve signal traffic. This structure has been interpreted to imply less control over breathing patterns than in modern humans, related to the absence of an ability for spoken language (but see unit 32 for a qualification).

The anatomy of the Turkana boy's pelvis provides an important insight into the changing patterns of behavior brought by this new species. The birth canal was smaller than in modern humans, but its absolute size suggests that humanlike infant development appeared for the first time. Alan Walker, who directed much of the study of the Turkana boy's skeleton, calculated from the birth canal size in the boy's pelvis that the brain size in *ergaster* neonates would have been 275 cm^3. An apelike pattern of development (a brain-size doubling from birth to maturity) would lead to an adult brain of less then 600 cm^3, which is significantly smaller than actually develops. Continued brain growth at a high rate for a time after birth would be necessary to achieve the observed adult brain capacity of at least 850 cm^3—the pattern seen in *Homo sapiens*. Infant helplessness and prolonged childhood would therefore have already begun in *Homo ergaster*, thus giving an opportunity for more cultural learning.

In an analysis of tooth development as an indicator of life history patterns, Holly Smith, of the University of Michigan, has also produced evidence for a shift to a life history pattern similar to that seen in modern humans (see unit 12). In apes, first molar eruption occurs at a little over 3 years, and lifespan is about 40 years; in humans, the corresponding figures are 5.9 years and 66 years, respectively. In other words, human life history patterns have slowed relative to those of the great apes, including factors such as age at weaning, age at sexual maturity, and effective gestation length. While late *Homo erectus* fit the modern human pattern, as do Neanderthals and other archaic *sapiens*, *Homo ergaster* was somewhat intermediate between humans and apes; its first molar eruption occurred at 4.6 years, and its lifespan averaged 52 years.

The accumulations of bones and stones that appear in the archeological record coincidentally with the origin of the genus *Homo* become more frequent through *ergaster* and *erectus* times, giving an increasingly clear putative signal of some hunting activity (see unit 26). Some investigators speculate that a more broadly based diet, which included a greater proportion of meat than was eaten by earlier hominine species, was a factor in the population expansion out of Africa. Whatever the niceties of taxonomy, the evolution of *ergaster/erectus* signals the appearance of a new grade of hominine evolution.

CHANGING PATTERNS OF BEHAVIOR

A number of important "firsts" were recorded in human prehistory with the appearance of *ergaster/erectus*:

- The first appearance of hominines outside Africa
- The first appearance of systematic hunting
- The first appearance of anything like "home bases"

- The first systematic tool making
- The first use of fire
- The first indication of extended childhood

Thus, these species were apparently capable of a life more complex and varied than had previously been possible. ❊

KEY QUESTIONS

- How would one explain the robusticity of the *Homo erectus/ergaster* skeleton?
- What factors might be important in the migration of *Homo ergaster* out of Africa?
- How could the notion of a cladistic separation between Asian and African *Homo erectus* hominines be further tested?
- What are the behavioral implications of a reduction in body size dimorphism in *Homo ergaster/erectus*?

KEY REFERENCES

Bräuer G, Mbua E. *Homo erectus* features used in cladistics and their variability in Asian and African hominids. *J Hum Evol* 1992;22:79–108.

Brown FH, *et al*. Early *Homo erectus* skeleton from west Lake Turkana, Kenya. *Nature* 1985;316:788–792.

McHenry HM. Behavior ecological implications of early hominid body size. *J Hum Evol* 1994;27:77–87.

Rightmire GP. *Homo erectus*: ancestor or evolutionary sidebranch? *Evol Anthropol* 1992;1:43–49.

Smith H. Growth and development and its significance for early hominid behavior. *Ossa* 1989;14:63–96.

Swisher CC, *et al*. Age of the earliest known hominids in Java, Indonesia. *Science* 1994;263:1118–1121.

Walker A, Leakey REF, eds. The Nariokotome *Homo erectus* skeleton. Cambridge, MA: Harvard University Press, 1993.

Wood B. Origin and evolution of the genus *Homo*. *Nature* 1992;355:783–790.

Wood B, Turner A. Out of Africa and into Asia. *Nature* 1995;378:239–240.

NEW TECHNOLOGIES

25

As we saw in unit 24, the evolution of *Homo ergaster* and subsequent appearance of *Homo erectus* brought many changes in the biology of our direct ancestors. Variations in life history factors, in social structure, and in subsistence patterns combined to make the species a great deal "more human" than earlier species of *Homo* or the contemporary species *Australopithecus*. In particular, the further development of meat as a significant component of diet (see unit 26) must have been very important, both in increasing the stability and richness of energy resources and in allowing new habitats to be exploited. *Homo ergaster/erectus* was the first hominine to move beyond the bounds of the African continent. It might be expected that these developments would be accompanied by significant enhancement of stone-tool technologies.

THE ACHEULEAN ASSEMBLAGE

A significant innovation is seen in the archeological record, with the appearance of the Acheulean assemblage. The earliest known example of this assemblage comes from Konso-Gardula, Ethiopia, and is 1.4 million years old. The name derives from the site of St. Acheul, in northern France, where many examples of handaxes were discovered in the last century. The innovation consisted of the introduction of larger tools—known as handaxes, picks, and cleavers—than appear in Oldowan assemblages. Although each of these tools is bifacially shaped, the teardrop-shaped handaxe is regarded as characterizing the new technology. Compared with Oldowan choppers, Acheulean handaxes required a higher level of cognitive ability in the conceptualization of the end-product and its manufacture.

The earliest known *Homo ergaster* fossils appear in the record close to 2 million years ago, while the earliest known Acheulean element occurs some half million years later. Several interpretations of this temporal gap have been suggested. For instance, the innovation may have been cultural, with later *Homo ergaster* populations inventing the new tool technology after having employed the simpler Oldowan technique for half a million years. Alternatively, the Acheulean may have been a *Homo erectus* innovation. This latter explanation seems less likely, as archeological assemblages of the appropriate age in eastern Asia lack characteristic Acheulean artifacts (as discussed later in this unit).

The precise path through which the Acheulean innovation emerged is not clear. Glynn Isaac argued, for instance, that it required the production of large ovoid flakes, greater than 10 cm long, which were then trimmed by a few or many repeated blows along both edges. Some large flakes were apparently functional without further trimming. The regular production of large flakes according to a preferred shape would have represented a punctuation in technological expression upon which other bifacial implements could be built. The emergence of the handaxe may, however, have been more gradual. The Developed Oldowan (see unit 23) included small bifaces, sometimes constructed from ovoid cobbles and sometimes derived from relatively large flakes. Acheulean bifaces may be envisioned as a further development of the technique that emerged earlier.

Once the large, bifacial handaxe appeared, it remained a characteristic of Acheulean assemblages for a very long time, both in Africa and Eurasia. Production became refined through the millennia, so that some late examples appear finely hewn compared with the crude earlier specimens. While no early handaxe was the product of long, careful flaking to yield an esthetically pleasing, perfectly symmetrical teardrop shape, many late examples appeared as crude as the earlier versions. Part of the development of the Acheulean included an increasing reliance on more detailed preparation of the core upon which the handaxe was then made. This core preparation, known as the Levallois technique (named after the site in France where the first examples of later prepared-core assemblages were found), became especially dominant in Middle Stone Age and Middle Paleolithic technologies. Overall, then, the Acheulean, like the Oldowan before it, was marked by a tremendous technological stasis maintained through a very long period of time.

Acheulean assemblages are known from many sites in Africa, some of which are spectacularly rich. At Olorgesailie (700,000 years old), for instance, discovered 50 miles south of Nairobi, Kenya, by Louis and Mary Leakey and excavated by Glynn Isaac, hundreds of handaxes were strewn over the land surface. This industry persisted until roughly 200,000 years ago, when it was superseded

REPRESENTATIVE EXAMPLES OF ACHEULEAN TOOLS: (*top row*) Ovate handaxe; pointed handaxe; cleaver; pick. (*bottom row*) Spheroid (quartz); flake scraper; biface trimming flake. (All artifacts, except the spheroid, are lava replicas made by Nicolas Toth.) (Courtesy of Nicholas Toth.)

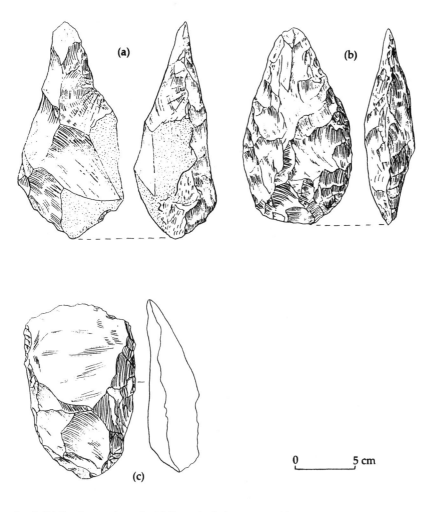

EARLY BIFACES FROM AFRICA: Drawings of (a) a pointed handaxe, (b) an ovate handaxe, and (c) a cleaver. The key innovation of the Acheulean industry was the ability to produce a large, ovoid shape from a core.

0   5 cm

by Middle Stone Age (Middle Paleolithic) assemblages. Chopping-tool assemblages (like the Oldowan) never completely disappeared during the 1.3 million years of the Acheulean period, for reasons that remain unclear. One interpretation is that this persistence simply reflects sites of different functional activities.

The earliest Acheulean site outside of Africa is Ubeidiya, west of the Sea of Galilee, in Israel. Dated at approximately 1 million years old, the site occurs along the natural migration route out of Africa into Asia. Migration into Europe may have followed the same route, or it may have moved across the narrow Straits of Gibraltar from northwestern Africa to Spain, or it may have involved island hopping across the Mediterranean; it may also have occurred via any combination of these paths. The dating of early sites in Europe is difficult because of the lack of volcanic rocks suitable for radiometric dating (see unit 7). Early sites include Isernia in Italy (700,000 years) and Vértesszöllös in Hungary and Arago in France (both somewhat older than 300,000 years). All three of these sites exhibit chopping-tool assemblages. Acheulean sites in Europe began to appear soon after 500,000 years ago. The many famous later sites include Terra Amata (France), Torralba and Ambrona (Spain), and Swanscombe and Hoxne (both in England).

Many Acheulean industries in Africa, Europe, and Asia

bear local names, implying local ethnic expressions of the same kind of technology. Overall, however, the continuity of form over a vast period of time and over a huge geographical area is more impressive than the local variation.

GEOGRAPHICAL DISTRIBUTION OF THE ACHEULEAN

The earliest Acheulean assemblages are located in Africa, but later sites are found in western Asia and Europe. They remain absent in eastern Asia, however—a curious pattern that was first emphasized by Hallam Movius in the 1940s. Stone-tool assemblages east of the so-called Movius line take on the chopping-tool form. Many hypotheses have been put forth in an attempt to explain this pattern. Movius, for example, considered the hominines in the east to be less evolutionarily developed than hominines elsewhere. In 1948, he claimed that the people of the east could not have "played a vital and dynamic role in early human evolution."

Some scholars suggest that the pattern is simply the result of an absence of suitable raw material for fashioning large bifaces east of the Movius line or that other material allowed the manufacture of tools that substituted for Acheulean handaxes. For instance, University of Illinois archeologist Geoffrey Pope suggests that bamboo may

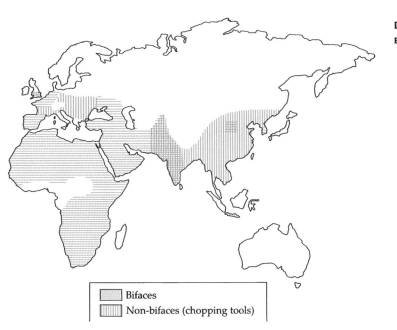

Bifaces
Non-bifaces (chopping tools)

DISTRIBUTION OF BIFACE AND NON-BIFACE INDUSTRIES: Biface assemblages are confined principally to Africa, western Asia, and Europe; they are absent in eastern and southern Asia, where chopping-tool industries are found. The dividing line between the two regions is called the Movius line.

have been used extensively by the Lower Paleolithic people east of this line. He points out that the region is rich in bamboo, an extremely versatile raw material that is used in the modern world for applications ranging from furniture to scaffolding in the building of skyscrapers. Simple, effective knives can be made from bamboo, which may have obviated the need to fashion handaxes; the latter tools require more work and a less abundant raw material.

Others suggest that the pattern reflects a division of cultural tradition, and has no functional or technological significance. Recent ideas about the evolution of the genus *Homo* and the redating of fossils in Java (see unit 24) offer a simpler alternative. If, as seems likely, *Homo ergaster* extended its range beyond Africa soon after it arose, then the first occupants of Asia would have long predated the first appearance of the Acheulean technology. Later incursions into eastern and southern Asia by Acheulean-bearing hominines might have been prevented if populations there were already well established.

THE FUNCTION OF ACHEULEAN HANDAXES

The function of Acheulean handaxes has long been a subject of speculation. A particularly unlikely explanation is that they were used as lethal projectiles, thrown like discuses as a means of killing prey. More prosaic suggestions hypothesize that they were used as axes or heavy-duty knives. In experimental studies, Indiana University archeologist Nicholas Toth found that handaxes (and cleavers) were highly effective at slicing tough hide, such as that of elephants. The combination of weight and relatively sharp edges gives them greater efficacy than the ubiquitous small, sharp flakes. Microwear studies by Lawrence Keeley, of the University of Illinois, reveal that handaxes were used for many functions, and for materials ranging from meat and bone to wood and hide. Thus, the Acheulean handaxe may have been the Swiss Army knife of the Lower Paleolithic.

The end of the Acheulean industries, which occurred from 300,000 to 200,000 years ago throughout some areas of the Old World, marked the end of these stone-tool assemblages that had few artifact types and enjoyed enormous longevity. Both the Oldowan and the Acheulean lasted at least 1 million years, and both produced a dozen or less identifiable implements. The end of the Acheulean brought the Lower Paleolithic (Early Stone Age) to a close and marked the beginning of the industries of the Middle Paleolithic (Middle Stone Age). This period lasted only from 300,000 years ago to roughly 40,000 years ago, and included many more identifiable tool types. Real technical innovation had begun, although even this development was overshadowed by what followed in the Upper Paleolithic (Later Stone Age) (see unit 30). ✳

KEY QUESTIONS

- What is the best explanation for the temporal gap between the first known appearance of *Homo ergaster* and the first known appearance of characteristic Acheulean artifacts?

- What is the best explanation for the geographic variation in the form of handaxes?
- How would you evaluate the range of proposals for explaining the distribution of different tool technologies across the Movius line?

- What kinds of evidence are most persuasive for understanding the function of handaxes?

KEY REFERENCES

Asfaw B, *et al*. The earliest Acheulean from Konso-Gardula. *Nature* 1992;360:732–735.

Gowlett J. Culture and conceptualization: the Oldowan-Acheulean gradient. In: Bailey G, Callow P, eds. Stone Age prehistory. Cambridge: Cambridge University Press, 1986:243–260.

Pope GG. Bamboo and human evolution. *Nat Hist* Oct 1989:49–56.

Schick KD, Toth N. Making silent stones speak. New York: Simon and Schuster, 1993.

———. Early Paleolithic of China and Eastern Asia. *Evol Anthropol* 1993;2:22–35.

Wynn T. Two developments in the mind of early *Homo*. *J Anthropol Archaeol* 1993;12:299–322.

HUNTER OR SCAVENGER?

26

Some time between the beginning of the hominine lineage and the evolution of *Homo sapiens*, an essentially apelike behavioral adaptation was replaced by what we would recognize as human behavior—namely, the hunter-gatherer way of life. How and when this development occurred is central to paleoanthropological concerns. As we have seen, fossil evidence reveals the fundamental anatomical changes during this period, but it is to archeology that one turns for direct evidence of behavior.

The earliest stone artifacts recognized in the record are dated to approximately 2.6 million years ago (see unit 23), which coincides closely with the earliest evidence of the genus *Homo*. From their earliest appearance in the record, stone tools occur both as isolated scatters and, significantly, in association with concentrations of animal bones. What this association between bones and stones means in terms of early hominine behavior has become the subject of heated debate among archeologists.

Until recently, some archeologists argued by analogy with modern hunter-gatherer societies that the associa-

tions represented remains of ancient campsites, or fossil home bases, to which meat and plant food were brought to be shared and consumed amidst a complex social environment. Others have countered by suggesting that these combinations merely indicated that hominines used the stones to scavenge for meat scraps and marrow bones at carnivores' kill sites; according to this hypothesis, the associations had no social implications. Hence the debate, which has often been characterized as "hunting versus scavenging," is being fought over how "human" was the behavior of hominines 2 million years ago.

EARLY HYPOTHESES AND RECENT DEVELOPMENTS

During the 1960s and early 1970s, paleoanthropologists considered hunting to be the primary human adaptation, a notion that has deep intellectual roots, reaching back as far as Darwin's *Descent of Man*. The apogee of the "hunting hypothesis" was marked by a Wenner-Gren Foundation conference in Chicago in 1966, titled "Man the Hunter." The conference not only stressed the idyllic nature of the hunter-gatherer existence—"the first affluent society" as one authority termed it—but also firmly identified the technical and organizational demands of hunting as the driving force of hominine evolution.

A shift of paradigms occurred in the mid- to late 1970s, when the late Glynn Isaac proposed the "food-sharing hypothesis." Cooperation was what made us human, argued Isaac—specifically, cooperation in the

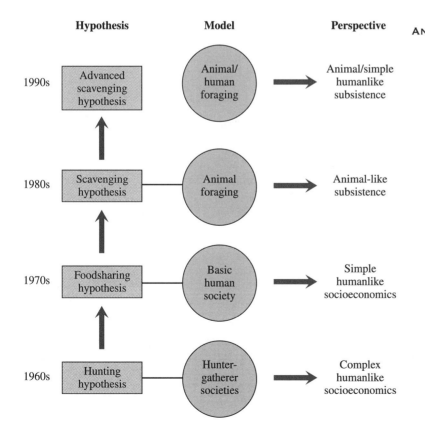

AN EVOLUTION OF HYPOTHESES: During the past four decades, ideas about the nature of early hominine subsistence (social and economic) activities have passed through several important stages. In the 1960s, anthropologists envisioned hominine evolution in terms of the impact of cooperative hunting. In the 1970s, the image shifted, with the focus emphasizing social and economic cooperation through a mixture of hunting and gathering in a protohuman context. This view changed again in the 1980s, effectively taking any "humanity" out of the picture and attributing a marginal scavenging behavior to hominines. The current position is that scavenging was probably a very important route of meat acquisition, but not the exclusive one; this view is taken within the context of a human/animal model.

sharing of meat and plant food resources that routinely were brought back to a social focus, the home base. In this system, the males did the hunting while the females were responsible for gathering plant foods. As for "Man the Hunter," Isaac claimed that it was not possible to evaluate the importance of hunting relative to that of scavenging. "For the present it seems less reasonable to assume that protohumans, armed primitively if at all, would be particularly effective hunters," he concluded in 1978.

Although the shift from the hunting hypothesis to the food-sharing hypothesis changed what was perceived to be the principal evolutionary force in early hominines, it nevertheless left them recognizably human. Specifically, the conclusion that the coexistence of bones and stones on Plio/Pleistocene landscapes implied a hominine home base immediately invoked a hunter-gatherer social package. Although the food-sharing hypothesis was often described by proponents as merely one of many possible candidates for explaining the evolution of human behavior, it proved very seductive. As Smithsonian Institution paleoanthropologist Richard Potts has observed: "The home base/food sharing hypothesis [was] a very attractive idea because it integrates many aspects of human behavior and social life [that] are important to anthropologists—reciprocity systems, exchange, kinship, subsistence, division of labor, technology, and language."

TESTING ASSUMPTIONS

Realizing that several assumptions were implicit in these interpretations, in the late 1970s Isaac initiated a program of research that would test the food-sharing hypothesis. Lewis Binford, of the University of New Mexico, independently embarked on a similar venture. Both studies addressed several basic issues. First, what processes brought concentrations of stone artifacts and animal bones together in particular sites? Second, if the bones and stones are causally associated at these sites, what behavioral implications are possible? For Isaac and his associates, these questions were addressed by reexamining fossil bones from several already excavated, 1.8-million-year-old sites at Olduvai Gorge and a newly excavated, 1.5-million-year-old site at Koobi Fora, known as site 50. For Binford, the exercise entailed the scrutiny of published material on the Olduvai sites.

In fact, bone fragments and stone artifacts might accumulate at the same site and yet be causally unrelated for several reasons. For instance, they might be independently washed along by a stream and then deposited together—a hydraulic jumble, as it is known. Alternatively, carnivores might use a particular site for feeding on carcasses, while hominines might use the same site for stone knapping and whittling wood, having no interest in the bones whatsoever. The first possibility can be tested by the detailed stratigraphy of the site. The second hypothesis would require some indication that the stones were used on the bones in a particular way.

Of the six major early bone and artifact sites at Olduvai bed 1, the most famous site is the Zinj "living floor,"

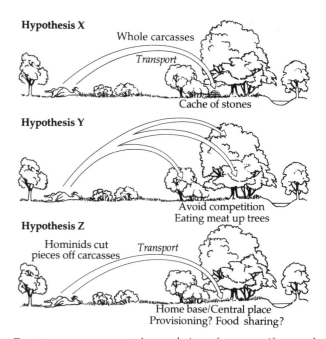

RIVAL HYPOTHESES: Accumulation of stone artifacts and broken animal bones in the same location forms an important element of the early archeological record. Traditionally interpreted as the remains of some kind of hominine home base (hypothesis Z), these accumulations are now subject to other interpretations. For instance, hypothesis Y suggests that the accumulation occurs at one location because hominines used trees there to escape competition from other carnivores while eating scavenged meat. Hypothesis X argues that hominines made caches of stones, to which they brought the more easily transported carcass fragments. Both cases produce the same result: an accumulation of bones and stones in one location. (Courtesy of Glynn Isaac.)

which includes an accumulation of more than 40,000 bones and 2647 stones. Geological analysis indicates that hydraulic processes probably had little or no influence in the formation of most of the bed 1 sites. Binford's analysis of the sites compared the pattern of bone composition with that of modern carnivore sites, using the assumption that any difference could be attributed to hominine activity—residual analysis, it is called. His conclusion was forthright: "The only clear picture obtained is that of a hominid scavenging the kill and death sites of other predator-scavengers for abandoned anatomical parts of low food utility, primarily for purposes of extracting bone marrow. . . . [There] is no evidence of 'carrying food home.'"

For Binford, therefore, the Plio/Pleistocene bone accumulations of the oldest archeological sites at Olduvai were principally the result of carnivore activity, with hominines playing the role of marginal scavengers. No humanlike social implications can be made for such species. "The famous Olduvai sites are not living floors," he concluded.

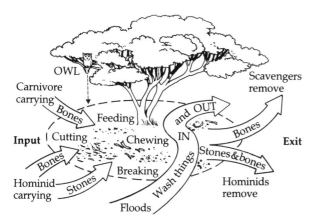

SITE DYNAMICS: Many factors influence the materials that might be brought to a locality and those that might be removed from it. Archeological excavations can recover only what remains at a site and what can be preserved (bones and stones, not plant and soft animal material). (Courtesy of Glynn Isaac.)

This last conclusion has also been reached by several of Isaac's associates, including Potts, Pat Shipman (of Pennsylvania State University), and Henry Bunn (of the University of Wisconsin). Their interpretations of the

bone accumulations, however, differ widely. Specifically, none of the three agrees with Binford that the accumulations are primarily the result of carnivore activity. All see the collections as the work of hominines, with carnivores visiting these sites only occasionally. The assessments made by Potts, Bunn, and Shipman differ in terms of how much of the accumulations are attributed to hunting and how much to scavenging.

Binford's analysis has been criticized on a number of grounds. For example, as Potts points out, this version of residual analysis makes the a priori assumption that hominines displayed no carnivore-like activity. If hominines hunted and consumed animals as other carnivores do, then the resulting bone fragment pattern would be subsumed under "carnivore activity," leaving no residual. Potts's own analysis of the Olduvai archeological sites indicates that the pattern of bone accumulation is more diverse than would be expected at exclusively carnivore sites. He concludes that the accumulations probably represent a mixture of scavenging and hunting, and argues that it is difficult—if not impossible—to distinguish between the bone accumulation patterns that would result from hunting and the patterns from what he terms "early scavenging." Early scavenging could occur when, for example, a hominine locates a dead animal that has not yet been partially eaten by a nonhuman carnivore.

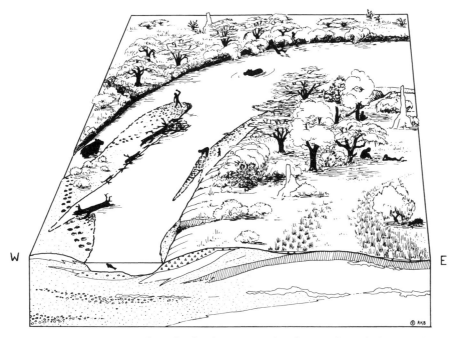

A 1.5 MILLION-YEAR-OLD SITE: Excavated on the floodplain east of Lake Turkana, site 50 has yielded 1405 stone fragments and 2100 pieces of animal bone. Nearly 1.5 million years ago, the site, which was located in the crook of a river course, was used for only a relatively short period of time. Stone fragments and debris struck during their manufacture could be reconstructed to form the original pebble used by the tool makers, and smashed animal bones could be conjoined to establish whole sections. Microscopic patterns on stone-tool edges indicate their use in cutting meat, soft plant material, and wood. This body of evidence invokes a picture of a rather humanlike subsistence behavior. (Courtesy of A. K. Behrensmeyer.)

CUTMARKS AND THEIR SIGNIFICANCE

In 1979, Potts, Shipman, and Bunn simultaneously discovered cutmarks on fossil bones at Olduvai, which apparently had been inflicted by stone flakes used to deflesh or disarticulate the bones. Cutmarks stand as perhaps the most direct evidence possible that hominines used the bones at the archeological sites. Once again, however, their interpretations of this phenomenon differ somewhat.

Shipman, for instance, sees little or no indication that the Olduvai hominines were disarticulating bones and therefore concludes that the bone accumulations were principally the fruits of scavenging from other carnivore kills. Both Potts and Bunn observe what they interpret as evidence of disarticulation of bones, which could indicate hunting or early scavenging. Of the two, Bunn more strongly favors hunting as an important aspect of the Olduvai hominines' behavior. Potts points out, incidentally, that nature includes very few pure hunters and pure scavengers, with most carnivores participating in both activities to some extent. "To ask whether early hominids were hunters or scavengers is therefore probably not an appropriate question," he says.

Nevertheless, whether they were hunted or scavenged, the remains of animals at the Olduvai sites could, in principle, serve as an indication of hominine home bases. This explanation seems unlikely, however. Typical hunter-gatherer home bases are places of intense social activity and havens of safety that are occupied for periods of a few weeks and then abandoned. In contrast, the Olduvai sites apparently accumulated over periods of between 5 and 10 years, and they were obviously visited by carnivores. The carnivores left their signatures on the sites in the form of tooth marks on certain bones. Some tooth marks overlap cutmarks, which seems to imply that hominines got to the bones first. Other tooth marks are overlapped by cutmarks, which appears to confirm that the hominines occasionally scavenged from carnivore kills.

CUTMARKS IN CLOSEUP: This fragment of bone, from a 1.5 million-year-old site in northern Kenya, bears characteristic marks that are left when a stone tool is used to deflesh a bone. The discovery of cutmarks provided an important method of testing the hypothesis that bones and stones at ancient sites were causally related.

A recent analysis by Robert Blumenschine, of Rutgers University, carried out on the Zinjanthropus site bones suggests that, although the Olduvai hominines were not the minimal scavengers of bones discarded by other carnivores (as Binford argues), they were principally scavengers and not significant hunters (as Bunn claims). This work was based on a comparison of tooth marks and percussion marks on fossil bones from Olduvai with marks produced experimentally.

If the Olduvai sites are not typical home bases, what were they? Potts has suggested that they formed around stone caches—places at which hominines accumulated raw material for making artifacts. Potts' computer simulations appeared to show that, on energetic grounds, forming stone caches and bringing carcasses to them would be an optimal strategy. In any case, the raw material for the artifacts at some sites apparently came from sources as far as 11 km away. Some of this raw material was never processed, but was left as lumps called manuports.

Thus, the Olduvai sites appear to have been formed by hominines transporting stone to particular localities; they probably also brought meat-bearing bones to these sites, the result most likely of scavenging but possibly of some hunting. Instead of home bases, these sites appear to have been meat-processing and consumption places. Not all early sites are identical, however. For instance, some locations at Koobi Fora, including site 50, are clearly not stone caches because the stone artifact raw material is sourced on the spot. Moreover, several of the stone flakes at site 50 show signs of wood whittling and processing of soft plant material, which might imply a more leisurely use of the site than might otherwise have been envisaged. Whether this development represents a change through

EXCAVATION IN PROGRESS: Site 50, on the eastern shore of Lake Turkana, has yielded important information with which to test the hypothesis that the co-occurrence of bones and stones resulted from hominine activity.

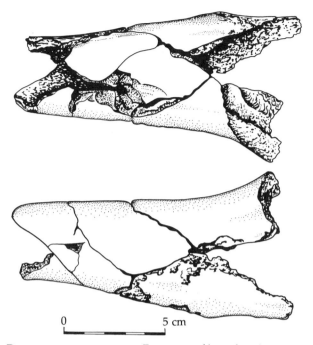

BONE JIGSAW PUZZLE: Fragments of bone found at site 50 were conjoined, producing these two ends of the humerus of a large, extinct antelope. The pattern of fracture indicates that it was the result of percussion by early hominines. Cutmarks were also present on the bone.

A SMALL CORE/UNIFACIAL CHOPPER: With hammerstone in position to knock off next flake (from site 50). (Courtesy of Glynn Isaac.)

time—site 50 is approximately 300,000 years younger than the Olduvai sites—or differences in ecological context remains unknown.

Isaac's response to the findings was to suggest that the food-sharing hypothesis be replaced by the central-place foraging hypothesis. "Conscious motivation for 'sharing' need not have been involved," he wrote in 1982. "My guess now is that in various ways, the behavior system was less human than I originally envisaged, but that it did involve food transport and *de facto*, if not purposive, food sharing and provisioning."

Here Isaac was touching on a difficult methodological issue—that of trying to imagine the lives of humanlike creatures in unhumanlike terms. Modern hunter-gatherers operate with sophisticated organization and (relatively speaking) technology. Lacking weapons to kill at a distance, as humans did until late in prehistory, hunters could achieve only very limited goals and might not qualify as hunters in the commonly understood sense. Scavenging, on the other hand, would have been both technologically and ecologically feasible.

It is worth noting the evidence, produced by Leslie Aiello, for a change of body proportion between *Australopithecus* and *Homo* that would be consistent with an adaptation for great routine activity (see unit 24). Taken together with the appearance of stone tools and archeological sites from 2.5 million years onward, plus shifts in dentition in *Homo* (reduced posterior, increased anterior), it might signal a significant shift in subsistence strategies. Unfortunately, the challenge of deciding to what degree scavenging and hunting contributed to this new adaptation may remain unmet.

The discovery of wooden spears at the site of Schöningen in Germany, dated about 400,000 years old, and their descriptions published in February 1997, implies that systematic hunting had been well developed by that date. ❊

KEY QUESTIONS

- How different are the patterns of bone accumulations at the Olduvai sites from those at pure carnivore sites?
- How are cutmarks distributed on Olduvai bones, and what does this pattern imply about the integrity of the bones that were transported to the sites?
- Can a distinction be made between evidence for hunting as against evidence for early scavenging?
- What kind of social organization might be implied by the central-place foraging hypothesis?

KEY REFERENCES

Binford L. Human ancestors: changing views of their behavior. *J Anthropol Archeol* 1985;4:292–327.

Blumenschine RJ. Percussion marks, tooth marks, and experimental determination of the timing of hominid and carnivore access to long bones at FLK Zinjanthropus, Olduvai Gorge, Tanzania. *J Hum Evol* 1995;29:21–51

————. Archeological predictions for hominid land use in the paleo-Olduvai Basin, Tanzania, during lowermost Bed II times. *J Hum Evol* 1998;34:565–607.

Blumenschine RJ, Cavallo JA. Scavenging and human evolution. *Sci Am* October 1992:90–96.

Bunn H, Kroll E. Systematic butchery by Plio/Pleistocene hominids at Olduvai Gorge, Tanzania. *Curr Anthropol* 1986;27:431–452.

Dennell R. The world's oldest spears. *Nature* 1997;385:767–768.

Hawkes K, et al. The behavioral ecology of hunter-gatherers, and human evolution. *Trends Ecol Evol* 1997;12:29–32.

Isaac G. The archeology of human origins. *World Archeol* 1982;3:1–87.

Lupo KD. Butchering marks and carcass acquisition strategies: distinguishing hunting from scavenging in archeological contexts. *J Archeol Sci* 1994;21:827–837.

Rosa L, Marshall F. Meat eating, hominid sociality, and home bases. Curr Anthropol 1996;37:307–338.

Shipman P. Scavenging or hunting in early hominids. *Am Anthropol* 1986;88:27–43.

Speth JD. Early hominid hunting and scavenging: the role of meat as an energy source. *J Hum Evol* 1989;18:329–343.

PART 7

ORIGIN OF MODERN HUMANS

THE NEANDERTHAL ENIGMA

27

Neanderthals, everyone's favorite "caveman," lived in much of Europe, part of Asia, and the Middle East between 150,000 to probably just less than 30,000 years ago (these last occurrences were observed in western Europe). The first fossil humans to be discovered, Neanderthals have long been the focus of anthropological investigation. More bones of Neanderthals are known than for any other fossil hominine group, including some 30 nearly complete skeletons, so this preoccupation within the anthropological profession is understandable.

In addition to questions about Neanderthals' daily life, two important issues have occupied anthropologists: Who were the Neanderthals' ancestors? And what was the Neanderthals' fate? Hypotheses about Neanderthals' evolutionary status (particularly their fate) have shifted back and forth in the century and a half since the first bones were unearthed. At times, they have been viewed as the direct ancestors of modern Europeans; at other times, they have been regarded as a side branch of the human evolutionary tree, with extinction as their fate. Today, the latter is the most widely supported hypothesis.

NEANDERTHAL ANATOMY

Neanderthal anatomy represents a mixture of primitive characters, derived characters that are shared with other hominines, and derived characters that are unique to Neanderthals (see accompanying figure). In general terms, Neanderthals may be described as being robustly built, heavily muscled, and short in stature. Evidence of the heavy musculature appears in the extremely large muscle attachments and the bowing of the long bones. This structure implies that, whatever the details of Neanderthal subsistence, this species' daily life involved routine, heavy work. The short, broad trunk is consistent with life in a cold environment (Bergmann's rule; see unit 11), as are the short forearm and lower leg relative to the humerus and femur (Allen's rule; see unit 11). For much of the time of the Neanderthals' existence (between 150,000 and just less than 30,000 years ago), Europe and the Middle East were indeed cold, reflecting the end of the Pleistocene Ice Age.

An aspect of the skull anatomy—the extreme protrusion of the upper face—has also been speculated to be related to cold adaptation. To see this relationship, imagine a normal human face, but made of rubber. Now hold the nose and pull it out several inches. This protrusion of the upper face and a broad nasal aperture combine to produce a large chamber in the nasal passage; according to University of New Mexico anthropologist Erik Trinkaus, this chamber would provide an effective way to warm frigid air before it enters the lungs. Two bony projections jut into the front of the nasal cavity from either side, an anatomical feature not seen in any other hominines.

Body weight for the Neanderthals is estimated at 140 pounds for males and 110 pounds for females; statures are estimated at 5 feet 6 inches for males and 5 feet 3 inches for females. Despite their short stature, Neanderthals had large brains, an average of 1450 cm³—some 100 cm³ larger than the modern average. The significance of the larger brain remains a matter of speculation.

Inevitably, brain size impinges on the question of Neanderthals' capacity for spoken language. Does a human-size brain imply a human level of language capacity? Nothing in the neuroanatomy of Neanderthals would deny this capability. These specimens share the overall shape of brains of earlier humans, with the brain appearing low and broadest near the base, with small frontal lobes and large occipital lobes (at the back of the brain). The discovery of a hyoid bone at the cave site of Kebara, Israel, similarly offers no characters that would rule out language capacity. The hyoid attaches to the base of the tongue, and thus is important in the mechanics of verbalization. The Kebara hyoid is modern in all respects, implying no mechanical barrier to spoken language in this respect. Evidence of limited verbal skills does appear in the structure of the larynx, which is inferred from the shape of the cranial base (see unit 32).

The Neanderthal pelvis is unique. In incomplete specimens, the pelvic canal appears to be unusually large, prompting speculation that the gestation was prolonged in this species and that the infant at birth was larger than in modern humans. When a more complete specimen came to light in 1987 (from Kebara), it revealed that the pelvic canal is not unusually large—just that the pubic bone is extraordinarily long.

NEANDERTHAL BEHAVIOR

Neanderthals lived by hunting and gathering, probably in small, nomadic groups, an existence that evidently required extraordinary strength. Their tool technology employed the Levallois technique (see unit 23) to produce flakes that were then further worked to yield as many as 60 different implements, according to François Bordes, a French archeologist. For the Neanderthals, this Middle Paleolithic technology is termed the **Mousterian** technology, with the name being derived from a cave at Le Moustier, France. The Mousterian flakes could be used for many purposes, including cutting flesh, scraping hides, and working wood. Mousterian assemblages show little use of bone, antler, or ivory.

Toward the end of the Neanderthals' tenure, a second, more refined tool assemblage appears in western Europe. Known as the **Chatelperronian**, after a cave site near Chatelperron in France, this technology was long a mystery to archeologists. While it is clearly Upper Paleolithic in character, having fine blades and artifacts made from bone, antler, and ivory, this assemblage also includes at least 50

NEANDERTHAL ANATOMY

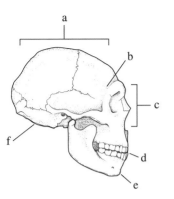

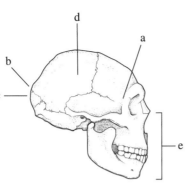

Primative features

a. Long, low cranial vault
b. Well developed supraorbital torus
c. Large face, broad nasal opening
d. Large dentition
e. Absence of chin
f. Broad cranial base

Shared, derived features

a. Lateral reduction of browridge
b. Reduced occipital torus
c. Rounder occipital profile
d. Large brain
e. Reduced facial prognathism

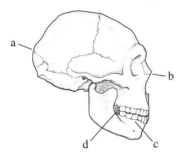

Unique, derived features

a. Spherical shape of cranial vault
   (seen in rear view)
b. Midfacial projection, large nose
c. Teeth positioned forward
d. Retromolar space

percent of flake tools, like those seen in the Mousterian. In 1979, Neanderthal remains were found in association with Chatelperronian tools at the cave site of Saint-Césaire in France, which is dated at 36,000 years old. In 1996, a second such association was demonstrated at the site of Arcy-sur-Cure, also in France. In this case, part of a 33,000-year-old temporal bone of a child was identified as Neanderthal on the basis of the structure of the bony labyrinth (inner ear), in which Neanderthals are derived with respect to *Homo erectus* and modern humans.

These two associations support the conclusion that the Chatelperronian was a Neanderthal industry, produced when Neanderthals and modern humans coexisted in western Europe. Upper Paleolithic technology first appeared in western Europe 40,000 years ago; called the Aurignacian (see unit 29), it represented the work of modern humans. Whether the Chatelperronian industry is a home-grown invention of the late Neanderthals or a result of cultural contact between Neanderthals and modern humans remains unknown.

Another tool that the Neanderthals used routinely was their front teeth. This dentition is often worn with characteristic shelving, perhaps through repeated biting or pulling on hide or other soft but tough material.

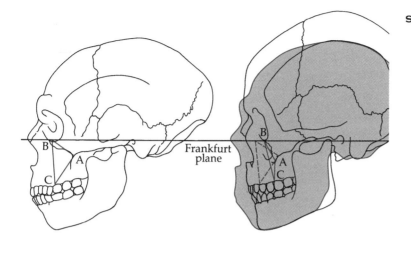

SKULL SHAPE: The triangle in the Neanderthal skull (*left*) shows the spatial relationships between the forward edge of the first molar (C), the lower edge of the cheekbone (A), and the upper edge of the cheekbone (B). A similar relationship drawn in a modern human skull (*right*, with a Neanderthal outline shaded in) produces a much flatter triangle, illustrating the significant forward protrusion in the Neanderthal face.

Remains of Neanderthals have often been found in caves, sometimes in circumstances suggesting deliberate burial, as at Shanidar. A 40,000-year-old skeleton discovered in a cave site near La Chapelle-aux-Saints, France, was found together with a bison leg, other animal bones, and some flint tools, for example; a woman's skeleton was also found in an exaggerated fetal position in the cave of La Ferrassie. Many other examples are described in the literature, often with the assumption that burial was deliberate and associated with ritual practice.

Some of the "burials" can probably be explained by natural events, such as the collapse of cave roofs on occupants or abandoned bodies, and thus are devoid of ritual. But chance would have to be invoked in too many other cases to explain associations of bodies and stone tools, of alignments of bodies, and so on. The evidence is convincing that Neanderthals, and probably other archaic *sapiens*, occasionally buried their dead with a degree of ritual that we recognize as human. The act of burial is

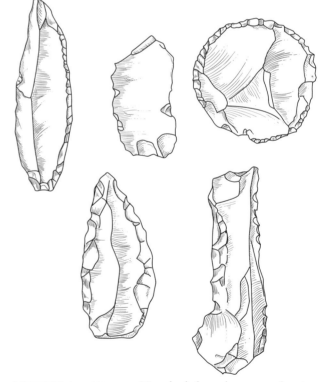

MOUSTERIAN TOOLS: Neanderthals made stone tools using the Levallois technique, which involves striking flakes from a prepared core and then fashioning tools from the flakes. The French archeologist Francois Bordes has identified 20 different artifact types in the Mousterian.

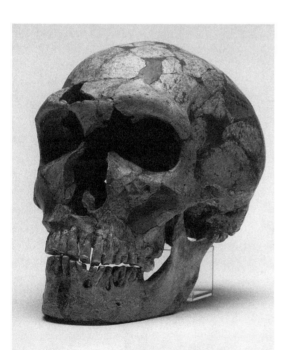

LA FERRASSIE: A 50,000-year-old Neanderthal from the site of La Ferrassie, France, discovered in 1908. (Courtesy of Margot Crabtree.)

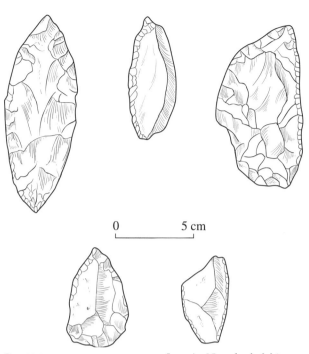

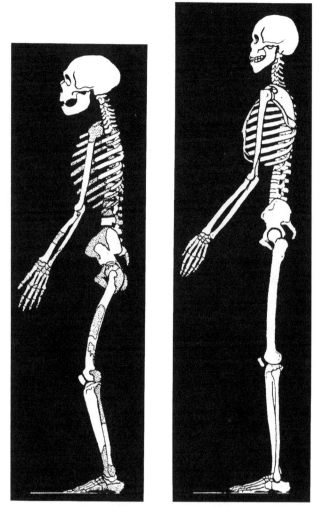

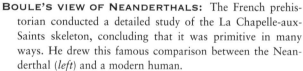

CHATELPERRONIAN TOOLS: Late in Neanderthal history, populations in western Europe manufactured tools that included many Upper Paleolithic elements, such as blades. This industry is called the Chatelperronian.

probably one reason why so many Neanderthal skeletons have been recovered.

A BRIEF HISTORY OF DISCOVERY AND INTERPRETATION

In August 1856, quarry workers in the Neander Valley, Germany, unearthed humanlike bones in a cave, Feldhofer Grotto, above the Düssel River. The fossilized remains—the top of a cranium, some leg and arm bones—were taken to Carl Fuhlrott, a mathematics

BOULE'S VIEW OF NEANDERTHALS: The French prehistorian conducted a detailed study of the La Chapelle-aux-Saints skeleton, concluding that it was primitive in many ways. He drew this famous comparison between the Neanderthal (*left*) and a modern human.

NEANDERTHAL BURIAL: This skeleton was recently recovered from the Kebara Cave, Israel, where it was evidently the subject of deliberate burial. (Courtesy of Ofar Bar-Yosef.)

teacher and local historian known to be interested in natural curiosities. Clearly the remains of a bulky and powerfully muscled individual, they were unlike anything Fuhlrott had seen before, so he sought the more informed viewpoint of Hermann Schaaffhausen, a professor of anatomy at the University of Bonn.

The anthropological community's reaction to the Neanderthal bones was mixed: some believed they represented a primitive race of human; others thought they belonged to a diseased individual; few considered the Neanderthal to be part of human ancestry. William King, an Irish anatomist, was unusual among his colleagues for regarding the Feldhofer specimen as different from *Homo sapiens*, and in 1864 he gave it the species name *Homo neanderthalensis*.

By the end of the century, the discovery of more fossil individuals with the same suite of curious anatomical

BOULE'S INFLUENCE: Marcellin Boule's conclusions about the primitiveness of Neanderthals influenced the profession's view, as seen here in a depiction of Neanderthal life drawn under the supervision of Henry Fairfield Osborn in 1915.

of Natural History in Paris, where Marcellin Boule undertook a detailed study, beginning in 1908. The picture Boule sketched of the Old Man of La Chapelle—and by implication all Neanderthals—was less than flattering. Effectively, he described a slouching, bent-kneed, bent-hipped semi-idiot. Very quickly the anthropological establishment accepted Boule's characterization of Neanderthals, and pronounced the species to be an evolutionary specialization that went nowhere.

Boule's description of Neanderthals as an evolutionary dead end left modern man without an ancestor. In 1912, Piltdown Man appeared on the scene and was accepted by many (but not all) anthropologists as our direct ancestor, filling this void in our evolutionary past. The so-called pre-*sapiens* theory was developed at this point, which argued that there had been an ancient split in the human lineage which led to the early appearance of a relatively modern skeletal form alongside a more archaic hominid, represented in the fossil record by the Neanderthals. The pre-*sapiens* theory effectively dominated anthropological thinking for almost 50 years, despite various vigorous efforts to dislodge it. For instance, the American anthropologist Alés Hrdlička tried, but failed, to resurrect the unilinear hypothesis in the 1920s, based on anatomical

characteristics effectively undermined the notion that pathology explained the appearance of the Feldhofer individual. More important, Eugene Dubois's discovery of *Pithecanthropus* in the early 1890s forced serious consideration of what ancestral forms of human might have looked like (see unit 24).

Gustav Schwalbe, a professor at the University of Strasbourg, suggested at the turn of the century that both *Pithecanthropus* and Neanderthal were part of a steady progression from primitive to modern human beings—a pattern known as unilinear evolution. Under this view, Neanderthals are usually given a subspecies attribution, *Homo sapiens neanderthalensis*, while modern humans would be *Homo sapiens sapiens*.

Neanderthal's status as a direct human ancestor did not last long, for two reasons: (1) the 1908 discovery, and subsequent misinterpretation, of the famous Neanderthal skeleton from the site of La Chapelle-aux-Saints, and (2) the 1912 discovery, and subsequent misinterpretation, of parts of the Piltdown Man, which turned out to be one of the biggest scientific hoaxes of all time (see unit 3).

The La Chapelle-aux-Saints Neanderthal was virtually complete, offering anthropologists the first opportunity to compare in detail Neanderthal anatomy with that of modern humans. The skeleton was sent to the Museum

NEANDERTHAL REHABILITATION: This depiction of Neanderthal, created by Carton Coon in 1939, was meant to show a "normal" Neanderthal, as evidenced by the street clothes.

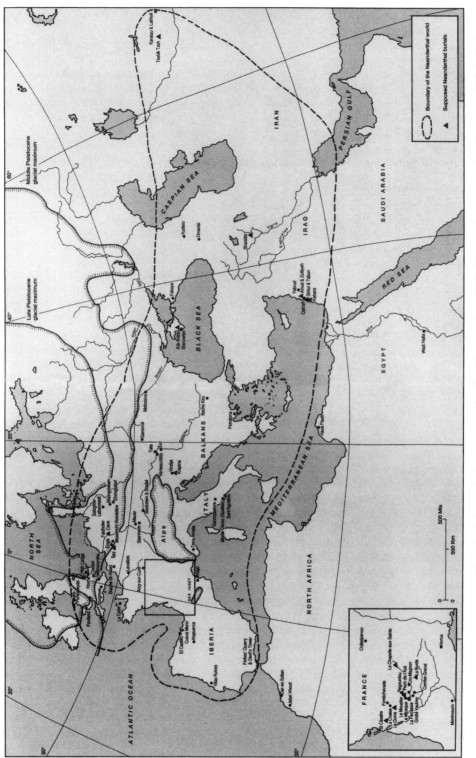

GEOGRAPHICAL DISTRIBUTION: Neanderthal populations were confined to Europe, the Middle East, and western Asia. (Courtesy of Chris Stringer.)

and archeological arguments. He called his hypothesis the "Neanderthal phase of Man."

New fossils discovered during the 1920s and 1930s in Europe and Asia initially failed to shake the pre-*sapiens* hypothesis, even though these specimens displayed Neanderthal-like features in these parts of the world at different times in prehistory. Later, the German anatomist Franz Weidenreich (see unit 24) invoked these fossils in a sophisticated elaboration of Hrdlička's Neanderthal-phase hypothesis. According to Weidenreich, the pithecanthropines gave rise to the Neanderthals, which were directly ancestral to modern humans—in broad outline, a scheme reminiscent of the hypothesis put forth by Gustav Schwalbe 40 years earlier. Moreover, Weidenreich envisaged parallel evolutionary lineages in various regions of the Old World, all leading through separate Neanderthal-like stages to the modern geographical variants of modern humans. Weidenreich's proposal was dubbed the candelabra model of modern human origins—drawn schematically, the long regional ancestries resemble an array of candles. This model, which was elaborated during the 1940s, is the precursor to a major position in the current debate (namely, the multiregional evolution model; see unit 28).

Despite Weidenreich's efforts, the unilinear point of view was slow to reemerge. Eventually, a confluence of events through the 1940s, 1950s, and 1960s overturned the dominance of the pre-*sapiens* theory; instead, it became just one of several competing theories, which included in their number the unilinear model. The first of these events was the development of the synthetic theory of evolution (see unit 4), which allows for anatomical variation within species. The second was the exposure of the Piltdown fossils as a hoax, which removed this pillar of support in a single stroke. The third event was the reevaluation of the La Chapelle-aux-Saints skeleton, which showed that it was not brutish, as Boule had concluded.

The rehabilitation of Neanderthal was effectively completed by Loring Brace, of the University of Michigan, whose 1964 paper, "The Fate of the 'Classic' Neanderthals," was highly influential. Brace reexamined the La Chapelle-aux-Saints skeleton and, like Straus and Cave before him, concluded that Boule had described anatomical features that simply were not present. By the late 1960s, Neanderthals had been restored, in many people's eyes, to their rightful place: as direct ancestors of modern humans. The unilinear theory was at last successfully revived, now as one of a handful of competing theories, including the pre-Neanderthal hypothesis and a version of the pre-*sapiens* hypothesis.

Brace added an extra stage to Schwalbe's original, three-stage scheme to transform it into "australopithecines, pithecanthropines, Neanderthals, and modern humans." According to Brace's so-called single-species hypothesis, only one species of hominine existed at any given period in human evolution—the ultimate expression of the unilinear pattern (see unit 3). Milford Wolpoff, also at Michigan, joined Brace as a vigorous supporter of this hypothesis. In the mid-1970s, the discovery of the coexistence at Koobi Fora of a small-brained, highly robust individual (KNM-ER 406, *Australopithecus boisei*) and a large-brained, nonrobust individual (KNM-ER 3733, *Homo ergaster*) demonstrated that the single-species hypothesis was invalid, at least for that period of human prehistory (close to 2 million years ago).

Wolpoff, now a major protagonist in the current debate on the origin of modern humans, nevertheless insists that the unilinear hypothesis holds for the later stages of human prehistory. A scientific tradition carrying the names of Schaaffhausen, Schwalbe, Hrdlička, Weidenreich, and Brace is therefore continued by Wolpoff. In unit 28, we will see how the modern version of this tradition measures up against the modern version of denying direct ancestry between Neanderthals and modern humans.

IMPLICATIONS OF NEANDERTHAL DNA

One of the more dramatic developments in the study of Neanderthal prehistory came in mid-1997, with the report of the extraction of mitochondrial DNA from the fossilized bones of the type specimen, discovered in the Neander Valley in 1856. Comparison of a short (328 base pair) sequence of Neanderthal mitochondrial DNA with that from a large selection of modern individuals showed it to be very different. For instance, the average number of nucleotide differences in this sequence among modern humans is eight; by contrast, the Neanderthal sequence differed in 28 nucleotide positions, implying that it was genetically very distant from modern humans, and could not have been ancestral to them. Based on the number of differences in nucleotide positions between modern humans and Neanderthals, the joint German/American team that did the work calculated that the last common ancestor of modern humans and Neanderthals lived about 600,000 years ago, and that modern humans originated in Africa, as the "out of Africa" hypothesis argues (see units 28 and 29). It is important to remember, of course, that this is a single sample, and needs to be replicated.

KEY QUESTIONS

- What aspects of Neanderthal anatomy imply an adaptation to cold environments?
- What is the most likely origin of the Chatelperronian tool industry?
- Why did so much resistance arise against accepting Neanderthals as a form of ancient human when they were first discovered?
- How is the current taxonomic status of Neanderthals best described?

KEY REFERENCES

Gargett R. Grave shortcomings: the evidence for Neanderthal burial. *Curr Anthropol* 1989;30:157–177.

Hublin J-J, *et al.* A late Neanderthal associated with Upper Paleolithic artifacts. *Nature* 1996;381:224–226.

Klein RG. Neanderthals and modern humans in West Asia. *Evol Anthropol* 1995/96;4:187–193.

Krings M, *et al.* Neanderthal DNA sequences and the origin of modern humans. *Cell* 1997;90:19–30.

Schwartz JH, Tattersall I. Significance of some previously unrecognized apomorphies in the nasal region of *Homo neanderthalensis*. *Proc Natl Acad Sci USA* 1996;93:10852–10854.

Smith FH. The Neanderthals: evolutionary dead ends or ancestors of modern people? *J Anthropol Res* 1991;47:219–238.

Spencer F. The Neanderthals and their evolutionary significance: a brief historical survey. In: Smith F, Spencer F, eds. The origins of modern humans. New York: Alan R. Liss, 1984:1–49.

Stringer CB, Gamble C. In search of the Neanderthals. London: Thames and Hudson, 1993.

Tattersall I. The last Neanderthal. New York: Macmillan, 1995.

Trinkaus E, Shipman P. The Neanderthals. New York: Alfred A. Knopf, 1993.

THE ORIGIN OF MODERN HUMANS: ANATOMICAL EVIDENCE

28

Since the 1980s, the question of the origin of **anatomically modern humans** has been among the most hotly debated issues in paleoanthropology, with very divergent opinions being vigorously expressed. One extreme hypothesis argues that the transformation occurred as a gradual change within all populations of *Homo erectus* wherever they existed, leading to the near-simultaneous appearance of multiple populations of modern humans in Africa and Eurasia. In this view, the genetic roots of modern geographical populations of *Homo sapiens* are deep, reaching back to the earliest populations of *Homo erectus* as they became established throughout much of the Old World (almost 2 million years in some cases). At the other extreme, an alternative hypothesis views modern humans as having a recent, single origin (in Africa), followed by population expansion into the rest of the Old World that replaced established nonmodern populations. In this scenario, the genetic roots of modern geographical populations of *Homo sapiens* are very shallow, going back perhaps 100,000 years. Other possibilities exist as intermediate positions between these two extremes.

COMPETING HYPOTHESES

The multiple-origins, or multiregional, hypothesis was the first comprehensive theory of the origin of modern humans. Its history stretches back more than 50 years, to Weidenreich's formulation (see unit 27). This hypothesis attempts to explain not only the origin of *Homo sapiens*, but also the existence of anatomical diversity in modern geographical populations. According to the multiregional hypothesis, this diversity resulted from the evolution of distinctive traits (through adaptation and genetic drift) in different geographical regions that became established in early populations of *Homo erectus* and persisted through to modern people. This persistence is known as **regional continuity**.

In its original formulation, the multiregional hypothesis posited limited gene flow (mating) between different geographical populations and was therefore dubbed the candelabra hypothesis. It has since been modified, with gene flow between populations now viewed as an important component. This most recent formulation, developed principally by Alan Thorne (of the Australian National University, Canberra) and Milford Wolpoff (of the University of Michigan, Ann Arbor), is now known as the **multiregional evolution hypothesis**. It views the *erectus*-

to-*sapiens* transformation as a balance between the maintenance of distinctive regional traits in anatomy through partial population isolation and the maintenance of a genetically coherent network of populations throughout the Old World through significant gene flow.

The recent, single-origin hypothesis has a shorter history, dating back to Louis Leakey's ideas developed in the 1960s. Leakey considered the Early and Middle Pleistocene hominines of Africa to be better candidates for modern human ancestry than the *Homo erectus* fossils of Asia; the latter, he said, were an evolutionary dead-end. Howells later dubbed the notion of a single origin the Noah's ark model. The most extreme form of this recent African origin (or **out of Africa**) hypothesis, which assumes substantial replacement of archaic populations by invading modern humans, is most closely associated with Chris Stringer, of the Natural History Museum, London. It accepts some interbreeding between archaic and early anatomically modern populations, but sees its long-term effects as minor. The hypothesis views the establishment of regional anatomical traits in today's geographic populations as the result of adaptation and genetic drift in local populations during the last 100,000 years.

PREDICTIONS OF DIFFERENT HYPOTHESES

Competing hypotheses are tested, of course, by assessing how accurately their predictions are proved in the fossil record. For the extreme hypotheses, the predictions are as follows. If the "out of Africa" model is correct, four principal predictions should hold:

- Anatomically modern humans should appear in one geographical region (Africa) significantly earlier than in others.

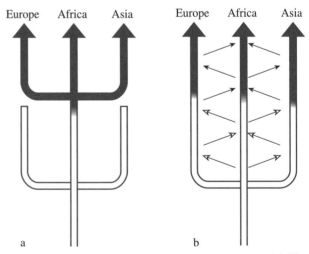

TWO MODELS FOR MODERN HUMAN ORIGINS: (a) The single, recent origin model, in which Africa serves as the source of modern humans, who then replaced established populations. (b) The multiregional evolution model, which balances gene flow between separate geographical populations and maintenance of regional anatomical integrity.

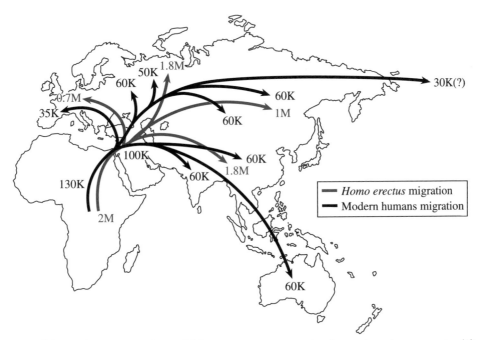

TWO MIGRATIONS: If the single, recent origin model is correct, then the original expansion of *Homo erectus* from Africa into the rest of the Old World would have been followed much later by a similar expansion of modern people. This presentation is certain to be an oversimplification because it implies two discrete events. In fact, multiple population movements must have occurred at different times and in different places.

- Transitional fossils from archaic to early modern anatomy should be found only in Africa.
- Because traits that distinguish modern geographical populations are recently developed, they will show no necessary links with those of earlier populations in the same region (note that this caveat also applies to Africa, because 100,000 years of evolution in diverse populations in that continent will lead to a variety of local traits).
- Little or no evidence should suggest hybridization between archaic and early anatomically modern populations.

In the multiregional evolution model, three expectations follow:

- Anatomically modern humans will appear throughout the Old World during a broadly similar period, although one area might see such populations earlier than the rest.
- Transitional fossils, from archaic to early modern anatomy, should be found in all parts of the Old World.
- In each region of the Old World, continuity of anatomy from ancient to modern populations should be apparent.

Hybridization between archaic and modern forms is not an issue with this hypothesis.

The two fundamental questions in testing the hypotheses against the fossil record are the *location* of the earliest anatomically modern humans and the issue of *regional continuity*. Apart from the Neanderthals, however, the relevant fossil record is frustratingly sparse.

WHERE WERE THE EARLIEST ANATOMICALLY MODERN HUMANS?

Anatomically modern humans are characterized by a reduction in skeletal robusticity and the development of modern bipedal locomotion. Nevertheless, these people were still more robust than modern-day humans. A general description of the skull would include a short, high, rounded cranium with a small face and the development of a chin.

Specimens of anatomically modern humans from Africa and the Middle East stand out as significantly older than those seen elsewhere in the Old World. For instance, the Omo 1 (Kibish) brain case and postcranial material, found in southern Ethiopia in 1967, are strikingly modern; they are estimated to be between 100,000 and 130,000 years old. (A second brain case, Omo 2 [Kibish], is slightly more primitive, but roughly the same age.) Slightly younger specimens—fragments of cranium, arm, and foot—with modern features come from the Klasies River Mouth Cave in South Africa. The dates for these fossils range between 70,000 and 120,000 years old. Border Cave, also in South Africa, has yielded modern-looking cranial and skeletal fragments that may be 100,000 years old. Provenance has been a concern in these cases, so that the true date may be substantially less than 100,000 years.

In the Middle East, the Israeli cave sites of Skhūl and Qafzeh have yielded extensive fossil material, including partial skeletons. Most anthropologists judge these specimens to be essentially modern, even though they have some archaic features. Recent dating efforts (with electron spin resonance and thermoluminescence techniques; see unit 7) give these specimens' ages as close to 100,000 years. Elsewhere in the Old World, the earliest modern remains come from the cave site of Liujiang, in southern China, with a date of 67,000 years, but possibly younger. In Southeast Asia (Java), modern humans appear to be late arrivals, with *Homo erectus* coexisting with early moderns in Africa and the Middle East. Similarly, the earliest modern people in Europe are latecomers, appearing some 40,000 years ago.

This pattern of the earliest appearance of modern humans more strongly supports predictions of the single-origin hypothesis than those of the multiregional hypothesis.

The roughly similar dates of the African and Middle Eastern fossils have led some anthropologists to suggest a north African origin for modern humans, with the Middle East as part of the same ecological zone. Others leave open the possibility that the Middle East itself was the region of origin. The strikingly modern form of the Omo 1 (Kibish) brain case dated at as much as 130,000 years old provides sub-Saharan Africa's strongest claim to being the region of origin.

THE QUESTION OF REGIONAL CONTINUITY

Regional continuity of anatomical traits from ancient to modern populations represents the cornerstone of the multiregional evolution hypothesis. The extreme form of the single-origin hypothesis denies such continuity, particularly through to the present day. The identification of such putative regional continuity in the Far East, in fact, led Weidenreich to formulate the multiregional hypothesis half a century ago. Modern proponents of the hypothesis claim to find such continuity in Asia, Africa, the Middle East, and Europe, as well as the Far East. The issue of regional continuity remains the most contentious aspect of the current debate, however, with little agreement between proponents of competing hypotheses over interpretation of relevant fossil anatomy in these geographical regions.

AUSTRALASIA

Proponents of multiregionalism argue that Australasia offers one of the strongest sets of evidence in favor of regional continuity. The argument is based on essentially three data points: the earliest inhabitants of Java, much more recent archaic forms in Java, and modern Australians. The earliest Javan inhabitants, *Homo erectus*, possessed especially thick skull bones, strong and continuous brow ridges, and a well-developed shelf of bones at the back of the skull. Their foreheads were flat and retreating, and the large, projecting faces sported massive cheek bones. Indeed, the teeth are the largest known in *Homo erectus*. As noted in unit 24, these people may have lived in Java as long as 1.8 million years ago.

The next data point is taken from a dozen brain cases found in 1936 at Ngandong, in western Java. Colloquially known as Solo Man, these specimens have many *Homo erectus* features. Multiregionalists see them as descendants of the earlier Javan *Homo erectus* people, displaying many of the same anatomical features mentioned above, but with enlarged brain cases. The age of the Ngandong fossils is surprising. Until recently, they had been estimated to have been more than 100,000 years old, but dates newly obtained at the Berkeley Geochronology Center place them between 27,000 and 53,000 years old. If correct, it means that the archaic Ngandong population lived long after modern humans had appeared elsewhere in the Old World and were contemporaries of the earliest *Homo sapiens* in the region. This development is parallel to the situation in Europe, where Neanderthals and modern humans coexisted for a while. The earliest inhabitants of Australia constitute the third data point. Archeological evidence indicates that humans first reached Australia approximately 60,000 years ago, although fossil evidence is considerably younger (see unit 34). According to multiregionalists, the earliest Australian fossils "show the Javan complex of features."

Can the features cited as evidence of regional continuity truly be traced from ancient Javan *Homo erectus* (1.8 million years old), to the Ngandong specimens (50,000 years), to modern Australians (60,000 years)? The very large time span over which these three points are distributed, and the clumping of the two most recent dates, makes the proposition unlikely. More particularly, are these features truly unique (that is, derived) to this region of the world?

A general anatomical similarity undoubtedly exists in these three populations, particularly in terms of their robusticity. Unfortunately, the comparison of facial and dental features cited as evidence of regional continuity cannot be tested with the Ngandong specimens because they comprise brain cases only. Two independent studies by Australian anthropologists Colin Groves and Phillip Hapgood in the late 1980s, however, questioned the phylogenetic validity of several of these features, concluding that they are retained primitive traits common to *Homo erectus* and archaic *Homo sapiens*, not derived features unique to the region. Indeed, many of these features occur with greater frequency in other Asian populations. More recently, Marta Lahr, of Cambridge University, England, reached similar conclusions based on an examination of cranial features. In a recent review of this evidence, Leslie Aiello, of University College, London, stated that "The only conclusion that can be drawn from this [evidence] is that the anatomical features used in support of continuity cannot be uncritically accepted as 'clade' features mirroring regional continuity in the Far East." This ambiguity does not *disprove* continuity, she noted, but merely indicates that the evidence currently adduced in its support is invalid.

The evolutionary history of the region was evidently complex. As Aiello observed, it "was characterized by a complicated mosaic of gene flow, population migration, and continuity throughout the Middle and Late Pleistocene periods and . . . this mosaic involved gene flow from east to west as well as from west to east." The bottom line, then, is that neither hypothesis is strongly supported or disproved in Australasia.

EAST ASIA

In a recent review, Wolpoff and fellow proponents of the multiregional hypothesis stated that East Asia provides "a continuous sequence of human fossil remains" from almost 1 million years ago to the present day. Although the East Asian fossil record is richer than that of the Far East, this statement is surely an exaggeration. Wolpoff and his colleagues also state that the fossils reveal "a smooth transition into the living peoples of East Asia."

The earliest known human fossil material in the region consists of a cranium from Lantian, in northeastern China; this *Homo erectus* specimen is dated at close to 1 million years. Next oldest is the collection of cranial parts from the main cave of Zoukoudian, which are also classic Asian *Homo erectus* (see unit 24) and span a period from more than 500,000 to 200,000 years ago. *Homo erectus* remains of similar age to the youngest fossil at Zoukoudian have also been found at Hexian. Fossils that display a mix of *erectus* and *sapiens* features have been discovered in China, including a partial skeleton from Jinniu Shan and a skull from Dali, both dated to approximately 200,000 years. These latter two specimens are generally known as archaic *sapiens*, and in the multire-

gionalists' scheme they represent forms transitional from *erectus* to modern *Homo sapiens*. They include the abundant fossil remains at the Upper Cave of Zoukoudian, dated at 20,000 years.

This picture became more complicated in the early 1990s, after the discovery of two crushed crania at Yunxian, in east-central China. Although their damaged condition makes anatomical analysis difficult, these specimens have been described as archaic *sapiens* and are dated to as long ago as 350,000 years. If their anatomical attribution and age are correct, they would be almost twice as old as some of the latest *Homo erectus* populations, undermining the pattern of regional transition as envisaged in the multiregional evolution hypothesis.

Multiregionalists argue that the fossils in eastern Asia differ from those in the Far East much in the same way that the modern populations vary. Eastern Asians, both ancient and modern, have smaller faces and teeth, flatter cheeks, and rounder foreheads; their noses are less prominent and are flattened on top. A feature that is particularly emphasized as reflecting regional continuity is the shovel-shaped upper incisors. Critics of multiregionalism point out that this supposed derived feature is also found in ancient populations elsewhere in the Old World, and therefore cannot be used to link Chinese *Homo erectus* to modern Chinese people. Lahr's recent analysis is also critical of this and other primitive traits that are used to infer regional continuity. Moreover, a study by J. Kamminga and Richard Wright strongly indicates that the Upper Cave population individuals do not resemble modern Chinese people, as they should if they were part of a gradual, regional transition from *erectus* to archaic *sapi-*

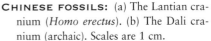

CHINESE FOSSILS: (a) The Lantian cranium (*Homo erectus*). (b) The Dali cranium (archaic). Scales are 1 cm.

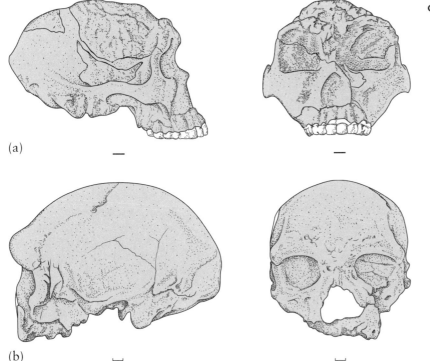

(a)

(b)

ens to modern *sapiens*; instead, this population is more closely allied with African morphology.

In Aiello's view, the evolutionary history of the region was complex, and neither the multiregional nor the single-origin hypotheses is strongly supported or refuted in East Asia.

THE MIDDLE EAST

The fossil record of the Middle East is rich, including several partial skeletons. The first excavations began in the 1930s at the cave sites of Skhūl and Tabūn on Mount Carmel, Israel, and produced partial skeletons that probably resulted from deliberate burial. The Tabūn individuals are Neanderthals, while those from Skhūl are primitive-looking moderns. Excavations conducted over the next five decades yielded additional human fossils from these and three more sites (Kebara, also on Mount Carmel; Amud, near the Sea of Galilee; and Qafzeh, near Nazareth). Kebara and Amud yielded Neanderthals and Qafzeh moderns.

Until a decade ago, the Neanderthals were thought to predate the modern population (with ages of 60,000 and 40,000 years old, respectively) and were assumed to be ancestral to them, in line with the multiregional hypothesis. Recent dating efforts, however, have revealed a more complicated picture that offers much less support for the multiregional hypothesis. Kebara and Amud appear to be nearly 60,000 years old, as believed earlier, but Tabūn is much older at approximately 120,000 years. A more significant redating affected the modern populations, with Skhūl and Qafzeh being placed at 100,000 years. Clearly, a simple ancestor/descendant relationship between Neanderthals and moderns is not possible, as the moderns are near contemporaries with the earliest Neanderthals of the region, and Neanderthals persist for at least 40,000 years after the first appearance of moderns.

The temporal overlap of the two populations is construed by proponents of a single-origin model as strong support for their hypothesis. The body proportions of the modern people more closely resemble those of warm-adapted Africans than those of cold-adapted Neanderthals (see unit 11), which provides additional support for the single-origin hypothesis.

Proponents of multiregionalism counter these conclusions by arguing that the region was occupied by one highly variable premodern group, not separate Neanderthals and moderns. (The fact that the two populations used identical stone-tool technologies, the Mousterian, is adduced in support of this notion; see unit 30.) Most observers find the claim for a single, variable population

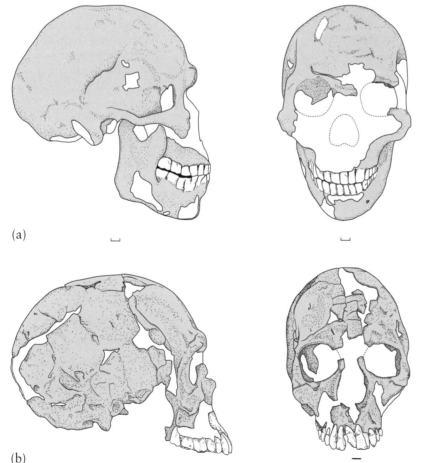

(a)

(b)

MIDDLE EASTERN MODERNS: Crania from (a) Skhūl and (b) Qafzeh. Both sites are dated at approximately 100,000 years, which means that modern humans moved out of Africa soon after they originated south of the Sahara; alternatively, it might indicate that north Africa or the Middle East was the site of origin. Scales are 1 cm.

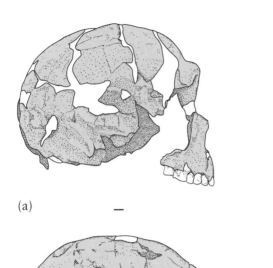

(a)

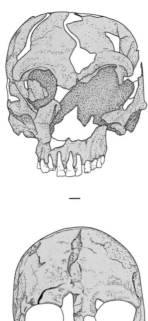

MIDDLE EASTERN NEANDERTHALS:
Crania from (a) Tabūn and (b) Amud. These Neanderthals were once thought to predate the modern population in the Middle East, suggesting an ancestor/descendant relationship. Although the Tabūn people predated the moderns, those from Amud did not, making such a relationship impossible. Scales are 1 cm.

(b)

unconvincing, noting that it would require a range of variation unknown in any other hominine population. Moreover, aspects of the Neanderthal postcranial anatomy show retention of certain primitive features (in the femur and pelvis) that the moderns lack.

Overall, the Middle East offers more support for the single-origin hypothesis than for the multiregional hypothesis, and may even refute the latter.

EUROPE

The Middle to Late Pleistocene hominine fossil record of Europe is dominated by the Neanderthals (see unit 27). For this discussion, the pertinent question involves the identity and fate of their ancestors. According to the multiregional evolution hypothesis, the Neanderthals were part of a gradually evolving lineage that eventually yielded anatomically modern humans in Europe. In contrast, the single-origin hypothesis purports that they represent a locally evolved species that became extinct approximately 30,000 years ago and that contributed nothing to modern European populations.

As noted in unit 24, no unequivocal fossil evidence exists to prove the presence of *Homo erectus* in Europe. Many examples of so-called archaic *sapiens* have been located, however, including some recent spectacular finds at Atapuerca, in northeast Spain. These remains of many individuals include some that may be 780,000 years old.

According to some proponents of the single-origin hypothesis, most of these specimens should be assigned to *Homo heidelbergensis*, which may have been ancestral to Neanderthals in Europe and to *Homo sapiens* in Africa. However, in May 1997, the discoverers of the fossils elected to name the fossils a new species *Homo antecessor*. Multiregionalists view this group as evidence of a transition toward modern *Homo sapiens*.

The Mauer mandible, found in 1907 and dated at roughly 500,000 years old, combines primitive features (robusticity) with modern features (molar size). It was given the species name *Homo heidelbergensis* in 1908. Other fossils with a similar mix of ancient and modern were found in the mid-1930s, such as a cranium at Steinheim, Germany, and skull fragments at Swanscombe, England. Both of these items date to between 200,000 and 300,000 years old. The Steinheim skull possesses heavy brow ridges and a low forehead that betray its primitive status—albeit one not equivalent to *Homo erectus*.

In 1960, Greece joined in the panoply of European archaic human sites, with the discovery of a robust but large cranium in a cave at Petralona. Dating this fossil has long posed a challenge, but most recently it has been estimated to be 200,000 years old. In the early 1970s, the face, forehead, and two jaws of an archaic form were found at the cave of Arago, near Tautavel, in south-

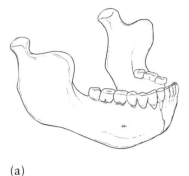

(a)

TWO GERMAN FOSSILS: (a) The Mauer mandible, found in 1907, combines archaic features (robusticity) with modern features (molar size). It is dated at 500,000 years and is the type specimen of *Homo heidelbergensis*. (b) The Steinheim cranium, found in 1933 and dated between 200,000 and 300,000 years, displays a mix of archaic features (heavy brow ridges) and modern features (large brain). Scales are 1 cm.

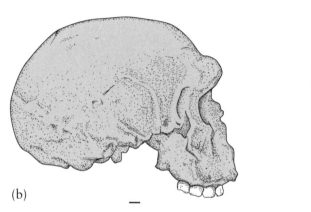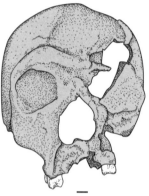

(b)

west France. The face protrudes forward, the brow ridges are heavy, the forehead is slanting, and the brain is smaller than the modern average. Overall, the Arago fossil is more primitive than the Steinheim model, and perhaps 100,000 years older. In 1993, a massive tibia, or shin bone, was found at Boxgrove, England, together with some Acheulean tools. Its age has been estimated at 500,000 years, or similar to that of the Mauer mandible.

The most spectacular finds of recent times, however, are those at the fossil-rich Atapuerca Hills in northern Spain. In 1993, a team of Spanish researchers reported the discovery of 1300 human fossil remains (representing 30 individuals) from a single site (Sima de los Huesos), dated at 300,000 years old. It represents the largest single collection of early human fossils bones anywhere in the world. Like other human fossils of this age in Europe, the specimens display a mix of ancient and modern features, and may be assigned to *Homo heidelbergensis*.

These various specimens represent potential ancestors for Neanderthals. What of their fate? Beginning some 40,000 years ago, classic Neanderthal anatomy disappeared in Europe, with an east-to-west progression that ended nearly 27,000 years ago. The latest evidence of Neanderthals is found at the site of Zafarraya, southern Spain. Fossil evidence indicating the presence of anatomi-

cally modern humans follows the same trajectory. For instance, modern jaw and tooth fragments from the cave of Bacho Kiro, Bulgaria, are dated at 43,000 years. A frontal bone with a high forehead and small brow ridges has been found at Velíka Pečina in Croatia and dated at 34,000 years. A similar specimen, but with more robust frontal bone, from Hahnöfersand, Germany, has been dated at 33,000 years. A large collection of somewhat robust modern human remains was found at Mladeč, Czechoslovakia. The age of the famous Cro-Magnon fossils, from France, is placed at approximately 30,000 years.

When Bräuer and his colleague K. W. Rimbach compared the crania of the early moderns of Europe, the early moderns of Africa, and the Neanderthals, they found a close morphological similarity between the first two but saw no link between early European moderns and Neanderthals. Similarly, Cro-Magnon skeletons exhibit a warm-adapted body stature, not the cold-adapted formula seen in Neanderthals. This character may be taken as strong evidence of the replacement of Neanderthals and supports the single, African origin hypothesis.

Although some proponents of the multiregional hypothesis accept that Neanderthals were replaced, at least in the west, most argue for continuity. As evidence, they adduce the size of the nose in Neanderthals and later

1985 view   **1990 view**

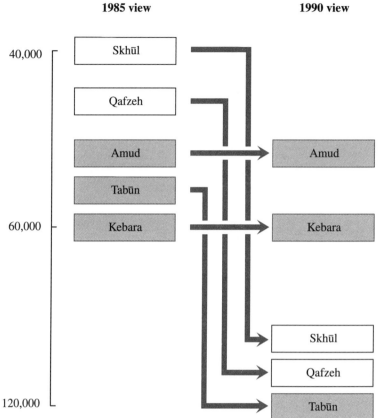

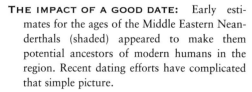

THE IMPACT OF A GOOD DATE: Early estimates for the ages of the Middle Eastern Neanderthals (shaded) appeared to make them potential ancestors of modern humans in the region. Recent dating efforts have complicated that simple picture.

Europeans, some details of the back of the skull, and, most particularly, the shape of the mandibular nerve canal. This opening is grooved in most living people, but it is surrounded by a bony ridge in 53 percent of Neanderthals. The incidence in later, modern Europeans is just 6 percent. According to multiregionalists, this incidence is 44 percent in early moderns in Europe, indicating continuity. Stringer and Bräuer have recently criticized this claim, saying that while it might indicate gene flow between Neanderthals and early moderns, it is just as likely to be a statistical fluke. The sample used by multiregion-alists comprises just four individuals, including one from Vindija, Croatia, that many consider to be Neanderthal. Given the small sample size of just three individuals, the inclusion of just one with an infrequent feature would produce an erroneously high incidence. Of course, the chances of this type of occurrence in a population with low incidence is not great.

Aiello's assessment is that the anatomical (and archeological) data of Europe "do not contradict an ultimate African origin for modern humans; however, they also do not clearly substantiate this hypothesis."

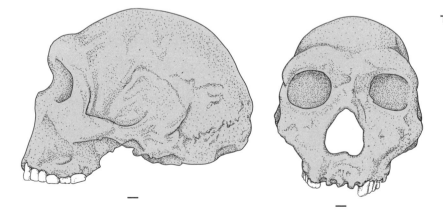

THE PETRALONA CRANIUM: Found in 1960, the cranium is robust but has a large brain case, thus combining archaic and modern features. It has recently been redated at 200,000 years. Scales are 1 cm.

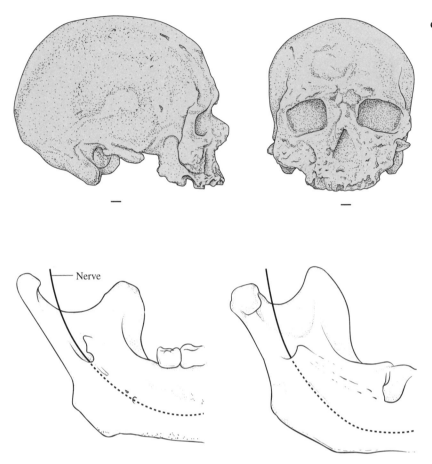

CRO-MAGNON: The famous cranium from Les Ezyies, in France, dated at 30,000 years, provides an example of early modern people in western Europe. Scales are 1 cm.

MANDIBULAR NERVE CANAL: In most living and fossil people, the rim around the nerve canal is grooved (*left*); in roughly half of all Neanderthals, it is surrounded by a bony ridge. Multiregionalists argue that early modern people in Europe also had a high incidence of the bony ridge, indicating important morphological continuity.

AFRICA

The Middle and Late Pleistocene human fossil record of Africa is not extensive, but a sufficient number of specimens has been found to prove a transition from primitive to modern humans. The first find was made in 1921 at a cave site at Kabwe (formerly Broken Hill), Zambia. The specimen, a cranium, was originally called Rhodesian Man, but is now more generally referred to as the Kabwe cranium. The cranium is large, having a capacity of 1280 cm³, and possesses a sloping forehead and prominent brow ridges reminiscent of Neanderthal. Associated limb bones are, however, straighter and more slender than those of Neanderthals. The specimen's age is estimated to be at least 200,000 years.

Similar archaic forms have been found at Elandsfontein, South Africa (dated at 300,000 years); Bodo, Ethiopia (of similar age); and near Lake Ndutu, Tanzania (perhaps 100,000 years older than the other two finds). The cranial shape of most of these African archaics is long, as in *Homo erectus*, but more elevated; from the rear, it appears to be wider at the top than at the base,

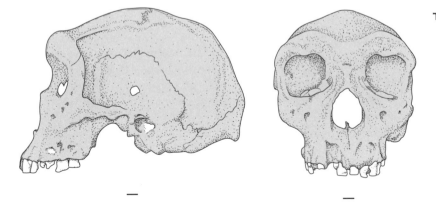

THE KABWE CRANIUM: Estimated to be at least 200,000 years old, this cranium was the first early human fossil found in Africa. Scales are 1 cm.

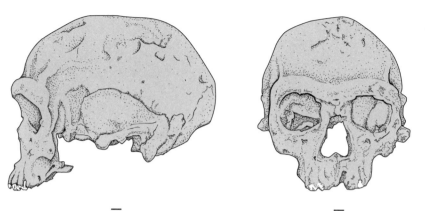

JEBEL IRHOUD: One of a group of later human fossils that were perhaps intermediate between *Homo heidelbergensis* and *Homo sapiens*. The species name of *Homo helmei* would be appropriate for this specimen, if a specific designation is justified. Scales are 1 cm.

unlike the structure in *Homo erectus*. The Ndutu cranium is shorter and less flattened. In northern Africa, archaic forms of Middle Pleistocene age have been found at Salé and the Thomas Quarries, in Morocco.

Some proponents of the single-origin hypothesis group these specimens in *Homo heidelbergensis*, which they claim evolved in Africa and then moved into other regions of the Old World. The species is held to be ancestral to modern humans, through a form represented by several specimens that are generally modern, but not yet fully modern. These remnants include the following: cranial fragments from Florisbad, South Africa; a cranium and

lower face from Ngaloba, Tanzania; a skull (KNM-ER 3884) from Koobi Fora, Kenya; the Omo 2 (Kibish) brain case from Ethiopia; and various cranial and postcranial fossils from Jebel Irhoud, Morocco.

The dates of some of these specimens remain somewhat uncertain, but they are generally later than the above group of *Homo heidelbergensis* specimens. Recent dating of the Florisbad cranium indicates that it may be as old as 300,000 years. Whether this group can be contained within *Homo heidelbergensis* or should be assigned to a separate species (*Homo helmei*) is a matter of debate. In any case, these individuals could represent a form transi-

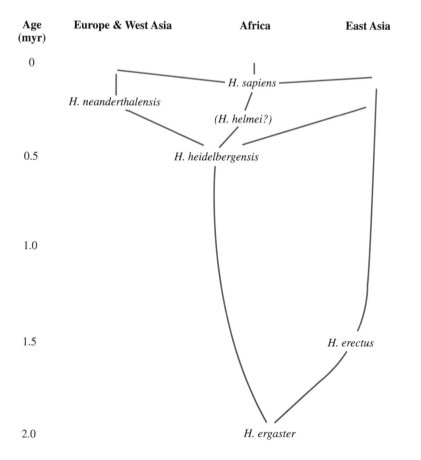

Age (myr)	Europe & West Asia	Africa	East Asia

A PHYLOGENETIC SCHEME, REFLECTING THE SINGLE ORIGIN MODEL

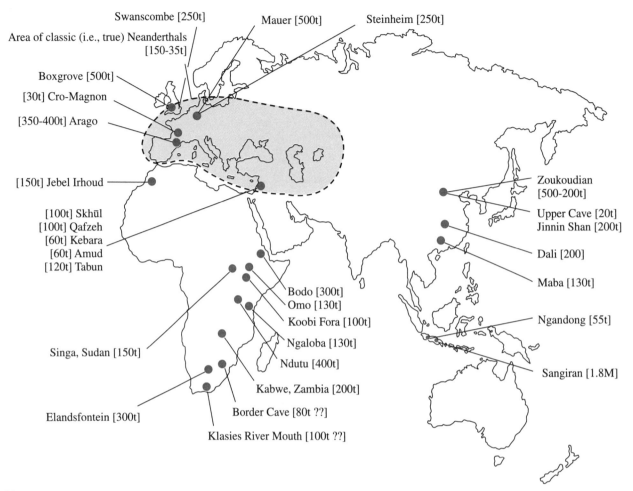

Swanscombe [250t]   Mauer [500t]   Steinheim [250t]

Area of classic (i.e., true) Neanderthals [150-35t]

Boxgrove [500t]

[30t] Cro-Magnon

[350-400t] Arago

[150t] Jebel Irhoud

[100t] Skhūl
[100t] Qafzeh
[60t] Kebara
[60t] Amud
[120t] Tabun

Singa, Sudan [150t]

Elandsfontein [300t]

Klasies River Mouth [100t ??]

Border Cave [80t ??]

Kabwe, Zambia [200t]

Ndutu [400t]

Ngaloba [130t]

Koobi Fora [100t]

Omo [130t]

Bodo [300t]

Zoukoudian [500-200t]

Upper Cave [20t]
Jinnin Shan [200t]

Dali [200]

Maba [130t]

Ngandong [55t]

Sangiran [1.8M]

MAP SHOWING SOME OF THE MOST IMPORTANT SITES

tional to modern humans, such as those found at Omo 1 (Kibish), Klasies River Mouth, and Border Cave.

This pattern of transitional forms from archaic to modern fits both the single-origin and multiregional evolution hypotheses, of course. The fact that it occurred earlier in African than elsewhere provides support for the former concept. In addition, the anatomical similarities between some of these African archaic forms and archaics elsewhere in the Old World supports the single, African origin hypothesis rather than the multiregional evolution hypothesis.

THE ANATOMICAL EVIDENCE AS A WHOLE

The origin of modern humans was undoubtedly complex, involving much population movement at different times, and local population expansions and extinctions. Lahr and her colleague Robert Foley have argued that multiple dispersals from a variable source population in Africa at different times and via different routes may help explain how morphological variability developed in the modern world. In any case, the weight of evidence offers more support for some form of single origin hypothesis than for the multiregional evolution hypothesis. ✳

KEY QUESTIONS

- How has the history of the interpretation of Neanderthals' place in human evolution influenced the modern debate over the origin of modern humans?
- Why is the same fossil evidence often interpreted differently by different anthropologists?

- What is the strongest evidence in favor of (1) the multiregional evolution hypothesis and (2) the single-origin hypothesis?
- What additional fossil evidence would help to resolve the current debate?

KEY REFERENCES

Aiello LC. The fossil evidence for modern human origins in Africa: a revised view. *Am Anthropol* 1993;95:73–96.

Ambrose SH. Late Pleistocene human population bottlenecks, volcanic winter, and differentiation of modern humans. *J Hum Evol* 1998;34:623–651.

Bermudez de Castro JM, *et al.* A hominid from the Lower Pleistocene of Atapuerca, Spain: possible ancestor to Neanderthals and modern humans. *Science* 1997;276:1392–1395.

Bar-Yosef O, Vandermeersch B. Modern humans in the Levant. *Sci Am* April 1993:64–70.

Frayer DW, *et al.* Theories of modern human origins: the paleontological test. *Am Anthropol* 1993;95: 14–50.

Howell FC. Some thoughts on the study and interpretation of the human fossil record. In: Eikle EM, *et al.*, eds., Current issues in human evolution. San Francisco: California Academy of Sciences, 1996: 1–38.

Lahr MM. The multiregional model of modern human origins. *J Hum Evol* 1994;26:23–56.

Lahr MM, Foley R. Multiple dispersals and modern human origins. *Evol Anthropol* 1994;3:48–60.

Nitecki M, Nitecki D, eds. Origins of anatomically modern humans. New York: Plenum Press, 1994.

Rightmire GP. Deep roots for the Neanderthals. *Nature* 1997;389:917–918.

Ruff CB, *et al.* Body mass and encephalization in Pleistocene *Homo*. *Nature* 1997;387:173–176.

Stringer CB. The emergence of modern humans. *Sci Am* Dec 1990:98–104.

Stringer CB, McKie R. African exodus. New York: Henry Holt, 1996.

Swisher CC, *et al.* Latest *Homo erectus* of Java: potential contemporaneity with *H. sapiens* in Southeast Asia. *Science* 1996;274:1870–1874.

Tattersall I. Out of Africa again . . . and again? *Sci Am* April 1997:60–67.

Thorne AG, Wolpoff MH. The multiregional evolution of humans. *Sci Am* April 1992:76–83.

THE ORIGIN OF MODERN HUMANS: GENETIC EVIDENCE

29

The first application of genetic data to the question of the origin of modern humans took place in the early 1980s, but not until 1987 did it become highly visible in this realm. The initial work, conducted first in Douglas Wallace's laboratory at Emory University and later in the University of California, Berkeley, laboratory of Allan Wilson, focused on mitochondrial DNA. It inspired the so-called **mitochondrial Eve hypothesis**, which posited that the mitochondrial DNA in all living people could be traced back to a single female who lived in Africa approximately 200,000 years ago (hence the inclusion of the name "Eve"). This female was a member of a population of an estimated 10,000 individuals, all of whom were related to the founding population of modern humans; descendants of this population spread into the rest of the Old World, and replaced existing populations of various species of archaic *sapiens* and *Homo erectus*. Thus, the mitochondrial Eve hypothesis was consistent with the recent, single-origin (out of Africa) model and gave no support for the multiregional evolution model (see unit 28).

THE MITOCHONDRIAL EVE STORY: BRIEFLY TOLD

Most of the DNA in our cells is packaged within the 23 pairs of chromosomes in the nucleus, which in total measures about 3 billion base pairs in length; this struc-

ture is known as the **nuclear genome**. The cell also contains many copies of a second, much smaller genome that consists of a circular molecule of DNA, 16,569 base pairs long, called the **mitochondrial genome**. Mitochondria are the organelles responsible for the cell's energy metabolism, and each cell contains several hundred of these structures.

Mitochondrial DNA is useful for tracking relatively recent evolutionary events for two reasons. First, the DNA, which codes for 37 genes, accumulates mutations on average 10 times faster than occurs in nuclear DNA. Even in short periods of time, therefore, the DNA will accumulate mutations that can be counted. As mutations represent the equivalent of information, mitochondrial DNA provides more information over the short term than does nuclear DNA. Second, unlike an individual's nuclear genome, which consists of a combination of genes from both parents, the mitochondrial genome comes only from the mother (except under unusual circumstances). Because of this maternal mode of inheritance, no recombination of maternal and paternal genes occurs; such a mixture may sometimes blur the history of the genome as read by geneticists. Potentially, therefore, mitochondrial DNA offers a powerful way of inferring population history, unhindered by the genetic fog of recombination.

One of the first significant observations to emerge from this work was that the amount of variation of mitochondrial DNA types in the modern human population throughout the world is surprisingly low—just one-tenth of that known among chimpanzees, for instance. One explanation is that modern humans evolved very recently, a view that Wallace and Wilson independently supported. A calculation based on the rate of accumulation of mutations of mitochondrial DNA gave a time of origin of 140,000 to 280,000 years ago. An alternative explanation holds that modern humans passed through a population bottleneck recently, which reduced genetic variation. These explanations are not mutually exclusive: modern humans may have evolved recently *and* experi-

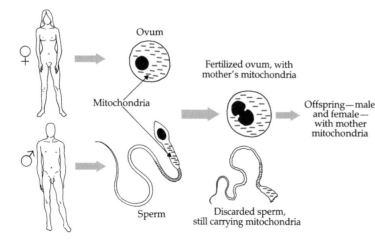

PATTERNS OF INHERITANCE: Unlike nuclear DNA, for which we inherit half from our mother and half from our father, mitochondrial DNA is passed on only by females. When the sperm fertilizes the egg, it leaves behind all of its mitochondria; the developing fetus therefore inherits mitochondria only from the mother's egg.

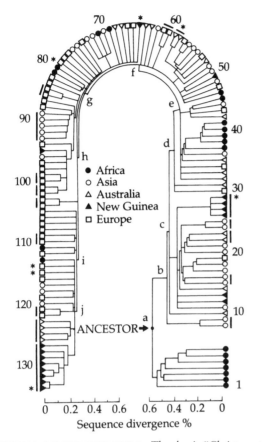

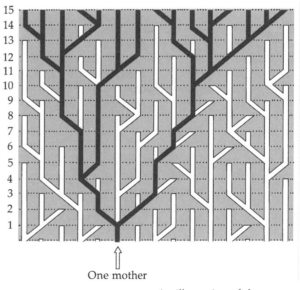

PATTERNS OF RELATEDNESS: The classic "Christmas tree" genealogy produced by Wilson and his colleagues in 1987 shows the genetic divergence among 147 individuals from different geographic populations, whose mitochondrial DNA was tested. The tree shows a split between African and non-African populations. The African population is the longest established, indicating the origin of modern humans in that continent. The different degrees of sequence divergence among the non-African populations give some indication of when different parts of the Old World were colonized. Recent analysis has shown that this tree, one of many possible, may not be the optimum interpretation. (Courtesy of Rebecca L. Cann *et al./Nature.*)

mitochondrial DNA types in today's human population can be traced back to a single female, not because she was the only woman living at the time, but because of the dynamics of loss of the DNA. This process is best explained by analogy. Imagine a population of 5000 mating pairs, each with a different family name. As time passes, the population remains stable (each couple produces only two offspring). In each generation, on average, one-fourth of the couples will have two boys, one-half will have a boy and a girl, and one-fourth will have two girls. If family names are passed only through males, one-fourth of the family names will be lost in the first generation. With each succeeding generation, more losses will occur, albeit at a slower rate. After approximately 10,000 generations (twice the number of original females), only one family name will remain (see diagram). The same pattern holds for the loss of mitochondrial DNA types, except that the transmission flows through the female line.

In the decade since the initial publication of the Berkeley results, a massive effort has been channeled into testing their validity. Two conclusions stand out. First, the claim for identifying an African origin of modern humans is not statistically significant, as was once stated, although it still remains the most likely case. Second, in the more than 5000 individual samples tested to date, not a single example of an ancient (that is, deriving from a deep *Homo erectus* lineage) mitochondrial DNA has been de-

LIFE OF A LUCKY MOTHER: An illustration of the concept that all maternal lineages in a population trace back to a single lineage in an ancestral population. At each generation one-fourth of the mothers will have two male offspring, one-fourth will have two female offspring, and one-half will have one female and one male offspring. The mitochondrial lineages of mothers bearing only male offspring will come to an end, leading eventually to one lineage dominating the entire population. (Courtesy of Allan Wilson.)

enced a population bottleneck. Another scenario would involve the evolution of modern humans in ancient times, followed by a recent population bottleneck.

A second finding from the early work was that Africans display the greatest degree of variation in their mitochondrial DNA. This discovery was taken to indicate that this population was oldest, and therefore represented the population of origin of modern humans. An alternative explanation, however, is that the early African population was larger than other populations, and its greater size promoted the accumulation of more extensive genetic variation.

Colorful though it is, the term "Eve" in the hypothesis title is misleading, and it originally led to widespread misunderstanding of the implication of the study. The

tected, which is contrary to what would be expected if the multiregional evolution hypothesis were correct.

The inability to wrest a statistically significant answer from the mitochondrial DNA data prompted two paths of research. The first involved a search for similar evidence from other genetic systems, such as genes and other elements in the nuclear genome. The second relied on extraction of a different kind of information from the mitochondrial data—namely, data relating to population history.

A SPECTRUM OF GENETIC EVIDENCE

The multiregional hypothesis suggests that the roots of all modern human populations go back to *Homo erectus*, which originated in Asia or Africa almost two million years ago. By contrast, the single-origin hypothesis states that modern humans originated less than 200,000 years ago, probably in Africa. Molecular evidence that indicates an African origin cannot differentiate between the hypotheses, because both claim an African origin. Instead, the *time of origin* distinguishes between them. The question is, How can the molecular data best be used to test the two hypotheses in terms of time of origin?

As we saw in unit 4, many genes accumulate mutations at a rather regular rate, giving a potential molecular clock. With a living population, the history of many different genetic variants of a gene, or alleles, can be traced by successive, inclusive steps, until a single ancestral type is reached. This ancestral type is known as the **coalescent**, and the time in history at which it is reached is called the **coalescence time**. If, when a new species is established, the population contains only a single allele of a particular gene, then the coalescence time for that gene may serve as a good indicator of the time of the speciation event. In other words, the gene tree is the same as the population (or species) tree (see unit 8). Frequently, however, the founding population of a new species will contain a subset of the existing genetic variation, so that the gene tree will show a more ancient divergence than the population tree. In this case, the coalescence time predates the time of the origin of the species. Under certain circumstances, the coalescence time may be substantially older than the time of origin of a species; in other (unusual) circumstances, the coalescence time may be younger.

For any particular species, a distribution of coalescence times of its various genes will exist. Some will coincide with the age of the species; many will be slightly older; some will be very much older; and a small number will be younger. Maryellen Ruvolo, of Harvard University, has recently proposed that hypotheses of the time of modern human origins may be tested by examining the distribution of coalescence times of a range of genes in modern populations. If the multiregional model is correct, then those times should cluster around 1.8 million years ago (close to the time of origin of *Homo erectus*); if the recent, single-origin model is correct, those times will cluster around, for example, 200,000 years ago (the coalescence time of modern mitochondrial lineages). Ruvolo points

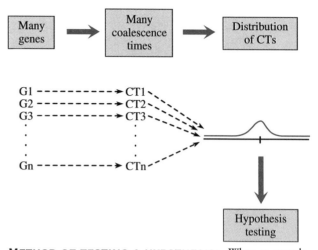

METHOD OF TESTING A HYPOTHESIS: When a population splits, it leads to a distribution of coalescence times from many genes (denoted as G1, G2, and so on). Coalescence times can be expected to cluster around the time of population division, thereby indicating the time of origin of new species. No single coalescence time is a reliable indicator because some genes will have an older coalescence time than the population split, while others will be younger. (Courtesy of Maryellen Ruvolo.)

out that, because only the *distribution* of coalescence times is informative, a single coalescence time cannot prove or disprove either hypothesis. Even with a recent origin, more ancient coalescence times are expected; likewise, a certain probability of recent coalescence times arises with an ancient origin as well.

As of early 1997, 14 coalescence times had been calculated for various genetic loci, including 4 different measures in mitochondrial DNA and 10 in different genes in nuclear DNA. If the 4 mitochondrial results are counted as a single data point (to reflect their common inheritance), then the remaining independent coalescence times are as follows: 6 cluster around 200,000 years ago, while the rest are scattered at 0.5, 1.2, 1.3, 3.0, and 35 million years ago. (Two independent studies on different regions of the Y chromosome, the male equivalent of mitochondrial DNA, gave coalescence times of 188,000 and 270,000 years.) Remember that *clustering* of coalescence times is the most important criterion—not the position of individual times. The results so far clearly favor the recent origin model.

New genetic data used in human origin analyses include two types that are particularly interesting: one is derived from microsatellite DNA and the other involves so-called *Alu* sequences. Although they may appear to represent arcane elements of modern molecular biology, these data sets offer important practical tools for anthropologists. Both of these new, ground-breaking investigations appear to favor the recent, single-origin model.

Microsatellites, which are short stretches of DNA that contain many repeats of two- to five-nucleotide segments, evolve very rapidly. Unlike the rates of mutation for most

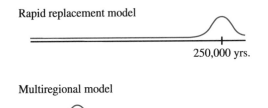

Rapid replacement model

250,000 yrs.

Multiregional model

1.8 my

COALESCENCE TIMES AND THE ORIGIN OF MODERN HUMANS: If modern humans originated close to 250,000 years ago, as implied by the mitochondrial DNA hypothesis, then the distribution of coalescence times would show a peak at that time (*top*). If the multiregional evolution model is correct, then coalescence times would cluster around 1.8 million years ago (*bottom*). (Courtesy of Maryellen Ruvolo.)

genetic elements, which often must be calculated by calibration against the fossil record, the rate of mutation of microsatellites can be determined by laboratory observation. This certainty adds some weight of confidence to the coalescence time calculated with this technique, which is 156,000 years.

Alu elements are sequences of DNA approximately 300 base pairs in length, which become inserted in large numbers over the nuclear genome. Once inserted, they are never removed (or at least not completely) and thus remain immune to the kinds of homoplastic changes that may obscure point mutations (see unit 4). A recent, multiauthored study on *Alu* elements in a large sample from around the world gave a coalescence time of 102,000 years.

Debate continues to swirl over the mutational dynamics of microsatellite sequences and *Alu* elements, just as the interpretation of coalescence times has inspired controversy. In particular, population history may influence

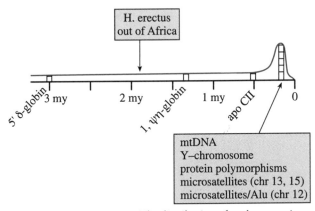

H. erectus
out of Africa

5' δ-globin   3 my   2 my   1, ψη-globin   1 my   apo CII   0

mtDNA
Y–chromosome
protein polymorphisms
microsatellites (chr 13, 15)
microsatellites/Alu (chr 12)

HYPOTHESES TESTED: The distribution of coalescence times from mitochondrial and nuclear genes supports the recent, single-origin model of modern humans. (Courtesy of Maryellen Ruvolo.)

coalescence times in ways unrelated to the establishment of a species, usually leading to an erroneously young date. The fact that the inferences drawn from the mitochondrial DNA data are matched closely by a significant proportion of those from nuclear data, however, encourages the view that they are collectively providing insight into species events rather than identifying population events. For example, population crashes and explosions would affect mitochondrial DNA variation to a greater extent than nuclear DNA variation. While most observers accept the apparent implications of this body of work, a minority of critics remain unconvinced. As always, more data are required.

And new data were forthcoming in mid-1997, in the form of mitochondrial DNA sequences from the fossilized bones of a Neanderthal specimen, which showed that this form of archaic human could not have been ancestral to modern human populations in Europe (unit 27).

A SECOND PATH OF INVESTIGATION: POPULATION HISTORY

Two factors play into the new line of investigation followed in population history analyses. The first stems from the difficulty that has been experienced in deriving an unequivocal phylogenetic tree from the mitochondrial DNA data. The low phylogenetic resolution in the data prompted certain researchers to seek other kinds of information that might be inferred from them, using a technique known as **mismatch distribution.** The insight gained with this technique can be applied to address the second factor—namely, the puzzle of the unusually low level of genetic diversity of mitochondrial DNA in modern populations.

The conclusion of this work is that, early in their history, the population of modern humans suffered a relatively severe bottleneck. Following that bottleneck, the population expanded explosively. These data imply that the multiregional evolution model cannot explain modern human origins. Rather, a modified form of the recent, single-origin model, known as the **weak Garden of Eden hypothesis,** is more likely to be correct.

Henry Harpending and his colleague Alan Rogers, of the University of Utah, developed a hypothetical model of a population that expanded within a brief period of time. Genetic data culled from the modern descendants of this population gave information about both the extent and timing of such an event (illustrated in the accompanying diagram).

In their model, Harpending and Rogers assumed that mutations accumulate regularly in all lineages (mutations are shown as crosses on the horizontal lines in the middle panel of the figure). They then compared DNA sequences between all pairs of lineages in a sample of this population, and counted the number of mutational differences between each pair (a sample of 50 individuals gives 435 pairs for comparison). The time scale is measured in terms of mutational time, in which one unit represents the time needed for a single mutational difference to accumulate between two lineages; two units are sufficient for two mutational differences; and so on. The rate at which mutations accumulate is determined by both the rate of mutation at all sites in the DNA and the generation time. In this case, one mutational unit equates to 8333 years, given the known rate of mutation of certain mitochondrial sequences in humans.

Because the population underwent expansion at seven mutational units of time in the past (see diagram), a large proportion of lineages in the current population will include seven mutational differences between them. Some lineages split after the expansion event of course, and these lineages will differ by fewer than seven mutations. When all pairs of lineages have been compared and mutational differences counted, these numbers are then arrayed on a histogram, with the horizontal axis representing the mutational time, going from zero in the present to ever-increasing numbers as one moves back in time. The histogram shows a peak at seven mutational differences, with fewer points at older and younger times, forming a wave pattern (see the bottom panel of the figure). Harpending and Rogers describe this pattern as "the signature of an ancient population expansion." The position of the crest of the wave indicates *when* population expansion occurred; the shape of the wave shows its magnitude (the sharper the peak, the more rapid was the expansion).

When Harpending and Rogers applied the mismatch distribution analysis to real mitochondrial DNA data from modern human populations from around the globe, they found the same wave pattern. This discovery implies that the modern human population underwent a rapid expansion of numbers, the timing of which was centered around 60,000 years ago. Further analysis revealed that the expansion took place at different times for different geographical populations. The African population expanded first, followed later by expansions in the European and Asian populations. This conclusion came from a mismatch distribution analysis conducted *within* each geographical population, followed by a similar analysis performed *between* pairs of populations (this latter technique is termed **intermatch distributions**).

Several possible scenarios exist to explain what happened here, the most persuasive of which is the weak Garden of Eden hypothesis. Remember that the recent, single-origin hypothesis posits that modern humans arose as a small, isolated population, and that descendants of this population spread throughout the Old World, replacing existing populations of archaic *sapiens*. This concept is also called the **Garden of Eden hypothesis**. The intermatch distribution analysis implies a little more complicated history. According to this hypothesis, once established (some 100,000 years ago), the founding

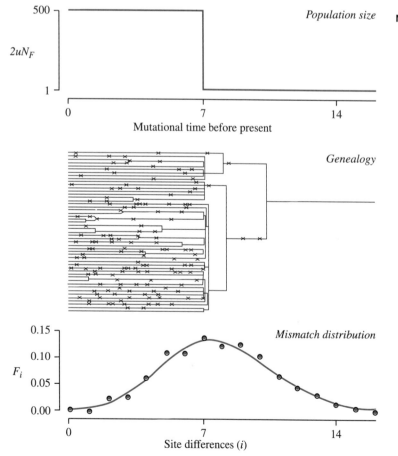

MISMATCH DISTRIBUTION: This method uses genetic variation in modern populations to infer population events in the past. (See the text for details.) (Courtesy of Alan Rogers and Lynn Jorde.)

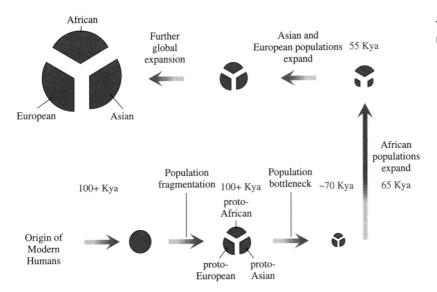

African

Further
global
expansion

Asian and
European populations   55 Kya
expand

European   Asian

100+ Kya

Population
fragmentation   100+ Kya

Population
bottleneck   ~70 Kya

African
populations
expand

65 Kya

proto-
African

Origin of
Modern
Humans

proto-   proto-
European   Asian

THE WEAK GARDEN OF EDEN HYPOTHESIS: Developed from mismatch distribution analysis, this hypothesis represents a variant of the single-origin model. It posits the origin of modern humans in Africa, prior to 100,000 years ago. This population fragmented (within Africa), and the separate populations subsequently developed genetic distinctiveness. A population bottleneck reduced population size and genetic variation within them. The African population was the first to expand, followed by the proto-Asian and proto-European populations, which migrated into these geographical regions. Population expansion then continued.

population of modern humans fragmented into separate populations; these groups later spread out geographically to form the modern populations of Africa, Europe, and Asia. The genetic distinctiveness of these populations was therefore established prior to the expansion; the mismatch and intermatch distribution data indicate that these separate expansions took place at different times. Thus, replacement of archaic *sapiens* populations would still have occurred, but would not have involved the same dynamics as envisaged with the original Garden of Eden hypothesis.

According to this new line of investigation, the low level of mitochondrial DNA diversity reflects a population bottleneck after the establishment of the modern human population; this bottleneck was followed by sequential population expansions in different parts of the world. Several questions arise here, the most important of which is, What was the severity of the bottleneck?

The complicated calculation required to answer this question is based on the current genetic diversity of mitochondrial diversity in the world and on the mutation rate of these DNA sequences. The simplest answer indicates the existence of some 3500 breeding females, which would give a total population of approximately 10,000 individuals. (Similar numbers have been obtained from other data, including nuclear DNA data.) In fact, population genetics equations show that if this population was distributed in discrete geographical populations over the Old World, as required by the multiregional hypothesis, the number of females would have been smaller—close to 1500. This figure creates a fatal problem for the hypothesis because, as Harpending and Rogers note, "It is difficult to imagine that a population this small could have populated all of Europe, Africa, and Asia. . . . Knowledge that Eve lived recently would imply that the human population was . . . too small to have populated three continents." In other words, the numbers that flow from this analysis (if correct) make the multiregional hypothesis untenable. Some form of a recent, single-origin model would seem much more reasonable. ✳

KEY QUESTIONS

- Why is mitochondrial DNA a potentially useful tool for tracking recent evolutionary and population events?
- What are the limitations of mitochondrial DNA in inferring phylogenetic history?
- What is the significance of the coincidence of coalescence times of mitochondrial and nuclear genes?
- What further genetic evidence might clarify the validity of competing hypotheses for the origin of modern humans?

KEY REFERENCES

Ayala FJ. The myth of Eve: molecular biology and human origins. *Science* 1995;270:1930–1936.

Erlich H, *et al.* HLA sequence polymorphism and the origin of modern humans. *Science* 1996;274:1552–1554.

Gibbons A. The mystery of humanity's missing mutations. *Science* 1995;267:35–36.

———. Y chromosome shows that Adam was an African. *Science* 1997;278:804–805.

———. Calibrating the mitochondrial clock. *Science* 1998;279:28–29.

Goldstein DB, *et al.* Genetic absolute dating based on microsatellites and the origin of modern humans. *Proc Natl Acad Sci USA* 1995;92:6723–6727.

Hammer MF, Zegura SL. The role of the Y chromosome in human evolutionary studies. *Evol Anthropol* 1996;5:116–134.

Harpending HC, *et al.* Genetic traces of ancient demography. *Proc Natl Acad Sci USA* 1998;95: 1961–1967.

Krings M, *et al.* Neanderthal DNA sequences and the origin of modern humans. *Cell* 1997;90:19–30.

Manderscheid EJ, Rogers AR. Genetic admixture in the late Pleistocene. *Am J Physical Anthropol* 1996;100:1–5.

Pritchard JK, Feldman MW. Genetic data and the African origin of humans. *Science* 1996;274: 1548–1549.

Rogers AR, Jorde LB. Genetic evidence on modern human origins. *Hum Biol* 1995;67:1–36.

Ruvolo M. A new approach to studying modern human origins. *Molec Phylogenet Evol* 1996;5: 202–219.

Stoneking M. DNA and recent human evolution. *Evol Anthropol* 1993;2:60–73.

———. In defense of "Eve"— a response to Templeton's critique. *Am Anthropol* 1994;96:131–141.

Takahata N. A genetic perspective on the origin and history of humans. *Annu Rev Ecol Systematics* 1995;26:343–372.

Templeton AR. The "Eve" hypothesis: a genetic critique and reanalysis. *Am Anthropol* 1993;95:51–72.

———. "Eve": hypothesis compatibility versus hypothesis testing. *Am Anthropol* 1994;96:141–155.

Tishkoff SA, *et al.* Global patterns of linkage disequilibrium at the CD4 locus and modern human origins. *Science* 1996;271:1380–1387.

Wilson AC, Cann RL. The recent African genesis of humans. *Sci Am* April 1992:68–73.

THE ORIGIN OF MODERN HUMANS: ARCHEOLOGICAL EVIDENCE

30

Although the archeological evidence related to the origin of modern humans is relatively good in Europe and western Asia, it is poor in East Asia and, unfortunately, in Africa. For instance, while more than 100 sites dating between 250,000 and 40,000 years old have been carefully excavated in southwestern France (and many more are known in less detail), only about a dozen such sites have been studied in East Africa, a region almost 100 times larger in geographical extent. This disparity has led inevitably to a distinctly Eurocentric interpretation of the archeological record, which gives the impression that pertinent behavioral changes principally took place in Europe. Several important discoveries have been made in Africa in recent years, however, and their interpretation is leading some archeologists to favor a different view of our behavioral evolution.

THE ARCHEOLOGICAL BACKGROUND

In looking for signs of modern human behavior, we are concerned with a shift from the Middle Stone Age (MSA) to the Later Stone Age (LSA) in Africa, dated at some 250,000 to 40,000 years and 40,000 to 10,000 years, respectively. The equivalent stages in Europe, Asia, and North Africa are known as the Middle Paleolithic and Upper Paleolithic.

The end of the Lower Paleolithic, 250,000 years ago, saw the end of innovation-poor, long-lasting stone-tool industries. With the beginning of the Middle Paleolithic, the number of identifiable tool types quadrupled, reaching perhaps 40. With the Upper Paleolithic, beginning 40,000 years ago, the number of tools more than doubled again, to as many as 100. Moreover, European tool industries cascade through at least four identifiable traditions in less than 30,000 years—a pace of innovation and change unprecedented in the archeological record. In addition to new forms of tools, raw materials that were only infrequently used earlier, such as bone, ivory, and antler, became very important in the Upper Paleolithic industries.

The Middle Paleolithic (mode III) and Middle Stone Age technologies were characterized by the predominance of the prepared core technique, such as the Levallois technique, which appeared earlier (see unit 25). Flakes produced by this method may then be further fashioned to give what some archeologists identify as approximately 40 different implements, each with its own putative cutting, scraping, or piercing function. Some variation exists in Middle Paleolithic assemblages throughout the Old World, which has encouraged the development of a plethora of local names. The most generally applied name, however, is Mousterian, after the Neanderthal site of Le Moustier, in the Perigord region of France.

Just as flakes from prepared cores characterize Mousterian (and Mousterian-like) industries in the Middle Paleolithic, blades produced from prepared cores constitute something of a signature for the many industries in the European Upper Paleolithic (mode IV). Blades are defined as flakes that are at least twice as long as they are wide. The preparation of the cores used for their manufacture requires great skill and time. Many blades may then be detached sequentially using a pointed object, such as the end of an antler, hammered by a hammerstone. The

MIDDLE PALEOLITHIC ARTIFACTS: These typically retouched flakes of various types were made between 250,000 and 40,000 years ago. (*top row, left to right*) Mousterian point; Levallois point; Levallois flake (tortoise); Levallois core; disc core. (*bottom row, left to right*) Mousterian handaxe; single convex side scraper; Quina scraper; limace; denticulate. (Scale bar: 5 cm.) (Courtesy of Roger Lewin and Bruce Bradley.)

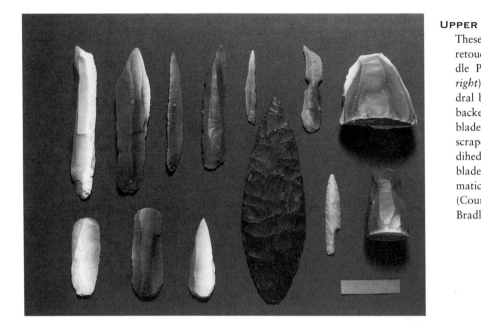

UPPER PALEOLITHIC ARTIFACTS These artifacts are typically formed from retouched blades and are finer than Middle Paleolithic tools. (*top row, left to right*) Burin on a truncated blade; dihedral burin; gravette point; backed knife; backed bladelet; strangulated blade; blade core. (*bottom row, left to right*) End scraper; double end scraper; end scraper/dihedral burin; Solutrean laurel leaf blade; Solutrean shouldered point; prismatic blade core. (Scale bar: 5 cm.) (Courtesy of Roger Lewin and Bruce Bradley.)

blades, often small and delicate, may be functional without further preparation, or they may merely serve as the starting point for specifically shaped implements. In addition to the signature blade, Upper Paleolithic tool makers also made extensive use of bone, ivory, and antler as raw material for some of the most delicate implements. Thus, a strong sense of directed design and elaborate use characterize Upper Paleolithic tool assemblages.

An important issue in the context of the origin of modern humans is the dynamics of the shift between the Middle Paleolithic (and MSA) and the Upper Paleolithic (and LSA). For Stanford University archeologist Richard Klein, the evidence reveals "the most dramatic behavioral shift that archeologists will ever detect." For this reason, the transition has been regarded as revolutionary, not gradual. If true, then it would imply that the evolution of modern morphology (which appeared more than 130,000 years ago) occurred separately from the evolution of modern behavior (40,000 years ago). Recent discoveries in Africa may raise questions about this interpretation, however.

EUROPEAN EVIDENCE

The European archeological evidence for the stages in question is extensive, and it does appear to give a clear signal of a revolutionary change some 40,000 years ago. For this reason, the transition in Europe has been dubbed the Upper Paleolithic revolution. It coincides

UPPER PALEOLITHIC RANGE OF FORMS: The French archeologist G. Laplace produced this typology of Upper Paleolithic tools in the late 1950s and early 1960s. It forms the basis of all Upper Paleolithic typologies. The intricacy as well as the variety of production can be seen.

with the first appearance of modern humans in the region, carrying the cultural tradition known as the Aurignacian. Aurignacian sites throughout Europe show the typical blade-based technology and use of bone, ivory, and antler, not only to make points but also to create beads as body ornamentation. The sites are also associated with other characteristics of the Upper Paleolithic: they are larger than those of the Middle Paleolithic; open-air (as opposed to rock shelter or cave) sites are more distinctive and organized; artifacts indicate the existence of long-distance contact and even trade (shells and exotic stone that must have come from afar); and musical instruments, specifically simple flutes made from bone, are present.

As the Upper Paleolithic progressed, substantial temporal and spatial variability of style developed in artifact assemblages; the sense of cultural traditions in the way we would mean today was strongly present for the first time. Although sculpting and engraving appeared from the Aurignacian onward, evidence of cave painting did not become strong until the Gravettian, some 30,000 years ago.

The contrast between the Middle Paleolithic in Europe (specifically, the Mousterian) and the Upper Paleolithic is striking. Although not every aspect of Upper Paleolithic culture, especially technological advances and artistic traditions, was present from the beginning, overall it surely

Chatelperronian			
Aurignacian			
	Gravettian		
		Solutrean	
			Magdalenian Azilian
40,000 years before present	30,000 years before present	20,000 years before present	10,000 years before present

TOOL INDUSTRIES OF THE UPPER PALEOLITHIC: The pace of change of tool technologies becomes almost hectic from 40,000 years onward. In addition, the tool industries themselves take on a complexity and refinement unmatched in earlier periods. A distinct sense of fashion and geographic variation is also well developed.

offers evidence of a revolutionary change. Agreement on this latter point—revolution or not—is divided, both in terms of its dynamics and its explanation.

The match between archeological and fossil evidence in Europe is quite good. For instance, wherever hominine remains have been found with Mousterian assemblages, they have been Neanderthal. Virtually all hominine fossils associated with Upper Paleolithic assemblages have been modern humans. Two exceptions to the latter generalization have been identified, at the French sites of Arcy-sur-Cure and Saint-Césaire. Although the fossil evidence at Arcy-sur-Cure is fragmentary, a classic Neanderthal partial skeleton has been found at Saint-Césaire. These sites are interesting because the tool assemblages represent an intermediate form between Mousterian and Aurignacian, termed Chatelperronian.

Some scholars have argued that the intermediate nature of the Chatelperronian technology indicates the presence of a population in biological transition—that is, changing from Neanderthal to modern humans. The anatomy of the Saint-Césaire individual shows no such characteristics, however (see unit 27). The age of the skeleton, recently dated at 36,000 years, leaves little or no time for an evolutionary transition to local modern human populations. In any case, the site postdates the earliest Aurignacian sites, which have no local precursors. One possible explanation of the Chatelperronian is that it was developed by late Neanderthal populations that had cultural contact with incoming modern human populations.

Although no consensus has been reached on the meaning of the European archeological evidence, a strong case can be made for its support of revolutionary change as a result of population replacement. It does not, however, address the issue of the *origin* of modern humans.

ASIAN EVIDENCE

The archeological evidence in Asia is open to even more diverse interpretation than in Europe, partly be-

SOLUTREAN LAUREL LEAF BLADE: Some examples of these blades are so thin as to be translucent. They were probably used in rituals rather than in practical affairs. (Scale bar: 5 cm.) (Courtesy of Roger Lewin and Bruce Bradley.)

cause the data are fewer and partly because some apparent paradoxes exist. Great differences are also noted between western Asia and eastern Asia, where the evidence is sparsest of all. Western Asia, which includes the Middle East, is closely allied to Africa geographically and provides a natural migration route out of Africa. Between 200,000 and 50,000 years ago, this region was variously occupied by Neanderthal and early modern humans, while the Far East was inhabited by populations that were neither Neanderthal nor modern.

The archeological transition from archaic to modern in the Middle East is typologically very similar to the Mousterian to Upper Paleolithic transition in Europe, and apparently occurs about the same time (40,000 years ago). If the transition tracks the migration of modern humans out of Africa, through the Middle East, and finally into western Europe, then the evidence for it in the Middle East might be expected to predate the evidence gleaned further west. Tentative confirmation of this movement might come from the site of Boker Tachtit in Israel, which dates to between 47,000 and 38,000 years ago. Evidence of Upper Paleolithic human remains in the Middle East is scarce, but is essentially that of modern humans.

Where western Asia differs from Europe is in the occurrence of anatomically modern humans with classic Mousterian assemblages, at the Israeli sites of Skhūl and Qafzeh (see unit 28), which have been dated to approximately 100,000 years. These fossil remains are either equal in age or predate Neanderthals of the region, and thus would seem to preclude an evolutionary transformation of Neanderthals into modern humans. Nevertheless, the occurrence of modern human anatomy with Mousterian assemblages some 60,000 years before Upper Paleolithic assemblages appear in the region represents a puzzle. It implies either that modern human anatomy evolved long before modern behavior or that the modernity of the Skhūl and Qafzeh remains has been overstated. Recent analyses have implied that the two populations used different hunting strategies, with modern humans being more efficient.

Klein points out that the Skhūl/Qafzeh specimens are extremely variable anatomically and that they possess some archaic features, such as prominent brow ridges and large teeth. "Both cranially and postcranially, they clearly make far better ancestors for later modern humans than the Neanderthals do," he says. "However, it seems reasonable to suppose that they were not yet fully modern biologically—perhaps, above all, neurologically." Clark and Lindly's reading of the evidence differs from Klein's interpretation, with the duo arguing for continuity between the archaic and the modern species, both in the fossils and the archeology.

The interpretation of eastern Asian evidence poses a challenge because of the scarcity of sites and uncertain dating. There does appear to be a continuity of chopping-tool assemblages from *Homo erectus* times through approximately 10,000 years ago, with no dramatic shift equivalent to that seen in the European Upper Paleolithic.

One site in Sri Lanka, Batadomba Iena cave, contains a microlithic tool assemblage that has been radiocarbon dated at 28,500 years old. In addition, sites in Siberia, dated between 35,000 and 20,000 years old, contain Upper Paleolithic-like artifacts and art objects, suggesting a more European-like pattern. The migration from Southeast Asia to Australia between 60,000 and 45,000 years ago implies the evolution of modern human behavior by at least this date (see unit 34).

The Asian evidence is therefore equivocal at best, but offers little to suggest the appearance of modern human behavior early in the record.

AFRICAN EVIDENCE

For the past two decades, the Middle Stone Age of Africa has been viewed as equivalent to the Middle Paleolithic in Europe, both chronologically and technologically. The prevailing view of the Middle to Later Stone Age transition was that it resembled the Middle to Upper Paleolithic transition—that is, it was revolutionary, reflecting the sudden appearance of modern behavior. This view is now being questioned by some prehistorians, particularly by Alison Brooks and Sally McBrearty, of George Washington University and the University of Connecticut, respectively.

Brooks and McBrearty point out that evidence of blade production, such as that found in Ethiopia, dated at 180,000 years ago, and South Africa (the Howieson's Poort industry), dated at 80,000 years ago, has been assigned too little importance. Recently, McBrearty has reported blade production at a site in central Kenya (the Kapthurin formation), which is some 240,000 years old. These tools are 125,000 years older than the oldest known blades from the European Middle Paleolithic and more than 200,000 years older than those from the European Upper Paleolithic. If the production of such blades represents a signature for modern human behavior, then evidence of this behavior clearly has a long history.

One explanation for this production could be that the earlier blades were made by a less sophisticated technique. According to this theory, by themselves the blades do not constitute an unequivocal signal of modern human behavior. Instead, other behaviors must be considered as well, such as production of tools made from materials other than stone, artistic behavior, and other complex social behavior, such as long-distance trade or exchange of objects.

For instance, tools made from bone are common in the Upper Paleolithic, but are almost unknown earlier. A striking exception is a collection of barbed bone points (like harpoon heads) found at the Katanda site in eastern Zaire, and reported by Brooks and her colleagues in 1995. These artifacts have been dated by thermoluminescence and electron spin resonance techniques at between 90,000 and 160,000 years old, or 135,000 years older than the previously oldest known artifacts of this kind. This discovery has encouraged archeologists to reconsider

Years ago (× 1000)	O-isotope stages and climate stratigraphy	Europe	Western Asia	Eastern Asia	Africa	Years ago (× 1000)
10	1 / Holocene	Neolithic, etc. Mesolithic	Neolithic, etc.	Neolithic, etc.	Neolithic, etc.	10
20	2	Later Upper Paleolithic — Aurignacian and modern *H. sapiens*; Chatelperronian and Neanderthal	"Upper and "Epi-" Paleolithic" — Upper Paleolithic and modern *H. sapiens*	"Late Paleolithic" — modern *H. sapiens*	Later Stone Age and Upper Paleolithic — modern *H. sapiens*	20
30				???		30
40	3		Mousterian and ???	???	Middle Stone Age / Mousterian and ? early modern *H. sapiens*	40
50				???		50
60	4			???		60
70				???		70
80	5a	Mousterian and Neanderthal	Mousterian and Neanderthal	flake/chopper industry and archaic *H. sapiens*	Howieson's Poort, Aterian and early modern *H. sapiens*	80
90	5b	Last interglaciation				90
100	5c					100
110	5d					110
120	5e	Mousterian and Neanderthal	Mousterian and early modern *H. sapiens*		Middle Stone Age / Mousterian and early modern *H. sapiens*	120
130						130
	6 / Penultimate glaciation	Acheulian, etc.	Acheulian, etc.			
190						190

CONTINENTS COMPARED: The picture of modern human origins derived from archeological evidence is at best incomplete. In Europe, where the evidence is most plentiful, the picture is quite sharp, showing a sharp transition approximately 40,000 years ago that reflects the inward migration of anatomically modern humans carrying modern cultural behavior. In Asia, the picture is less clear. In Africa, new evidence suggests that modern human behavior begins to appear early in the Middle Stone Age, congruent with the early appearance of anatomically modern humans in that continent. (Courtesy of Richard Klein/*Evolutionary Anthropology*.)

claims for other bone tools at several Middle Stone Age sites, though none is said to be as old as those found at Katanda.

Nothing discovered in Africa has matched the artistic expression for which the Upper Paleolithic of western Europe is so famous. The oldest, reliably dated rock painting in Africa appears in the Apollo cave, Namibia, dated at 27,000 years, which is equivalent to the oldest examples of art in Europe. In contrast, pigments and grinding stones for processing pigments have been found in many regions of Africa, dating from at least 80,000 years ago. If such pigments were used for body decoration, for example, rather than treating hide, it would be significant in the context of the current question. It is impossible to prove which of these possibilities is correct, however. Evidence of personal adornment, such as ostrich eggshell beads, appears in the record relatively late, about 60,000 years ago. Are these artifacts to be taken as evidence of absence of early symbolic behavior that is so often considered as reflecting the modern human mind at work? Not necessarily so, argues Brooks, given the very unfavorable conditions of preservation in

the African environment and the paucity of sites investigated.

For an increasing number of archeologists, these separate lines of evidence tell us something about a gradual emergence of modern human behavior. Once it passed a certain threshold, that behavior appears to have exploded, producing the rich fabric of social complexity associated with the Upper Paleolithic and Later Stone Age. That explosion was a cultural change, however, not a biological one. By contrast, Klein and others have argued that only with a critical biological change—such as facilitation of linguistic ability—did modern human behavior become possible; they define modern human behavior as including the ability to produce the entire range of activities, not just one of them at different times and different places. Undoubtedly this issue will continue to inspire debate for some time to come.

HYPOTHESES TESTED

As a test of competing hypotheses—the "out of Africa" and multiregional evolution hypotheses—the archeological evidence is equivocal, and certainly not as

188   HUMAN EVOLUTION: AN ILLUSTRATED INTRODUCTION

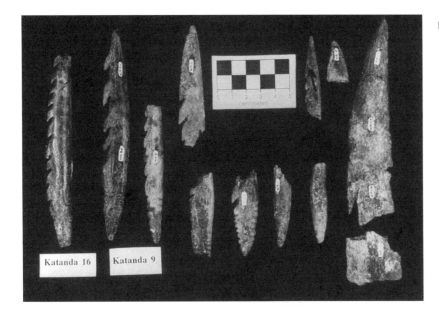

MIDDLE STONE AGE BONE TOOLS: Discovered recently in Zaire, these harpoon-like bone points are the earliest known examples of worked bone, dated at between 90,000 and 160,000 years old. (Courtesy of Alison Brooks and John Yellen.)

strong as the anatomical and genetic evidence. Nevertheless, it can be argued that a signal of modernity appears first in Africa, representing a chronological precursor of what later appears in Eurasia. The appearance of modern cultural activities in Europe seems to coincide with the first appearance of anatomically modern humans there—a culture brought by migrants, not developed locally. Thus, the "out of Africa" model is more strongly supported than the multiregional evolution model. ✳

KEY QUESTIONS

- Which model of modern human origins does the present archeological evidence most strongly support?
- Under what circumstances might the Chatelperronian industry have arisen?
- Is it reasonable to assume a tight coupling between modern morphology and modern behavior?
- What kind of archeological evidence would be most valuable in addressing the question of the tempo and mode of the origin of modern humans?

KEY REFERENCES

Brooks AS. Behavior and human evolution. In: Meikle WE, *et al.*, eds. Contemporary issues in human evolution. San Francisco: California Academy of Sciences, Memoir 21, 1996.

Clark GA, Lindly JM. The case for continuity: observations on the biocultural transition in Europe and western Asia. In: Mellars P, Stringer CB, eds. The human revolution. Princeton, NJ: Princeton University Press, 1989:626–676.

Foley RA, Lahr MM. Mode 3 technology and the evolution of modern humans. *Cambr J Archeol* 1997;7:3–32.

Harrold FB. Mousterian, Chatelperronian, and early Aurignacian: continuity or discontinuity? In: Mellars P, Stringer CB, eds. The human revolution. Princeton, NJ: Princeton University Press, 1989:677–713.

Klein RG. The archeology of modern humans. *Evol Anthropol* 1992;1:5–14.

———. Anatomy, behavior, and modern human origins. *J World Prehistory* 1995;9:167–198.

Lieberman DE, Sea JJ. Behavioral differences between archaic and modern humans in the Levantine Mousterian. *Am Anthropol* 1994;96:300–332.

McBrearty S, *et al.* Variability of Middle Pleistocene hominid behavior in the Kapthurin Formation, Baringo, Kenya. *J Hum Evol* 1996;30:563–580.

Mellars P. Major issues in the emergence of modern humans. *Curr Anthropol* 1989;30:349–385.

Straus LB. The Upper Paleolithic of Europe: an overview. *Evol Anthropol* 1995;4:4–16.

Thieme H. Lower Pleistocene hunting spears from Germany. *Nature* 1997;385:807–810.

Yellen JE, *et al.* A Middle Stone Age worked bone industry from Katanda, Upper Semliki Valley, Zaire. *Science* 1995;268:553–556.

PART 8

THE HUMAN MILIEU

Evolution of the Brain, Intelligence, and Consciousness
The Evolution of Language
Art in Prehistory

EVOLUTION OF THE BRAIN, INTELLIGENCE, AND CONSCIOUSNESS

31

The brain is a very expensive organ to maintain. In adult humans, for instance, even though it represents just 2 percent of the total body weight, the brain consumes some 18 percent of the energy budget. Given the fact that the human brain is three times larger than it would be if humans were apes, we have to ask, Why and how did brain expansion occur in the human lineage?

As we saw in unit 12, life history factors—gestation length, metabolic rate, precociality versus altriciality, and so on—have an important impact on the size of brain that a species can develop. In this context, two major ideas have been advanced in recent years that bear on the special problem faced by hominines in brain expansion.

The first, proposed by Robert Martin, of the Anthropological Institute in Zurich, is that the mother's metabolic rate is the key to the size of brain a species can afford—the higher the metabolic rate, the bigger the relative brain size. The second, proposed by Mark Pagel and Paul Harvey of Oxford University, is that gestation time and litter size represent the determining factors—long gestation, with a litter of one is optimal for a large-brained species. Although both hypotheses are said by

their authors to have empirical support, debate continues as to which is the more germane. Whichever case proves to be correct, both pathways require the same kind of environmental context: a stable, high-energy food supply, with minimum predation pressure.

In being well endowed mentally, humans and other primates are a part of a very clear pattern among vertebrates as a whole. Depending somewhat on the measure used, mammals are approximately 10 times "brainier" than reptiles and amphibians. Underlying this stepwise progression, which takes into account successive major evolutionary innovations and radiations, is the building of more and more sophisticated "reality" in species' heads.

By being mammals, primates are therefore better equipped mentally than any reptile. Two orders of mammal have significantly larger brains than the rest of mammalian life: Primates and Cetaceans (toothed whales). And among primates, the anthropoids (monkeys and apes) are brainier still. Only humans are outliers from the monkey/ape axis: the brain of *Homo sapiens* is three times bigger than that of an ape of the same body size.

The need to grow such a large brain has distorted several basic life history characteristics seen in other primates. For instance, the adult ape brain is nearly 2.3 times bigger than the brain in the newborn (neonate); in humans, this difference is 3.5 times. More dramatic, however, is the size of the human neonate compared with ape newborns. Even though humans are of similar body size to apes (57 kilograms for humans, compared with 30 to 100 kilograms for apes) and have a similar gestation period (270 days versus 245 to 270 days), human neonates are approximately twice as large and have brains twice as large as ape newborns. "From this it can be concluded that human mothers devote a relatively greater quantity of energy and other resources to fetal brain and

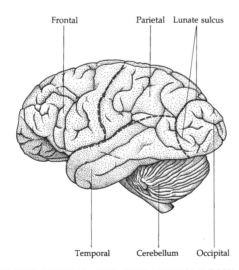

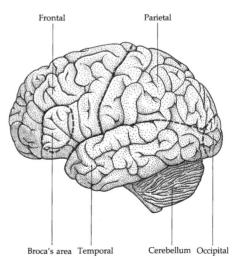

DIAGRAM OF THE TYPICAL APE AND HUMAN BRAIN PATTERN: The large human brain (*right*) compared with that of the chimpanzee is also distinguished by its relatively small

occipital lobe and large parietal lobe. The human brain is three times the size of the ape brain. (Courtesy of Ralph Holloway/ *Scientific American*, 1974, all rights reserved.)

early australopithecine brain is often said to be roughly the same size as modern gorilla and chimpanzee brains. This interpretation is misleading, however, for two reasons: (1) early australopithecines were smaller in body size than modern gorillas, and (2) modern ape brains almost certainly are larger than those of their 3 million-year-old ancestors. It is therefore safe to say that brain expansion had already been established by the time *Australopithecus afarensis* appeared.

Marked brain expansion is seen with the origin of the genus *Homo*, specifically *Homo habilis/rudolfensis*, which existed from 2.5 to 1.8 million years ago and had a range of brain size of 650 to 800 cm³. The size range for *Homo ergaster/erectus*, dated at 1.8 million to 300,000 years ago, is 850 to slightly more than 1000 cm³, although the concomitant increase in body size means that encephalization was not commensurately increased. The comparable measurements for archaic *Homo sapiens*, including Neanderthals, range from 1100 to more than 1400 cm³, or larger than in modern humans. Using the **encephalization quotient** (E.Q.), a measure of brain size in relation to body size, this progression can be discerned more objectively. The australopithecine species have E.Q.s in the region of 2.5, compared with 2 for the common chimpanzee, 3.1 for early *Homo*; 3.3 for early *Homo ergaster/erectus*; and 5.8 for modern humans.

By looking at overall brain structure as revealed in endocasts, it is possible to differentiate between an apelike and a humanlike brain organization. Each hemisphere contains four lobes: frontal, temporal, parietal, and oc-

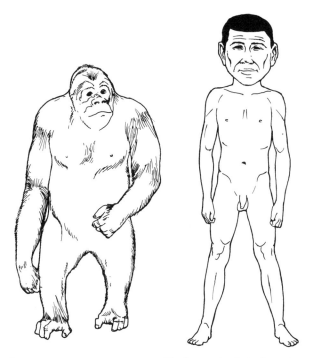

EXPANDED HUMAN BRAIN: The human brain is three times bigger than an ape's brain would be, given the same body size.

body development over a standard time than do our closest relative, the great apes," notes Martin.

Another major difference is the pattern of growth. In mammals with precocial young—which includes primates—brain growth proceeds rapidly until birth, whereupon a slower phase ensues for roughly a year. In humans, the prenatal phase of rapid brain growth continues for a longer period after birth, a pattern that is seen in altricial species. Compared with other altricial species, however, the rapid postnatal phase (at a fetal rate) of brain growth continues for a relatively longer period in humans. This extension effectively gives humans the equivalent of a 21-month gestation period (9 months in the uterus, and 12 months outside). This unique pattern of development has been called secondary altriciality. One important consequence is that human infants are far more helpless, and for a much longer time, than the young of the great apes. This extended period of infant care and subsequent "schooling" must have had a major impact on the social life of hominines.

FOSSIL EVIDENCE

Two types of fossil evidence are related to brain evolution: indications of absolute size, and information about the surface features—convolutions and fissures—of the brain. Both pieces of evidence can be obtained from either natural or man-made endocasts, which show the convolutions of the brain as they became impressed on the inner surface of the cranium.

Brain size is the first and most obvious piece of information to be gleaned, and it can often be gained even with partial crania. Measured at a little less than 400 cm³, the

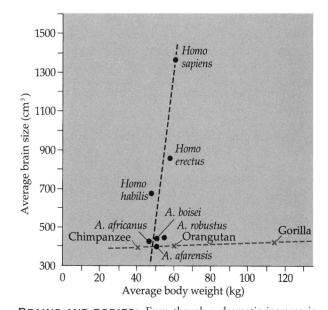

BRAINS AND BODIES: Even though a dramatic increase in body size did not occur in the *Homo* lineage, absolute (and therefore relative) brain size expanded significantly from *habilis* to *erectus* to *sapiens*. Brain size did not change significantly among the australopithecines or the modern apes, despite a large body size difference in the latter.

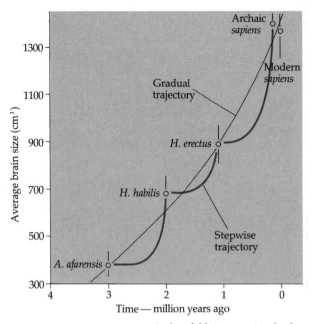

BRAINS THROUGH TIME: A threefold increase in absolute brain size occurred during the past 3 million years. Whether this increase took place gradually (as indicated by the smooth slope) or episodically (as indicated by the steps) is a matter that will be settled only with the discovery of additional, accurately identified fossils.

cipital. Very briefly, a brain in which the parietal and temporal lobes predominate is considered humanlike, whereas apelike brains contain much smaller parietal and temporal lobes. In addition, human frontal lobes are considerably more convoluted than apes.

Anthropologists find it very helpful to know when a human brain organization emerged in hominine history. Ralph Holloway, of Columbia University, examined in detail a wide range of hominine fossil endocasts, including *Australopithecus afarensis*, and concluded that brain organization was very humanlike. His analysis included the position of the lunate sulcus, a short groove that lies at the margin between the occipital and temporal lobes. In humans, the sulcus lies relatively further back than in apes. According to Holloway, in all fossil hominine endocasts in which the lunate sulcus could be discerned, this structure lies in the human position.

In 1980, Dean Falk, of the State University of New York at Albany, challenged this view after a study of the hominine endocasts in South Africa. The two researchers have since exchanged more than a dozen papers, each defending his/her position, but no resolution has been reached. Falk's position has recently received support independently from two researchers, Este Armstrong and Harry Jerison.

If brain reorganization toward the human configuration began only with the origin of *Homo*, while the australopithecine brain remained essentially apelike, then it would be consistent with other events in human prehistory, including the shift toward a humanlike life history

pattern with the origin of *Homo,* the evolution of human-like body proportions, the reduction of body size dimorphism, and the first appearance of stone-tool technology. Falk has argued that an important anatomical feature in the expansion of the brain in *Homo* was the distributed structure of the blood vessels, which permits efficient cooling; this concept is known as the radiator hypothesis.

MEASURES OF INTELLIGENCE

It is relatively easy to plot brain expansion through hominine history, but how are we to measure the rise of intelligence through time? The archeological record is notoriously lacking in tangible indications of the working of the mind. Thus, we are left with stone tools and other clues to subsistence activity as measures of intelligence. As we saw in units 23, 25, and 30, the imposition of standardization and expansion of complexity emerged very slowly in prehistoric stone-tool industries. The earliest stone-tool-making hominines, however, apparently possessed greater cognitive skills than modern chimpanzees (see unit 23). Apparently something changed in the brains of the earliest hominine tool makers to permit the development of this ability.

One other insight into how fossil evidence might show expanding brain size concerns the impact of brain expansion on social organization, specifically in infant care. Once hominines shifted from the basic primate pattern of brain growth, producing a much more helpless infant whose brain continued to grow at the fetal rate, then greater allocation of time and resources would be needed for rearing offspring. This had occurred by the time of the evolution of *Homo ergaster* (see unit 24).

POSSIBLE CAUSES OF BRAIN EXPANSION

A long popular notion was the hypothesis that the very obvious difference between hominines and apes—that humans made and used stone tools—was the most likely cause of brain expansion: the tripling of hominine brain size was seen as being accompanied by an ever-increasing complexity of tool technology. "Man the tool maker" was the encapsulation of this approach in the 1950s, followed a decade later by "Man the hunter." In either case, the emphasis was placed on the mastering of practical affairs as the engine of hominine brain expansion.

New ideas have emerged more recently that might be described by the phrase "Man the social animal." The new insight begins with a paradox: Laboratory tests have demonstrated that monkeys and apes are extraordinarily intelligent, and yet field studies have revealed that the daily lives of these creatures are relatively undemanding, in the realm of subsistence at least. Why, then, did this high degree of intelligence develop?

The answer may lie in the realm of primate social life. Although, superficially, a primate's social environment does not appear to be more demanding than that of other mammals—the size and composition of social groups is matched among antelope species, for example—the *interactions* within the group are far more complex. In other words, for a nonhuman primate in the wild, learning the

THE SOCIAL MILIEU: Socializing has become an important part of primate life. Making alliances and exploiting knowledge of others' alliances are key to an individual's reproductive success. Biologists now believe that the intellectual demands of complex social interaction were an important force of natural selection in the expansion of primate—and ultimately, human—brains.

distribution and probable time of ripening of food sources in the environment is intellectual child's play compared with predicting—and manipulating—the behavior of other individuals in the group. But why should social interactions be so complex—so Machiavellian—in primate societies?

When one observes other mammal species and sees instances of conflict between two individuals, it is usually easy to predict which animal will triumph: the larger one, or the one with bigger canines or bigger antlers (or whatever is the appropriate weapon for combat). Not so in monkeys and apes. Individuals devote much time to establishing networks of "friendships" and ob-

serving the alliances of others. As a result, a physically inferior individual can triumph over a stronger individual, provided the challenge is timed so that friends are at hand to help the challenger and that the victim's allies are absent.

In a survey of much of the field data relevant to primate social intelligence, Dorothy Cheney, Robert Seyfarth, both of the University of Pennsylvania, and Barbara Smutts, of the University of Michigan, posed the following question: "Are [primates] capable of some of the higher cognitive processes that are central to human social interactions?" This question is important, because if anthropoid intellect, honed by complex social interaction, is merely sharper than that of the average mammal and more adept at solving psychologist's puzzles, then it does not qualify as *creative* intelligence.

Cheney and her colleagues had no difficulty in finding many examples of primate behavior that appear to reflect humanlike social cognition. The researchers conclude that "primates can predict the consequences of their behavior for others and they understand enough about the motives of others to be able to be capable of deceit and other subtle forms of manipulation." Supporting this hypothesis, the British anthropologist Robin Dunbar has found that primate species with more complex social interaction have larger cerebral cortexes.

If nonhuman primate intellect has truly been honed, not in the realm of practical affairs, but in the hard school of social interaction, one is still left to explain why this situation has arisen. Why have primates found it advantageous to indulge in alliance building and manipulation? The answer, again gleaned from field studies, is that individuals that are adept at building and maintaining alliances are also reproductively more successful: making alliances opens up potential mating opportunities.

Once a lineage takes the evolutionary step of using social alliances to bolster reproductive success, it finds itself in what Nicholas Humphrey, a Cambridge University psychologist, calls an evolutionary ratchet. "Once a society has reached a certain level of complexity, then new internal pressures must arise which act to increase its complexity still further," he explains. "For, in a society [of

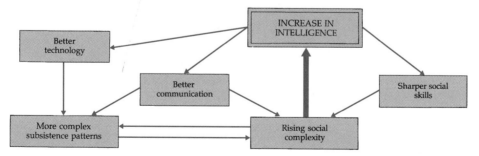

SOCIAL COMPLEXITY AND INCREASED INTELLIGENCE:
The need to cope with rising social complexity—including increasingly demanding subsistence patterns but particularly

a more ramified social structure and unpredictable social interactions—may have represented a key selection pressure for increased intelligence.

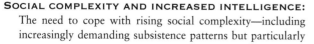

this kind], an animal's intellectual 'adversaries' are members of his own breeding community. And in these circumstances there can be no going back."

Where does consciousness fit into this mix? Humphrey describes it as an "inner eye," with pun intended. Consciousness is a tool—the ultimate tool—of the social animal. By being able to look into one's own mind and "see" one's reactions to things and other individuals, one can more precisely predict how others will react to those same things and individuals. Consciousness builds a better reality—one that is attuned to the highly social world that humans inhabit. ✳

KEY QUESTIONS

- What limitations arise when measuring differences in intelligence from differences in brain size and overall organization?
- How might one infer levels of intelligence from different stone-tool technologies?
- What key pieces of information might lend support to the "Man the social animal" hypothesis?
- How would one test whether nonhuman primates possessed a humanlike consciousness?

KEY REFERENCES

Aiello LC, Wheeler P. The expensive-tissue hypothesis. *Curr Anthropol* 1995;36:199–211.

Byrne R. The thinking ape. Oxford: Oxford University Press, 1995.

Calvin WH. The emergence of intelligence. *Sci Am* Oct 1994:101–107.

Dunbar RIM. Neocortex size as a constraint on group size in primates. *J Hum Evol* 1992;22:469–493.

———. The social brain hypothesis. *Evol Anthropol* 1998;6:178–190.

Falk D. 3.5 million years of hominid brain evolution. *Semin Neurosci* 1991;3:409–416.

Mithen S. The prehistory of the mind. London: Thames and Hudson, 1996.

Noble W, Davidson I. Human evolution, language and mind. Cambridge: Cambridge University Press, 1996.

Pagel MD, Harvey PH. How mammals produce large-brained offspring. *Evolution* 1988;42:948–957.

Seyfarth RM, Cheney DL. Meaning and mind in monkeys. *Sci Am* Dec 1992:122–128.

Toth N, *et al.* Pan the tool-maker: investigations into the stone tool-making and tool-using capabilities of a bonobo. *J Archeol Sci* 1993;20:81–91.

Wills C. The runaway brain. New York: Basic Books, 1993.

Wright K. The Tarzan syndrome. *Discover* Nov 1996:88–102.

Wynn T. Archeological evidence for modern intelligence. In: Foley RA, ed. The origin of human behavior. London: Unwin Hyman, 1991:52–66.

THE EVOLUTION OF LANGUAGE

32

One great frustration for anthropologists is that, by its nature, language is virtually invisible in the archeological record. Clues must therefore be sought from indirect sources: in stone tools, among indications of social and economic organization, in the content and context of paintings and other forms of artistic expression, and in the fossil remains themselves.

One general question about the evolution of human language relates to the dynamics of its emergence. Was it a slow, gradual process, beginning early in hominine history and becoming fully modern only recently? Or was it a rapid process, beginning recently in hominine history? This unit will examine several lines of evidence, taken from fossils and aspects of behavior identified in the archeological record.

FOSSIL EVIDENCE

In recent years, researchers have pursued two kinds of evidence from fossil hominines. First, information is gleaned from endocasts, those crude maps of the surface features of the brain. Second, indications of the structure of the voice-producing structures in the neck (the larynx and pharynx) provide clues as to language ability.

The major neural machinery for language functions is located in the left hemisphere in the great majority of modern humans, even in most left-handed people. As with many complex mental functions, however, language capabilities cannot be pinpointed precisely to particular centers. Traditionally, Broca's area, visible as a small lump on the left side of the brain toward the front, has been associated with language, particularly with the production of sound. A second center, Wernicke's area, located somewhat behind Broca's area, is involved in the perception of sound. Recent PET scan studies, however, have shown that this concept oversimplifies the situation. Many aspects of language—for instance, the lexicon, or vocabulary with which we work—defy precise localization.

Consequently, paleoneurologists can obtain few definite signs of language capacities from fossil endocasts. Signs of Broca's area have been found in *Homo rudolfensis* and later species of *Homo*, but not in australopithecines. For this reason, paleoneurologist Dean Falk believes that language capacity was already to some degree developed at the beginning of the *Homo* lineage. She disagrees with Ralph Holloway, however, who argues that language capacity began to develop earlier, among australopithecine species. His conclusion is based on the humanlike brain reorganization he detects in australopithecines. In contrast, Falk sees no reorganization in the human direction until *Homo* evolves (see unit 31).

If the fossil brains provide only tantalizing hints of verbal skills in our ancestors, what can we learn from the voice-producing apparatus? A number of researchers have pursued this question in recent years—in particular, Edmund Crelin, Philip Lieberman, and Jeffrey Laitman. The human vocal tract is unique in the animal world. In mammals, the position of the larynx in the neck assumes one of two basic patterns. One location is high up, which allows the animal simultaneously to swallow (food or liquid) and breathe. The second pattern places the larynx

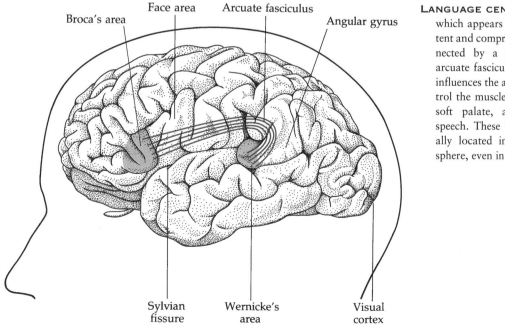

LANGUAGE CENTERS: Wernicke's area, which appears to be responsible for content and comprehension of speech, is connected by a nerve bundle called the arcuate fasciculus to Broca's area, which influences the areas of the brain that control the muscles of the lips, jaw, tongue, soft palate, and vocal cords during speech. These language centers are usually located in the left cerebral hemisphere, even in many left-handers.

Broca's area • Face area • Arcuate fasciculus • Angular gyrus • Sylvian fissure • Wernicke's area • Visual cortex

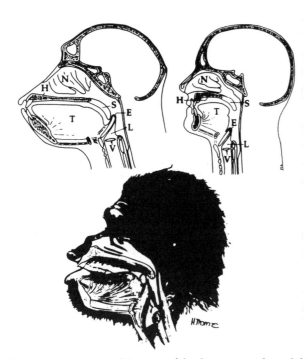

THE VOCAL TRACT: Diagrams of the chimpanzee (*above, left*) and human (*right*) vocal tract: N = nasal cavity; S = soft palate; T = tongue; P = pharynx; L = larynx; E = epiglottis; V = vocal fold. In the chimpanzee—as in all mammals—the larynx is high in the neck, enabling simultaneous breathing and swallowing. In mature humans, the larynx appears lower in the next, making simultaneous breathing and swallowing impossible, but increasing the size of the pharynx and scope of vocal production. Below is sketch of the australopithecine vocal tract, which resembles that of the chimpanzee. (Courtesy of J. Laitman, Patrick Gannon, and Hugh Thomas.)

low in the neck, requiring temporary closing of the air passage during swallowing; otherwise solids or liquids will block it and cause choking. Adult humans have the second pattern, while all other mammals, and infant humans, possess the first. The low position of the larynx greatly enlarges the space above it, which allows the sounds emitted from the it to be modified to a great degree. Nonhuman mammals are limited to modifying laryngeal sounds by altering the shape of the oral cavity and the lips. Human newborns maintain the basic mammalian pattern until about 1.5 to 2 years; the larynx then begins to migrate lower in the neck, achieving the adult configuration at approximately age 14 years.

Laitman and his colleagues discovered that the position of the larynx is reflected in the shape of the bottom of the skull, the basicranium. In adult humans, this structure is arched; in other mammals, and in human infants, it is much flatter. By looking at this feature in the fossil record, it should therefore be possible to discern something about the verbal skills of extinct hominine species. What does the fossil record indicate?

"In sum," says Laitman, "we find that the australopithecines probably had vocal tracts much like those of

living monkeys or apes. . . . The high position of their larynges would have made it impossible for them to produce some of the universal vowel sounds found in human speech." Unfortunately, the fossil record for *Homo rudolfensis/habilis* is poor as far as indications of the basicranium are concerned. Laitman and his colleagues have found that, in its putative evolutionary successor, *Homo ergaster*, "the larynx . . . may have begun to descend into the neck, increasing the area available to modify laryngeal sounds." The position of the larynx appears to be equivalent to that found in an 8-year-old human. Only with the origin of archaic *Homo sapiens*, some 300,000 years ago, does the fully modern pattern appear, indicating at least the mechanical potential for the full range of sounds produced by people today.

THE QUESTION OF NEANDERTHALS

A continuing controversy concerns Neanderthals' language abilities. Because they appeared 150,000 years after the fully arched basicranium evolved in archaic *sapiens*, implying fully developed speech potential in that species, Neanderthals might be expected to be similarly developed. However, basicranial flexion is less than that observed in earlier archaic *sapiens*. It looks as if the direction of evolution had been reversed, depriving Neanderthals of fully articulate speech. Laitman notes that the degree of basicranial flexion differs among different geographic specimens of Neanderthals, but suggests that their collective reduction in flexion may be related to their unusual upper respiratory tract anatomy, a possible adaptation to cold climes.

The notion that Neanderthals had poorly developed language abilities has become the majority position among anthropologists, and that this may have contributed to the extinction of the species. This conclusion has been challenged, however. In 1989, a team of researchers led by Baruch Arensburg, of Tel Aviv University, reported the discovery of a hyoid bone from a Neanderthal partial skeleton, at Kebara. This small, U-shaped bone lies between the root of the tongue and the larynx, and is connected to muscles of the jaw, larynx, and tongue. In size and shape, the Kebara hyoid is virtually identical to the modern bone. Arensburg and his colleagues claim this feature as proof that Neanderthals' language capacity resembled that of modern humans. Laitman challenges this conclusion, saying that the anatomy of the hyoid bone is insufficient evidence for inferring the overall shape of the vocal tract. No other hominine fossil hyoid bone has been found that would permit comparisons.

A second challenge to the accepted view comes from David Frayer, of the University of Kansas. He points to a new reconstruction of the famous La Chapelle-aux-Saints Neanderthal cranium, which, he says, indicates much more flexion in the basicranium than has been assumed. Frayer also argues that basicranial flexion in other Neanderthals falls within the range of other Upper Paleolithic and Mesolithic European populations. Laitman questions whether the new reconstruction is necessarily better than the earlier one. In any case, he says, measurements from the new reconstruction still imply a relatively undeveloped vocal tract for Neanderthals. This matter remains unresolved.

Overall, fossil endocasts and laryngeal structure indicate a rather gradual acquisition of language capabilities through hominine history, possibly beginning with the origin of the genus *Homo*. Holloway would put language origins further back in time.

It should be remembered that higher primates are able to produce a wide range of sounds, which they use to subtle effect. For instance, when juvenile monkeys are threatened by an older opponent, they scream, which usually brings help. This scream differs subtly, depending on the intensity of the threat and the dominance rank and kinship of the aggressor. Experiments with tape-recorded screams show that mothers' responses to the screams vary according to the indicated danger. In addition, some higher primates give different alarm calls for different predators (leopard, snake, and so on). Although the different calls are not "words," they do appear to be labels.

In thinking about the acquisition of spoken language by hominines, one must therefore imagine the buildup of an ever-greater range of primate sounds, and their eventual conjunction as words. Terrence Deacon, of Harvard University, suggests that neurological evidence supports such a scenario, and that language origins began with the genus *Homo* and developed gradually. For some researchers, however, the structured use of words—syntax—that characterizes human speech differs so dramatically from primate vocalization that it is seen as disjunct. In other words, these researchers argue that human language is part of a continuum with primate vocalization.

ARCHEOLOGICAL EVIDENCE: TOOLS

Some anthropologists have argued that the pattern of tool manufacture and language production—essentially, a series of individual steps—implies a common cognitive basis. If true, then following the trajectory of the complexity of stone-tool technology through time should reveal something about the change in language capabilities.

Thomas Wynn, of the University of Colorado, has used psychological theory to examine the validity of this argument. "It is true," he says, "that language and tool making are sequential behaviors, but the relationship is more likely to be one of analogy rather than homology." In other words, only a superficial similarity connects the two, and their cognitive underpinnings remain quite separate. Thus, one cannot look at the complexity of a tool assemblage on one hand and learn anything *directly* about language abilities on the other.

Glynn Isaac has also searched for indications of language function in ancient tool technologies, albeit via a different approach. He has argued that the complexity of a tool assemblage might provide some information about social complexity, not cognitive complexity, relating to mechanical or verbal processes. Beyond a certain degree of social complexity there is an arbitrary imposition of standards and patterns. Discerning such a relationship is to some extent an abstract exercise, which would be impossible in the complete absence of language.

As we saw in units 23, 25, and 30, the trajectory of technological change through hominine history falls into two phases: an incredibly slow phase leading from the earliest artifacts some 2.5 million years ago to approximately

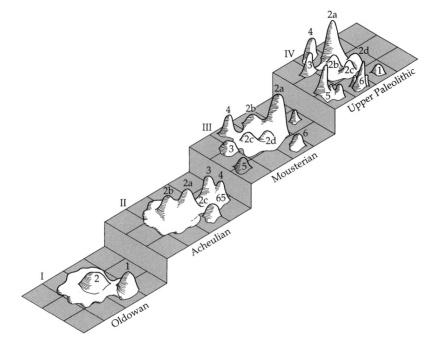

SHARPENING THE MIND, SHARPENING THE TONGUE: With the passage of time and the emergence of new species along the *Homo* lineage, stone-tool making became even more systematic and orderly. Peaks in the diagram represent identifiable artifact modes, with tall, narrow peaks implying highly standardized products. The increased orderliness in stone-tool manufacture must, argued archeologist Glynn Isaac, reflect an increasingly ordered set of cognitive processes that eventually involved spoken language. (I) Oldowan: 1 = core choppers; 2 = casual scrapers. (II) Acheulian (Olorgesailie): 2a = scrapers; 2b = nosed scrapers; 2c = large scrapers; 3 = handaxes; 4 = cleavers; 5 = picks; 6 = discoids. (III) Mousterian: 2a = racloir; 2b = grattoir; 2c = r. convergent; 3 = percoir; 4 = point; 5 = burin; 6 = biface. (IV) Upper Paleolithic: 2a = grattoir; 2b = nosed scraper; 2c = raclette; 3 = percoir; 4 = point; 5 = burins; 6 = backed blades. (Courtesy of Glynn Isaac.)

250,000 years ago, following by an ever-accelerating phase.

What lessons do we learn from this basic archeological evidence, in relation to origins of language? Writ on the large scale, it seems reasonable to infer that a language complex enough to conjure the abstract elements of social rules, myths, and ritual is a rather late development in hominine history; that is, it began only with archaic *Homo sapiens*, and became fully expressed only with anatomically modern humans. If one adds the economic and social organization necessary in hunting and gathering activities, which ultimately would involve the need for efficient verbal communication, then the archeological record shows the same pattern. Only in the later stages of hominine history does this organization take on a degree of sophistication that would seem to demand language skills.

ARCHEOLOGICAL EVIDENCE: ART

Australian scholars Iain Davidson and William Noble argue that spoken language is a very recent evolutionary development, closely tied to the cognitive processes of the development of imagery and art.

Painting or engraving an image of, for example, a bison does not necessarily imply anything mystical about the motives in the artist's mind. Nevertheless, the creation of art represents an abstraction of the real world into a different form, a process that demands highly refined cognitive skills. But the art created in the Ice Age was not simply a series of simple abstractions of images to be seen in the real world (see unit 33); rather, it was a highly selective abstraction. Whether it represented hunting magic or an encapsulation of social structure, this art speaks of a world created by introspective consciousness and complex language. It was, in fact, a world like ours, just technologically more primitive.

If artistic expression can inform us about the possession of complex language, the question is, How far back in prehistory did it stretch? Not very far, it seems. Although claims of some form of abstract artistic expression date back to 300,000 years ago, it is not until a little more than 30,000 years ago that artistic expression really began to blossom (see unit 33). Earlier than about 32,000 years ago, however, very little art has been recovered. Two pendants—one from reindeer bone, the other from a fox tooth—were discovered at the 35,000-year-old Neanderthal site of La Quina, France; an antelope shoulder blade etched with geometric pattern was also found at another French site, La Ferrassie. Elsewhere in Europe, bones and elephant teeth with distinct zigzag markings have been discovered that were carved by Neanderthals at least 50,000 years ago.

Bearing in mind the probable imperfections in the archeological record—in Europe, but especially in Africa—the inference to be drawn from artistic, abstract expression is that something important happened in the cultural milieu of hominines late in their history. The late British anthropologist Kenneth Oakley was one of the first to suggest, in 1951, that this "something important" was best explained by a quantum jump in the evolution of language. This development occurred, suggest Davidson and Noble, some 50,000 years ago.

Thus, the line of evidence from artistic expression suggests that the dynamic of language evolution was rapid and recent.

WHAT CAUSED THE EVOLUTION OF LANGUAGE?

The most obvious cause for the evolution of language was its development within the context in which it is so obviously proficient: communication. For a long time, this line of argument was pursued by a variety of anthropologists. The shift from the essentially individualistic subsistence activities of higher primates to the complex, cooperative venture of hunting and gathering surely demanded proficient communication. A popular hypothesis of language evolution included the notion that a first stage would have been a gesture language—gesturing, remember, is something humans do frequently, especially when lost for words.

In recent years, however, the explanatory emphasis has shifted, paralleling the shift in explanation for the evolution of intelligence. From the practical world of communication, explanation of language origins now turns to the inner mental world and social context.

"The role of language in communication first evolved as a side effect of its basic role in the construction of reality," argues Harry Jerison. "We can think of language as being an expression of another neural contribution to the construction of mental imagery.... We need language more to tell stories than to direct actions." As we saw in unit 30, anthropologists are beginning to recognize the importance of social interaction as the engine of the evo-

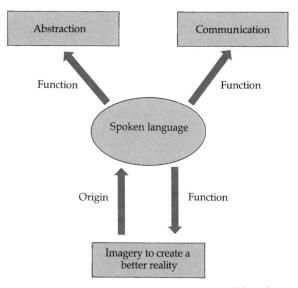

ORIGIN AND FUNCTION OF LANGUAGE: Although communication is clearly an important function of spoken language, its origins (and continued functions) probably centered on creating a better image of our ancestors' social and material worlds.

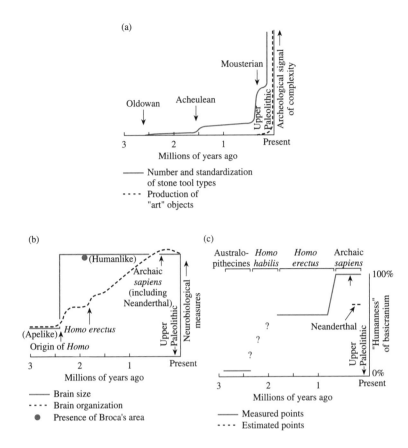

(a)

Mousterian

Oldowan   Acheulean

Upper Paleolithic

Archeological signal of complexity

Present

3   2   1

Millions of years ago

——— Number and standardization
of stone tool types
---- Production of
"art" objects

(b)

(Humanlike)

Archaic
sapiens
(including
Neanderthal)

Homo erectus

(Apelike)

Origin of Homo

Upper Paleolithic

Neurobiological measures

Present

3   2   1

Millions of years ago

——— Brain size
---- Brain organization
● Presence of Broca's area

(c)

Australo-   Homo   Homo   Archaic
pithecines   habilis erectus sapiens

100%

?

Neanderthal

?

?

Upper Paleolithic

"Humanness" of basicranium

0%

3   2   1   Present

Millions of years ago

——— Measured points
---- Estimated points

LINES OF EVIDENCE COMPARED: Evidence from archeology (a), brain size and brain organization (b), and indications of the structure of the larynx (c) is thought to be informative about the trajectory of the evolution of language. Archeological evidence indicates a recent, rapid evolution, whereas evidence about the brain and vocal tract implies an early, gradual evolution. (Courtesy of the Scientific American Library.)

lution of hominine intelligence. Consciousness and language go hand in hand with that view.

More recently, Robin Dunbar has suggested that language may have evolved as a way of facilitating social interaction in human groups, the equivalent of grooming in nonhuman primates. Beyond a certain group size, he argues, grooming becomes inefficient for maintaining social ties. Language is powerful because it can include individuals who are not present. These lines of investigation—the inner mental world and the social world—support an early, gradual dynamic of language evolution.

CONCLUSION

We have seen that different lines of evidence, as currently interpreted, lead to different conclusions about the dynamic of language evolution. Fossil evidence suggests a gradual trajectory, beginning early, as does certain cognitive evidence, such as internal mental worlds and the social context. Archeological evidence, ranging from stone-tool manufacture and artistic expression, is read to imply a recent, rapid evolution. The obvious conclusion is that one of these sources of evidence is being misread.

Most of the expansion of hominine brain size occurred before material and abstract expressions of culture became really vibrant. This incremental expansion might be taken to imply an incremental buildup of consciousness and language in our ancestors, rather than a final, sudden bound, as might be assumed in the Upper Paleolithic. Many examples in biology, however, include dramatic emergent effects as thresholds are passed. The origin of complex language and introspective consciousness might fit into this category. ✵

KEY QUESTIONS

- What is the relative importance of the different lines of fossil evidence in revealing past language capabilities?
- How would one test the idea that conformity of stone-tool production implies the impositions of social rules, and therefore the existence of language?
- What type of artistic expression provides the most persuasive evidence of the existence of language?
- If human language is discontinuous with primate vocalizations and communications, how might it have arisen?

KEY REFERENCES

Chazan M. The language hypothesis for the Middle-to-Upper Paleolithic transition. *Curr Anthropol* 1995;36:749–769.

Davidson I, Noble W. The archeology of depiction and language. *Curr Anthropol* 1989;30:125–156.

Deacon TW. The symbolic species: the coevolution of language and the brain. New York: Norton, 1997.

Dunbar R. Grooming, gossip, and the evolution of language. London: Faber, 1996.

Foley RA. Language origins: the silence of the past. *Nature* 1991;353:114–115.

Gannon, *et al.* Asymmetry of chimpanzee planum temporale: humanlike pattern of Wernicke's brain language area homolog. *Science* 1998;279:220–222.

Gibson K, Ingold T, eds. Tools, language, and intelligence. Cambridge: Cambridge University Press, 1992.

Jerison HJ. Brain size and the evolution of mind. 59th James Arthur Lecture, American Museum of Natural History, 1991.

Laitman JT. The anatomy of human speech. *Nat Hist* Aug 1983:20–27.

Noble W, Davidson I. Human evolution, language and mind. Cambridge: Cambridge University Press, 1996.

Pinker S, Bloom P. Natural language and natural selection. *Behav Brain Sci* 1990;13:707–784.

Raichle ME. Visualizing the mind. *Sci Am* April 1994:58–64.

ART IN PREHISTORY

33

Traditionally, the study of prehistoric art meant the study of prehistoric art in Europe, specifically in southwest France and northern Spain, created during the period 35,000 to 10,000 years ago (the Upper Paleolithic), the end of the Pleistocene Ice Age. Artistic expression undoubtedly flowed elsewhere in the Old World at this time—in Africa and Australia—but accidents of history and preservation have endowed Europe with a rich record of painted, engraved, and carved images that, properly interpreted, might give some insight into the workings of the human mind at this point in our history.

Recent years have witnessed a number of important developments in the study of prehistoric art, including discoveries beyond Europe, such as an engraved antler from Longgu Cave in China, the first prehistoric art object to be found there. The most spectacular new finds, however, have occurred in France, with the discovery of Chauvet Cave in the Ardeche, southern France,

and Cosquer Cave, on the southern coast near to Marseilles.

FEATURES OF UPPER PALEOLITHIC ART

The discovery of Chauvet Cave has upset some of the generalities that could be adduced for Upper Paleolithic art. For instance, carved and engraved images were thought to have preceded painted images by at least 10,000 years. Dated by radiocarbon analysis at 32,410 years old, Chauvet, however, is as old as some of the oldest-known carved objects, such as the ivory animal figures from Vogelherd, Germany, that date to a little more than 30,000 years. Moreover, the painted wall art consists mainly of large mammals, such as bison, aurochs, deer, horses, mammoth, ibex, and so on; carnivores are rare and usually sequestered in the deepest recesses of caves. This latter fact was interpreted as signaling prehistoric people's fear and respect for a fellow predator. At Chauvet, however, carnivores are prominent among the painted images, and they include a hyena and a leopard, animals not previously seen in prehistoric art.

Birds, plants, and humans are only infrequently represented in Upper Paleolithic art, and the latter are often depicted quite schematically when they do appear. The painted images are often very good, naturalistic representations of single animals or small groups of individuals, but they convey little sense of natural scenes. Again,

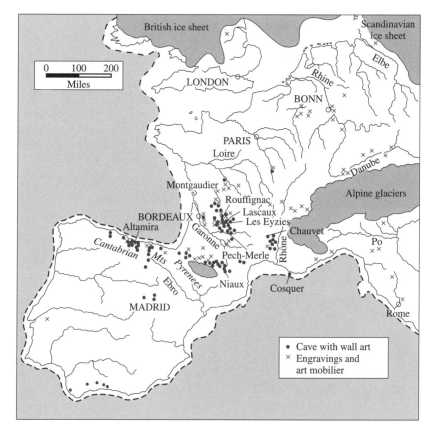

DISTRIBUTION OF ART SITES IN EUROPE: The limestone caves of Ice Age Europe have preserved a rich legacy of Paleolithic art. Although a certain stylistic continuity characterizes cave painting, motifs in art mobilier display much more variability.

(a)

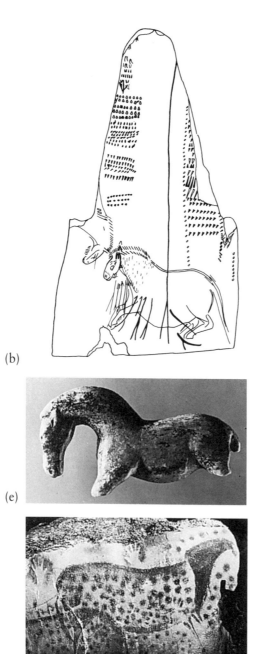

(b)

(c)

(d)

(e)

(f)

EXAMPLES OF PALEOLITHIC ART: (a) Fragment of rein-deer antler from La Marche, France, approximately 12,000 years old. Apparently used as an implement for shaping flint tools, the antler fragment is engraved with a pregnant mare, which seems to have been symbolically killed by a series of engraved arrows. Above the horse is a set of notches that have been interpreted by Alexander Marshack as documenting the passing lunar cycles.
(b) A drawing of the surface of the antler, "unrolled."
(c) An engraved antler baton from Montgaudier, France, dated at approximately 10,000 years old. Perhaps used in straightening the shafts of arrows or even spears, the baton's collection of engraved items suggests a representation of spring.

(d) A drawing of the antler-baton "unrolled."
(e) Vogelherd horse, carved from mammoth ivory some 30,000 years ago and worn smooth by frequent handling over a long period of time. The horse, which is the oldest known animal carving, measures 5 centimeters.
(f) The black outline of this horse was painted on the wall of a cave, Peche-Merle, France, approximately 15,000 years ago. Infrared analysis indicates that the mixture of black and red dots was added over a period of time. The black hand stencils are also later additions. Does the Peche-Merle horse, one of two in the cave, indicate the "use" of art? (Courtesy of Alexander Marshack.)

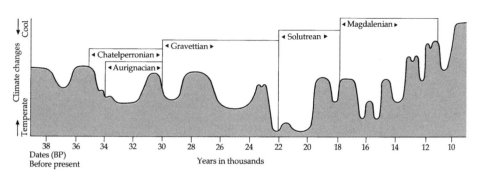

CLIMATIC FLUCTUATIONS: Although European prehistoric art was the product of the Ice Age, temperatures fluctuated somewhat throughout this period, driving dramatic shifts in ecological patterns. The most frigid period, from 22,000 to about 18,000 years ago, preceded the high point of prehistoric art, the Magdalenian. (Courtesy of the Randall White/American Museum of Natural History.)

Chauvet has a scene of two rhinos fighting, a unique depiction of an aggressive scene. Hand stencils—produced by brushing or blowing pigment around the hand while placed on a rock surface—are relatively common, often revealing what appears to be missing fingers. Some archeologists believe that, rather than representing mutilation, these stencils were produced by curling a finger under the palm, perhaps as a signature.

Painted images are usually scattered on rock surfaces in a seemingly random manner, often with one image superimposed partially or wholly on another. Sometimes interspersed among the animal images are simple geometric figures—some as simple as dots, others resembling grids and crescents.

Engraved or carved images, particularly on portable objects such as spear throwers, batons, pendants, and blade punches, often contain more detail in their execution. Overall, they give a sense of a wider representation of nature, including the large mammals seen in wall art (although in different proportions). For instance, birds, fish, and plants are often depicted, sometimes in rich combination; again, this illustration seems not to be the representation of a scene so much as an idea, such as a season. Interestingly, carnivore teeth are present in very high proportion in body ornamentation such as necklaces and pendants, in a striking contrast to most wall art.

The human image occurs more frequently in carved and engraved images than in painting. Here again, these depictions are often schematic in nature, as in the famous "Venuses." However, one site, La Marche in the French Pyrenees, contained a cache of more than 200 small engraved human faces, completely lifelike and individualistic—a portrait gallery from 20,000 years ago.

When the Ice Age finally came to a close, 10,000 years ago, the art ended as well, at least in the generally naturalistic, representational style that had persisted for 25,000 years. Geometric patterns became predominant, and people apparently no longer sought out deep caves in which to paint. It is quite possible, of course, that people painted just as much as before, but on open-air surfaces from which the images have disappeared.

INTERPRETATIONS OF PREHISTORIC ART

The first systematic study of Ice Age art was undertaken by the great French archeologist, the Abbé Henri Breuil. Throughout the first half of this century, he carefully copied images from many sites and attempted a chronology based on artistic style. He, and later scholars, believed that the art would grow more sophisticated through time—hence the notion that the famous Lascaux Cave (dated at 17,000 years old) was the high point of prehistoric art, given its brilliance in color and incorporation of perspective. The discovery of Chauvet has upset this simple idea of progress in execution of images, because it is Lascaux's equal in these respects and is twice as old.

Breuil developed the hypothesis that prehistoric art was also "hunting magic"—that is, a way of ensuring fruitful hunts and propitiating the victims. Supporting this idea is the presence among the images in many caves of animals apparently impaled by arrows or spears. Even the absence of such weapons does not militate against the idea, because an animal's image might be impaled symbolically during a ritual performance in front of it. The hunting magic hypothesis does face a problem in that the images painted in the caves very often depicted animals not included in the painters' diet, as indicated by bones found at living sites. In many cases, these bones show that reindeer were important as food—yet reindeer images are few. The reverse was true for horses and bison. As the French philosopher Claude Lévi-Strauss once observed, certain animals are depicted frequently, not because they were "good to eat" but because they were "good to think."

Breuil's hunting magic explanation persisted until his death in the 1960s, when it was replaced by the notion that the art somehow reflected the society that produced it. This thesis was developed independently by French archeologists André Leroi-Gourhan and Annette Laming-Emperaire. They noted that the inventory of animals depicted was comparable throughout Europe and described the presentation as remaining remarkably stable through

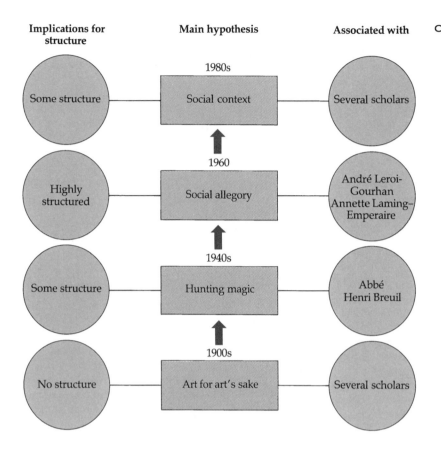

Implications for structure	Main hypothesis	Associated with

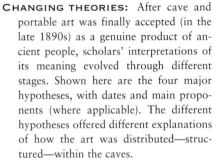

CHANGING THEORIES: After cave and portable art was finally accepted (in the late 1890s) as a genuine product of ancient people, scholars' interpretations of its meaning evolved through different stages. Shown here are the four major hypotheses, with dates and main proponents (where applicable). The different hypotheses offered different explanations of how the art was distributed—structured—within the caves.

time, an observation that contrasts with the much more locally idiosyncratic nature of portable art.

For Leroi-Gourhan and Laming-Emperaire, wall art reflected the duality of maleness and femaleness in society. Certain images were said to represent maleness, while others were female. The cave images were arranged so that female representations occurred at the center, with male representation located around the periphery, thereby reflecting a certain type of social structure. Although the two researchers did not fully agree on which images represented maleness and which femaleness, their work had the important effect of emphasizing social context in interpreting Paleolithic art.

Thus, where Breuil's explanation required no overall structure of the images within the caves, Leroi-Gourhan and Laming-Emperaire's very clearly did. Both explanations, however, were essentially monolithic. In recent years, this concept has changed as well. "We are beginning to see a great deal more diversity and complexity in Upper Paleolithic art," explains Randall White of New York University. "And this affects the way we envisage what was going on during this important stage of human evolution."

The Upper Paleolithic is divided into different cultural periods, based upon the tool technologies of the time (see unit 30). Throughout these different cultures, different aspects of the art changed in various ways, as Breuil noted

in his chronology. "It is important not to get the idea this pattern of change advanced on a broad front," cautions White. "In addition to differences through time, there are differences between regions, real geographic variations." These spatial and temporal variations in tool cultures are matched by similar variations in the art, although no precise correlation exists between a culture's technology and its art. Thus, a monolithic explanation of the meaning of the art is impossible.

Hunting magic may well explain some of the images. Rituals of other kinds almost certainly centered on the art as well. Something other than practicality drove Upper Paleolithic people to seek out and decorate deep caves, which appear to be otherwise unused. South African archeologists Davis Lewis-Williams and Thomas Dowson have suggested that the art is shamanistic—that is, produced by shamans in or after a state of trance. They base their conclusion on a study of San (Bushman) art of South Africa, which is known to be shamanistic, and on a survey of psychological studies on the hallucinatory images produced during trance.

During trance-induced hallucination, the subject experiences a small set of so-called entoptic ("within the nervous system") images, such as grids, zigzags, dots, spirals, and curves. In deeper stages of trance, these images may be manipulated into recognizable objects, and subjects may eventually come to see therianthropes, or chimeras of

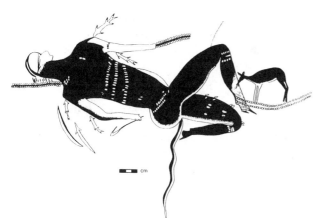

SHAMANISTIC ART: Supine therianthrope with fish. A small antelope, bleeding from the nose and therefore dying, stands on a double line of white dots. Human/animal chimeras are a feature of shamanistic art. This image is from the site of Maclear, in the eastern Cape, South Africa. (Courtesy of David Lewis-Williams.)

animal and human forms. Images that reflect these trance experiences are common in shamanistic art, in South Africa and elsewhere; Lewis-Williams suggests that they may have been part of Upper Paleolithic art, too.

The power of accurate dating in testing hypotheses was demonstrated in 1992, when a team of French and Spanish scientists published radiocarbon dates on images taken from two caves in Spain and one cave in France. Remember that Breuil had suggested that chronology could be inferred from style, given that style was held to change and improve over time. The researchers derived dates for certain images from Altamira and El Castillo, in Spain. The images were stylistically similar; thus, under Breuil's scheme, they should have been the same age. In fact, they were separated by more than 1000 years. A third image, from the Niaux Cave in the French Pyrenees, differed stylistically from the Spanish images; under Breuil's scheme, it would be expected to have been made at a different time from those in the Spanish caves. In fact, it is almost identical in age to the images found in El Castillo. Clearly, age and style do not always coincide.

Many portable art objects are decorated with geometric patterns. Some carry pictures of animals, fish, and plants; others include series of seemingly random notches. Alexander Marshack, an associate of the Peabody Museum at Harvard University, performed detailed studies of such objects. He suggested that some of the image combinations might represent seasons of the year: the images of a male and a female seal, a male salmon, two coiled snakes, and a flower in bloom, all engraved on a reindeer antler baton, is one such example.

In recent independent investigations, Denis Vialou, of the Musée de l'Homme in Paris, and Henri Delport, of the Musée des Antiquites Nationales, near Paris, conclude that less overall uniformity of structure connects the painted caves than originally envisaged by Leroi-Gourhan

and Laming-Emperaire. The discovery of Chauvet reinforces this point. Vialou and Delport acknowledge that most of the caves follow some kind of structure, but caution that each cave should be viewed as a separate expression.

Diversity, then, begins to emerge as a more realistic interpretative lens through which to view the Upper Paleolithic—a diversity of people, a diversity of cultures, and a diversity of the art. Paleoanthropologists have now shifted from trying to understand what an individual image or set of images might mean to attempting to understand the social context in which those images were produced. Most of all, an attempt is being made to divest modern interpretations of the bias inherent in modern eyes and minds. As Conkey says, "Perhaps we have closed off certain lines of inquiry, simply by using the label 'art'."

PRECURSORS TO UPPER PALEOLITHIC ART

A persistent question in archeology relates to the dynamic of the origin of symbolic image making: were hominines less advanced than *Homo sapiens* also capable of symbolic expression? Archeologists remain divided over the evidence and over its interpretation. (As we saw in unit 30, this issue is intimately tied to the question of the origin of modern humans.)

A decade ago, two anthropologists at the University of Pennsylvania, Philip Chase and Harold Dibble, surveyed the evidence for artistic and symbolic expression in the Middle to Upper Paleolithic transition, with the expressed purpose of determining the mode of the transition. Their conclusion was quite firm: "The most striking difference between the Middle and Upper Paleolithic is the contrast between the rich and highly developed art found in the latter period and the almost complete lack of it in the former." John Lindly and Geoffrey Clark, of Arizona State University, strongly disagree. In their examination of the archeological record, Lindly and Chase see that the Middle to Upper Paleolithic transition, as far as artistic expression is concerned, is a gradual, not a punctuational, event. According to the two researchers, the complexity of artistic expression in the Upper Paleolithic increases with time, with the Magdalenian being more developed than the Aurignacian.

Randall White disputes Lindly and Chase's contention that the Aurignacian is somehow poorer artistically than later periods in the Upper Paleolithic. "I have been struggling to understand the rich body of Aurignacian and Gravettian evidence, especially body ornamentation, from Western, Central, and Eastern Europe," he says. "The quantity of material is staggering." Others, including Paul Mellars, of Cambridge University, support White's view that the origin of symbolic art was punctuational.

Some evidence has been gathered to indicate the existence of image making earlier than the Upper Paleolithic, but it is very limited: a fragment of bone marked with a zigzag motif, from the Bacho Kiro site in Bulgaria, somewhat earlier than 35,000 years ago, for example, and a carved mammoth tooth, worn smooth with use and marked with red ocher, from the 50,000-year-old site of Tata, Hungary. Oldest of all is an ox rib engraved with a series of double arcs, from the French site of Peche de l'Azé, dated as being some 300,000 years old. Ocher has been found at several ancient living sites, including the campsite of Terra Amata, in southern France, which is dated to approximately 250,000 years ago. Nevertheless, argue Chase and Dibble, none of this art betrays modern human symbolism at work, merely weak glimmerings of its eventual development. They deem many of the supposed elements of evidence of Neanderthal mythology, such as the Cult of Skulls, to be the products of the overinterpretation of equivocal evidence by eager investigators.

More recently, Robert Bednarik, of the Australian Rock Art Association, has been promulgating the cause of pre-Upper Paleolithic art, arguing that it has not been recognized because archeologists believed it to be nonexistent (but see unit 34). Marshack has been applying microscopic analysis to incised flint pieces from the 54,000-year-old site of Quenitra, Israel, and a shaped piece of volcanic tuff from the Acheulean site of Berekhat Ram, which is between 233,000 and 800,000 years old. He has concluded that the incisions and the shaping represent the work of human hands. Although his findings may well be correct, many archeologists remain resistant to the notion that non-utilitarian artifacts prior to the Upper Paleolithic in Europe signify substantial symbolic, or abstract, expression. ✺

KEY QUESTIONS

- In what ways are modern interpretations of paleolithic art most likely to be biased?
- How would one test the hypothesis that, in some cases at least, paleolithic art is a form of hunting magic?
- What possible interpretations are there for the relative rarity of carnivore images in wall art compared with the extensive use of carnivore teeth in body ornamentation?
- Can the art of another culture ever be completely understood by those outside it?

KEY REFERENCES

Bahn P. New developments in Pleistocene art. *Evol Anthropol* 1996;4:204–215.

Bednarik RG. Concept–mediated marking in the Lower Paleolithic. *Curr Anthropol* 1995;36: 605–616.

Chase PG, Dibble HL. Middle Paleolithic: a review of current evidence and interpretations. *J Anthropol Archeol* 1987;6:263–296.

Clottes J. Rhinos and lions and bears. *Nat Hist* May 1995:30–34.

Davidson I. The power of pictures. In: Conkey M, *et al.*, eds. Beyond art: Pleistocene image and symbol. San Francisco: California Academy of Sciences, 1997.

Davidson I, Noble W. The archeology of depiction and language. *Curr Anthropol* 1989;30:125–156.

D'Errico F. Technology, motion, and the meaning of epipaleolithic art. *Curr Anthropol* 1992;33:94–109.

D'Errico F, Villa P. Holes and grooves: the contribution of microscopy and taphonomy to the problem of art origin. *J Hum Evol* 1997;33:1–31.

Lewis-Williams JD. Cognitive and optical illusions in San rock art research. *Curr Anthropol* 1986;27: 171–177.

Lindly JM, Clark GA. Symbolism and modern human origins. *Curr Anthropol* 1991;31:233–262.

Marshack A. A Middle Paleolithic symbolic composition from the Golan Heights: the earliest known depictive image. *Curr Anthropol* 1996;37:357–365.

Valladas H, *et al.* Direct radiocarbon dates for prehistoric paintings at the Altamira, El Castillo and Niaux caves. *Nature* 1992;357:68–70.

White R. Visual thinking in the Ice Age. *Sci Am* July 1989:92–99.

PART 9

NEW WORLDS

New Worlds
The First Villagers

NEW WORLDS

34

Following the origin of modern humans and their establishment throughout Africa and Eurasia, two major population dispersals occurred: one into the Americas, and another into Australia. Although paleontological, archeological, linguistic, and genetic evidence has been sifted to clarify the issue, the dates and modes at which these dispersals occurred remain uncertain.

Researchers have often displayed a tendency to contemplate aspects of human history in isolation from that of other groups of animals. Of course, in some respects the path of human history has been determined solely by the rather special behavioral repertoire displayed by the genus *Homo*. Equally, however, the human lineage on occasions must have responded to ecological changes in ways parallel to the responses produced by other animals.

For example, as Alan Turner, of Liverpool University, has argued, the initial dispersal from Africa and the later migration to North America can be viewed as territorial expansions in concert with other large predators. Rather than answering some inward spirit's urge for new lands, our ancestors were simply tracking their subsistence potential through new prey populations, as were other predators. One can only speculate, however, about the precise motivations of the first Australian colonists when they struck out in small boats for a land unseen. Whatever their goal, it was not simply taking part in a more general spread of other animals.

THE AMERICAS

Although population source (Asia) and the route (across the Bering Strait that separates Alaska and Siberia)

are undisputed, no consensus has been reached over the timing of this migration. One school of thought argues for a date close to 12,000 years ago. By 11,500 years ago, the Americas had clearly been peopled, as evidenced by the extensive archeological remains of Clovis and then Folsom cultures, evidence of which was first unearthed in the 1930s. But were the Clovis people the first Americans? Not according to the second school of thought, which argues for a date in the region of 30,000 years ago. Now genetic evidence points to there having been just one migration, with population influx perhaps extending over a period of a thousand years.

Whenever they arrived, the first Americans found a land very different from the one we know today. Between 75,000 and 10,000 years ago, the Earth was held in the pulsating grip of the Ice Age, its frigid grasp being tightest at 65,000 and 21,000 years ago. Throughout this time, at least part of North America was mantled with ice. The Laurentide ice sheet, 2 miles thick in places, buried much of Canada and the northern United States from the Atlantic coast to just east of the Rockies. The Cordilleran ice sheet ran ribbon-like up the Pacific coast from Washington State toward Alaska, submerging all but the highest peaks of the Rockies and the mountains of western Canada.

Except during a period between 20,000 and 13,000 years ago, an ice-free corridor appears to have linked southern North America with the ice-free regions of Alaska and Canada's Yukon and Northwest Territories, providing a potential migration route for people coming from Siberia. These individuals could have made the intercontinental crossing dry-shod or by island hopping, because the Beringia land bridge, which linked Siberia with Alaska, was fully or partly exposed for much of that time as the result of a drop in sea level; this fall in sea level measured as much as 100 meters at the glacial maxima, with the water being locked up in the greatly expanded polar ice caps. The time range for pos-

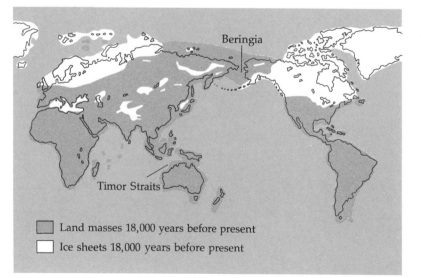

Land masses 18,000 years before present

Ice sheets 18,000 years before present

MIGRATION ROUTES TO AUSTRALIA AND AMERICA: Eighteen thousand years before present was the apogee of the last glaciation (75,000–10,000 BP). Expanded glacial cover (white areas) lowered sea levels to expose the shallow continental shelf (shaded areas over current coastlines). Although glaciation occurred less than shown 40,000 years ago, the Timor Straits were still considerably narrowed, facilitating the migration into Australia (and Tasmania). The reduced glaciation some 20,000–30,000 years ago might also have left an ice-free corridor linking North America and Siberia.

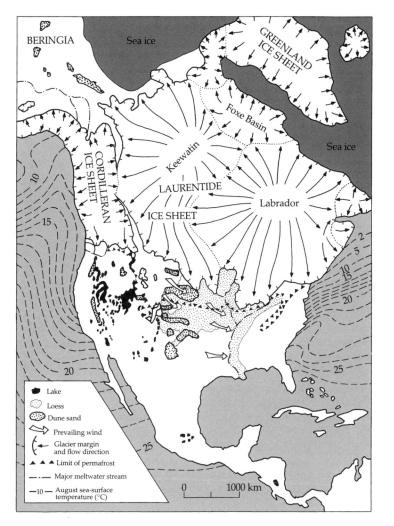

IN THE GRIP OF THE ICE: At the peak of the last glaciation, some 18,000 years ago, much of North America was covered by thick ice sheets. To the west was the Cordilleran ice sheet; in the center and east, the Laurentide ice sheet covered the land. There is still dispute as to whether an ice-free corridor existed throughout the period or was temporarily closed. (Courtesy of Stephen C. Porter.)

sible migration can probably be narrowed somewhat, because archeological evidence gathered to date appears to show that Siberia remained uninhabited until some 40,000 years ago.

Archeologists are faced with a perplexing question: who—if anybody—preceded the Clovis people into the Americas? Over the past few decades, many claims have been made for archeological evidence earlier than 11,500 years ago south of the area that was submerged under the ice sheets. Most have these claims have been viewed skeptically, with only a few being accepted as valid. Nevertheless, some people preceded the Clovis culture in the Americas, but the paucity of reliable sites suggests that this population was small. The explosion of sites from 11,500 years ago onward presumably reflects an explosion of populations, either from people already present in the continents or from a new migration.

Recently, some of the more famous "old" sites have lost their claims at predating Clovis. Calico Hills, California, which its proponents claim yields stone artifacts dating between 100,000 and 200,000 years, is no longer

taken seriously by most authorities. Del Mar Man, a collection of skulls once dated at 70,000 years, have been redated at approximately 8000 years. And the famous bone deflesher from Old Crow in the Yukon Territories, found in 1966 and dated at 27,000 years, was redated in 1987 at just 1400 years. Nevertheless, Richard Morlan, of the University of Toronto, believes that another Yukon site, Bluefish Caves, may prove to be in the vicinity of 25,000 years old. This last site relates to people north of the ice sheets, however.

The serious pre-Clovis contenders south of the ice are mostly in South America:

- Los Toldos Cave in the Argentine Patagonia, dated at 12,600 years
- The site of Tagua-Tagua in central Chile, dated at 11,380 years
- Also in central Chile, the site of Monte Verde, dated at 12,500 years
- Taima-Taima in northwestern Venezuela, dated at 13,000 years

CLOVIS AND AFTER: Although their skeletal remains are few, the Clovis people left their trademark—the Clovis point (*far left*)—spread widely over North America. The Clovis point, which usually measured about 7 centimeters in length, was apparently inserted into the split end of a spear shaft and bound in place by hide. Following in close succession after Clovis were the (*second-left to right*) Folsom, Scottsbluff, and Hell Gap cultures.

The evidence for Monte Verde's early date has recently become particularly strong, and most skeptics became convinced of its authenticity during a site visit in late 1996. The rock shelter site of Pedra Furada, in northeastern Brazil, has been claimed to have been inhabited as early as 50,000 years ago, which would make it by far the oldest pre-Clovis site in the Americas. Many archeologists remain skeptical that the stone artifacts on which the

claim is based are truly man-made; they may actually represent the result of natural stone breakage.

The most important site in North America, and among the strongest pre-Clovis contenders in all of the Americas, is the Meadowcroft cave shelter near Pittsburgh, Pennsylvania, a site that is said to have been occupied repeatedly since 19,600 years ago. Skeptics point out the possibility that the site's material has suffered contamination with carbon from nearby coal deposits, which would corrupt the radiocarbon dating used at the site. James Adovasio of the University of Pittsburgh, the site's principal investigator, counters by noting that the dates run from oldest to the youngest in the deposits from the bottom to the top in the site, just as they should if they were uncontaminated. This dating issue remains to be resolved.

Archeologists now agree that a pre-Clovis people existed in the Americas, perhaps as early as 30,000 years ago. If population growth was small, then the number of archeological sites to be discovered would be correspondingly small. As David Meltzer, of Southern Methodist University, recently observed, "Clovis, in that situation, may reflect the visible portion of a population curve that began much earlier."

When Columbus arrived in the Americas in the fifteenth century, 1000 different languages were spoken among the native Indian peoples. Stanford University linguist Joseph Greenberg has analyzed the 600 languages that survive, tracing them back to just three source languages: Amerind, the most widespread and diversified; Na-Dene, less widespread and diversified than Amerind; and Aleut-Eskimo, an even less widespread and diversified language than Na-Dene. It is possible, says Greenberg, that these three linguistic groups signal three separate migrations, with the Amerind group being the first arrivals.

Several molecular biology laboratories are conducting mitochondrial DNA analysis, so far without reaching an agreement as to whether the present population descends from a small founder population or from a large population. Several different mitochondrial DNA lineages have been identified in the modern population, all of Asian origin. The amount of genetic diversity among the lineages has been estimated variously to indicate separation as long as 78,000 years ago. Although humans could have been in the Americas that long, it is more likely that the mtDNA lineages diverged in Asian populations and were already established in the founding American population at a later date. That later date has yet to be determined, although several estimates close to 30,000 years ago have been made. Douglas Wallace and his colleagues at Emory University have tentatively indicated that the mtDNA evidence might lend support to Greenberg's three-wave migration hypothesis. Meanwhile, similar work at Oxford University has led to the conclusion of a single migration; in Japan, researchers have inferred four migrations from mitochondrial DNA data. Consensus has recently moved to support a single migration.

THE TIME OF CLOVIS: Clovis sites are scattered over much of North America (specifically the United States, as most of Canada was under ice at the time). As this diagram shows, dating of the sites lies in a tight range between 11,500 and just less than 11,000 years ago. Folsom sites follow close on behind, but again are confined to North America.

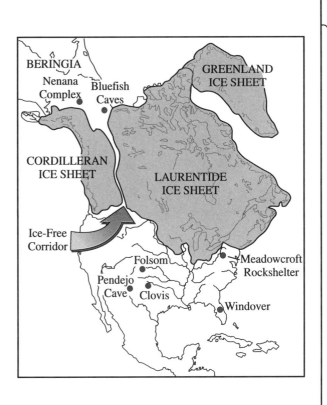

PUTATIVE PRE-CLOVIS SITES: The maps show the distribution of the sites in North America *(left)* and South America *(right)* that have the strongest claims for dating to pre-Clovis times. Pedra Furada, in Brazil, is the least likely candidate of those shown.

HUMAN IMPACTS OF THE ENTRY INTO THE AMERICAS

The Americas of the Ice Age differed dramatically from today's world. They teemed with large mammal species, including mammoth, mastodon, giant ground sloth, steppe bison, elk, yak, and lion—75 species in all, many of which were immigrants from Eurasia. Huge freshwater lakes ponded in the Great Basin. The great equatorial forests of Central and South America survived in sheltered "refuges," having largely been replaced by open grassland and woodland.

Clovis people, who manufactured a characteristic "fluted" projectile point (an American invention), lived in the narrow archeological window between 11,500 and 10,900 years ago. They were replaced by Folsom people, who produced smaller, more finely crafted projectile points. The Clovis and Folsom worlds were vastly different places, however. Clovis people hunted mammoth and mastodon. By Folsom times, none remained. Gone, too, were the great majority of large mammals, with some 75 species eventually going extinct and a few becoming restricted to South America.

One of the great debates over the peopling of the Americas has centered on this rapid extinction. Some authorities—Paul Martin of the University of Arizona being the most prominent—argue that the animals had been wiped out by a wave of Clovis and then Folsom hunters, advancing north to south for a millennium. Others—with Ernest Lundelius of the University of Texas most prominent—point to the dramatic climatic shift at the end of the Ice Age as the culprit.

Invasion of new lands by humans has been known to cause significant extinctions in relatively recent history. Climate change can also drive species to extinction—particularly a change as dramatic as that marked by the Pleistocene/Holocene transition. Thus, while both explanations are plausible, neither has been demonstrated beyond reasonable doubt in this case.

AUSTRALIA

The first Australians had to make a water journey to their New World. Even with sea levels at their lowest during glacial maxima, the journey from Sunda Land (the combined landmass of Southeast Asia and much of Indonesia) to the Sahul landmass (Australia, Tasmania, and New Guinea) would have required eight sea voyages, the last covering 52 miles. So far, no archeological evidence has been recovered from Australian sites of ves-

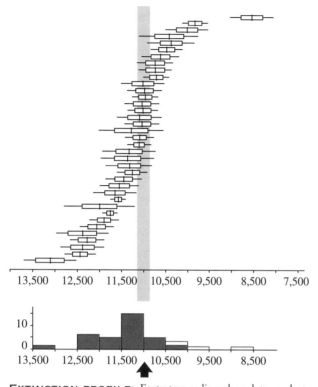

EXTINCTION PROFILE: Forty-two radiocarbon dates on last-appearing Shasta ground sloth dung from various sites in the U.S. Southwest. The arrow and shaded column above it indicate the approximate time of activity of Clovis hunters in the region. Last appearance dates cluster at this time.

sels that could have made such a journey. Coastal sites during the Ice Age are mostly now submerged beneath the sea, however. In any case, the ability to construct sea-going craft that could make the required journey to Australia may be taken as proof of modern human behavior.

Although hominines have been present in Southeast Asia for almost 2 million years, the first evidence of occupation in the Sahul outside of Australia is just 40,000 years old, taking the form of an archeological site on the northeast coast of New Guinea. Within Australia itself, the principal questions are, When and how was the continent first populated by humans? Fossil and archeological evidence is not extensive, and existing artifacts are often subject to differing interpretations; adding to the uncertainty is the difficulty of dating prehistoric material.

Until recently, the earliest known archeological sites—Malakunanja and Nauwalabila, in Arnhem Land, northern Australia—were approximately 60,000 years old. These dates, obtained with thermoluminescence analysis, are not universally accepted as valid. Even more contention surrounds claims for the archeological site of Jinmium, also in northern Australia; in late 1996, it was reported to be at least 60,000 years old and perhaps as much as 176,000 years old. To develop these dates, thermoluminescence analysis was used on sediments that con-

tained simple stone artifacts. More intriguing is a series of several thousand small depressions in rock surfaces at the site. If the depressions represent the work of human hands, as is suggested, and if the dates are correct, then archeologists' picture of the colonization of the continent must be dramatically revised. The notion of when humans first indulged in abstract symbolism would also have to undergo a vast reassessment. Both the dates for the site and the human agency for the depressions were critically questioned, however, and in mid-1998 new dating tests (based on optimally stimulated luminescence) showed the Jinmium site to be no more than 10,000 years old.

The oldest human fossils come from Lake Mungo, in southern Australia, with a secure date of 25,000 years. Several skeletons have been unearthed from this site, one of which appears to have been cremated, making it the oldest example of this form of ritual behavior in the world. A badly distorted cranium was discovered in the mid-1980s at Wilandra Lakes, near Lake Mungo, with a suggested date somewhat older than the Lake Mungo material. Details of the cranium have yet to be published. The Lake Mungo fossils are relatively gracile, with thin cranial bone, well-rounded foreheads, weak brow ridges, and small mandibles.

Standing in contrast to these fossils is a collection of crania from Kow Swamp, also in southern Australia, which date to approximately 12,000 years old. These specimens are more robust than the Lake Mungo people, having thick cranial bone, large and projecting faces, prominent brow ridges, and large mandibles.

The anatomical differences between these populations have prompted some anthropologists to propose that Australia was colonized at least twice (the multiple-source hypothesis). These researchers suggest that the gracile people came from China, while the robust colonists migrated from Indonesia. Interbreeding would have blurred the distinctions in later generations and produced the great anatomical variability present in modern aborigines. In fact, the division of the earliest fossils into gracile and robust is somewhat artificial, argues Phillip Habgood, of the University of Sydney. Both Habgood and a growing number of Australian scholars suggest that the early colonists were more anatomically homogeneous, with the variable morphology of the modern aborigines being the result of "genetic (and cultural) processes acting upon a small founding population." This concept is known as the homogeneity hypothesis. Evidence from mitochondrial DNA indicates that the island continent was colonized at least 15 times, not just once. The source population (or populations) for the earliest Australians remains unresolved (see unit 28).

HUMAN IMPACTS OF THE ENTRY INTO AUSTRALIA

As with the Americas, the first human inhabitants of Australia have been suggested to have caused extensive extinctions of giant mammals (marsupials in this case) through hunting. Extinctions did occur that were apparently clustered around 50,000 years ago, close to the time

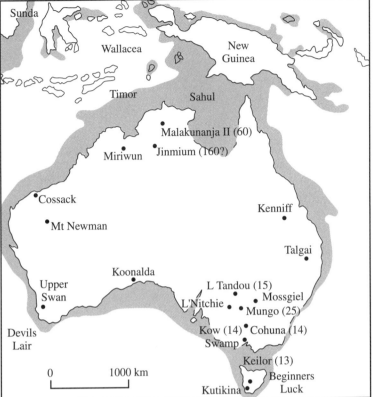

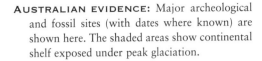

AUSTRALIAN EVIDENCE: Major archeological and fossil sites (with dates where known) are shown here. The shaded areas show continental shelf exposed under peak glaciation.

when many anthropologists believe Australia was colonized (60,000 years ago). Difficulties in dating the remains of these extinct forms casts uncertainty on this proposal. Recent dating efforts suggest that extinctions continued until some 20,000 years ago. As Iain Davidson has noted, "Under these circumstances, any argument for a direct human role in the extinction of megafauna seems premature." ❋

KEY QUESTIONS

- Why has it proven so difficult to establish a pre-Clovis presence in the Americas?
- What factors might lead to the conflicting conclusions that are being reached with genetic evidence on the peopling of the Americas?
- What population factors might lead to a highly variable population among the early Australians?
- What is the likelihood of an entry date into Australia that exceeds 100,000 years ago?

KEY REFERENCES

Bahn P. 50,000-year-old Americans of Pedra Furada. *Nature* 1993;362:114–115.

———. Further back down under. *Nature* 1996;383:577–578.

Brown P. Recent human evolution in East Asia and Australasia. *Phil Trans Roy Soc B* 1992;337:235–242.

Dillehay T, Meltzer DJ, eds. The first Americans: search and research. New York: CRC Press, 1991.

Flood J. Archaeology of the dreamtime. New Haven, CT: Yale University Press, 1990.

Gibbons A. The people of the Americas. *Science* 1996;274:31–32.

———. Young ages for Australian rock art. *Science* 1998;280:1351.

Greenberg JH, *et al*. The settlement of the Americas. *Curr Anthropol* 1986;27:477–497.

Greenberg JH, Ruhlen M. Linguistic origins of Native Americans. *Sci Am* Nov 1992:94–99.

Habgood P. The origin of anatomically-modern humans in Australasia. In: Mellars P, Stringer C, eds. The human revolution. Princeton, NJ: Princeton University Press, 1989:232–244.

Meltzer DJ. Clocking the first Americans. *Annu Rev Anthropol* 1995 24:21–45.

———. Monte Verde and the Pleistocene peopling of the Americas. *Science* 1997;276:754–755.

Merriweather DA, Ferrell RE. The four founding lineage hypothesis for the New World: a critical re-evaluation. *Mol Phylogenet Evol* 1996;5:241–246.

Morwood MJ, Smith CE. Rock art research in Australia, 1974–94. *Aust Archeol* 1994;39:19–38.

O'Connell JF, Allen J. When did humans first arrive in Australia and why is it important to know. *Evol Anthropol* 1998;6:132–146.

Roberts RG, *et al*. The human colonization of Australia. *Quat Sci Rev* 1994;13:575–583.

Weiss KM. American origins. *Proc Natl Acad Sci USA* 1994;91:833–835.

THE FIRST VILLAGERS

35

The date of 12,000 years before present (BP) is usually given as the beginning of what has been called the Agricultural (or **Neolithic**) Revolution. Prior to this date, human populations subsisted by various forms of hunting and gathering. After 12,000 BP, however, a shift toward plant and animal domestication occurred independently in several different parts of the world—first in the Fertile Crescent of the Near East, then in Meso America, and lastly in Southeast Asia. The adoption of agriculture was extremely rapid as measured against the established time scale of human prehistory and was accompanied by an escalation of the population size, rising from approximately 10 million at the outset of the Neolithic to 100 million some 4000 years ago. The tremendous changes wrought during the Neolithic period can be seen as a prelude to the emergence of cities and city states and, of course, to a further rise in population (which now totals 6 billion).

Until relatively recently, the Agricultural Revolution was viewed as a rather straightforward—if dramatic—transition. Responding to some kind of stimulus, hunters and gatherers, who were assumed to have lived in small nomadic bands of approximately 25 individuals, developed plant and animal domestication as a way of intensifying food production. As a result, these people began living in large, settled communities, whose social and political complexity far exceeded anything achieved earlier in history. In other words, sedentism and social complexity were explained as the *consequences* of the adoption of agriculture, and the Neolithic transition was characterized as a shift from the simple to the complex.

NEW INTERPRETATIONS

Given the discovery of new archeological and ethnographic evidence, and with a reassessment of some existing evidence, the Neolithic transition is now viewed in a different light. Most importantly, it is now clear that many populations established sedentary communities and elaborated complex social systems *prior* to the advent of agriculture. Hunters and gatherers of the late Pleistocene, it is now realized, were not necessarily living the simple, nomadic lifeway that anthropologists had imagined. Although debate persists about what triggered the Neolithic transition, it is not unreasonable to view some facets of agriculture as a consequence, not the cause, of social complexity.

The traditional characterization of the Neolithic transition as an Agricultural Revolution rested on two kinds of evidence: archeological and ethnographic. The former was seen as indicating an explosive change in economic organization; the latter was viewed as revealing a shift from simple to complex social organization. The phrase "Agricultural Revolution" seemed apt for a number of reasons—not least of which was the limited amount of archeological data with which to sketch this crucial period in human history. The few major sites, such as the early farming and trading community of Jericho, with its impressive tower and wall, seemed to burst out of an archeological void with dramatic suddenness. Indeed, the remains of Catal Huyuk, which was occupied by farming people between 8500 and 7800 BP, has been described as an archeological supernova. Excavated in the 1960s, this Turkish town covering some 30 acres boasted elaborate architecture and beautiful, symbolic wall paintings and carvings. A British team of archeologists, led by Ian Hodder of Cambridge University, began new excavations at this site in 1994.

In the two decades since the initial discovery of Catal Huyuk, further excavations in the Fertile Crescent have uncovered the remains of villages and towns, which collectively make clear that the adoption of agriculture was a much more gradual process than had been envisaged. Such sites include 'Ain Ghazal in Jordan and Abu Hureyra in northern Syria. In particular, a transition is now evident, in which settled communities based entirely on hunting and gathering gave way to a mixed economy of hunting and gathering combined with some domestication, and then to fully committed agriculture. Examination of this more complete archeological record reveals that the Neolithic transition was a step-by-step introduction of domestication, not an overnight revolution.

One of the most informative sites is that of Abu Hureyra, which was occupied from 11,500 to 7000 BP, with one major break from 10,100 to 9600 BP. Emergency excavation in 1974 showed that the first period of settlement, Abu Hureyra I, was a hunting and gathering community of 50 to 300 individuals who exploited the rich steppe flora (including many wild cereals) and the annually migrating Persian gazelle. A year-round settlement of simple yet substantial single-family houses, Abu Hureyra I confounds the traditional view of hunter-gatherer existence, which posits the existence of small, nomadic bands.

Perhaps because of overexploitation of local resources and an increasingly unfavorable climate, Abu Hureyra I was abandoned in 10,100 BP. It was reoccupied half a millennium later, this time by people who included plant—but not animal—domestication in their economy. For a millennium, the people of Abu Hureyra continued to hunt gazelle as their sole source of meat, after which time they turned to the domestication of sheep and goats. The overall pattern, therefore, is "a step by step introduction of domesticated plants and animals," explains Yale University's Andrew Moore, who led the 1974 excavation. "This is a pattern I see across southwestern Asia."

It should have come as no surprise that late Pleistocene hunters and gatherers led socially complex lives—indications of this lifeway have been known from the archeological record for some time. Most notable among this evidence was the art of the European Ice Age (see unit 33).

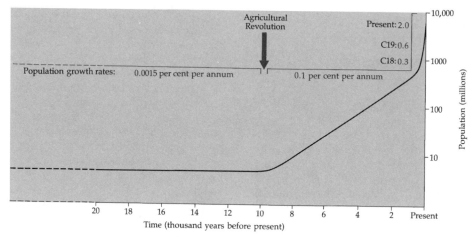

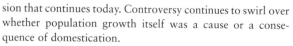

POPULATION CHANGE SINCE THE NEOLITHIC: The beginnings of substantial population growth coincided with the origin of plant and animal domestication, igniting an explosion that continues today. Controversy continues to swirl over whether population growth itself was a cause or a consequence of domestication.

"If one is looking for a single archeological reflection of sociocultural complexity, then presumably attention will continue to focus on the unique and impressive manifestations of Upper Paleolithic cave art from the Franco-Cantabrian region," notes Paul Mellars of Cambridge University. This period of wall and portable art began approximately 35,000 years ago and ended 10,000 years ago, with the termination of the Ice Age.

More tangible evidence of late Pleistocene social and economic complexity comes from the Central Russian Plain—specifically, a site near the town of Mezhirich, 1100 kilometers southwest of Moscow. Approximately 15,000 years ago, a settlement of some 50 people lived in a "village" consisting of at least five substantial dwellings,

each constructed from mammoth bones. "We are beginning to find evidence of semipermanent dwellings in the Central Russian Plain dating back to nearly 30,000 years ago," notes Olga Soffer of the University of Wisconsin.

Given this and other evidence, it is perhaps surprising that, until relatively recently, late Pleistocene humans were almost universally regarded as simple nomads who wandered endlessly from camp to camp in bands of no more than 25 individuals. This characterization was based on a very important and influential study during the 1960s of the !Kung San (Bushmen) of the Kalahari. Organized by Harvard University by anthropologists Richard Lee and Irven DeVore, the !Kung project examined in great detail the socioeconomic life of these people.

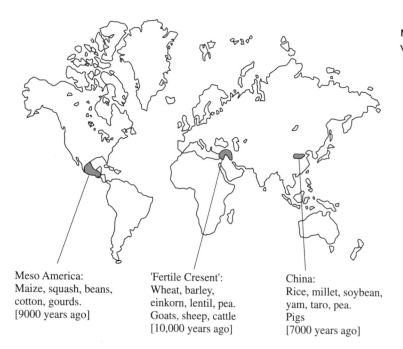

MAJOR CENTERS OF AGRICULTURAL INNOVATION: Plant and animal domestication apparently occurred independently and at different times in many different parts of the world. Three major centers of origin existed, whose influence spread geographically, eventually coming to dominate local innovations.

Meso America:
Maize, squash, beans, cotton, gourds.
[9000 years ago]

'Fertile Cresent':
Wheat, barley, einkorn, lentil, pea.
Goats, sheep, cattle
[10,000 years ago]

China:
Rice, millet, soybean, yam, taro, pea.
Pigs
[7000 years ago]

The project revealed that, despite living in a marginal environment, the !Kung were able to subsist on simple hunting and gathering, with the expenditure of just a few hours' work each day. In addition, !Kung social life was characterized as an egalitarian, harmonious, sharing environment. The collective results of the Harvard project were presented at a landmark meeting, titled "Man the Hunter," held at the University of Chicago in 1966. For several reasons—including the fact that no other ethnographic project had been so thoroughly and scientifically conducted—the Harvard team's portrayal of the !Kung became *the* image of the hunting and gathering lifeway, both in the modern world and in prehistory, despite existing archeological and ethnographic evidence to the contrary.

For more than a decade, the !Kung model of the hunter-gatherer lifeway dominated anthropological thought. By the early 1980s, however, its shortcomings had been gradually exposed. This shift in perception was driven by new historical, archeological, and behavioral ecology evidence. It indicated that a great deal more variability existed in the hunting and gathering lifeway of prehistoric peoples than had been allowed for in the !Kung model; this variability included a degree of social and economic complexity that hitherto had been associated exclusively with agricultural societies. "Many characteristics previously associated solely with farmers—sedentism, elaborate burial and substantial tombs, social inequality, occupational specialization, long-distance exchange, technological innovation, warfare—are to be found among many foraging societies," concluded anthropologists James Brown and T. Douglas Price in 1984, in a classic reassessment of hunters and gatherers.

In other words, the Agricultural Revolution was recognized to be neither a revolution nor a movement primarily focused on the adoption of agriculture. Instead, the Neolithic transition involved increasing sedentism and social complexity, which was usually followed by the gradual adoption of plant and animal domestication. In some cases, however, plant domestication preceded sedentism, particularly in the New World. For instance, Kent Flannery of the University of Michigan has shown that the first plant domesticated in the New World, the bottle gourd, which was grown about 9000 years BP in the southern highlands of Mexico, preceded sedentism by at least 1000 years. Clearly, the Neolithic was a complex period, and must have been influenced substantially by both local and global factors.

One long-standing question of interest in Europe, for instance, has been the mode by which agriculture spread. Was it carried by farmers moving into the region from the Middle East? Or did it develop locally, with the idea spreading throughout the continent, not the farming-oriented people? This question is amenable to genetic as well as archeological research. Work with classic genetic markers and, more recently, DNA sequences from nuclear genes suggested that population migration was important in the spread of agriculture. This conclusion, known as the demic expansion model, has been challenged by a recent survey of mitochondrial DNA patterns throughout the continent. This work implies that it was principally the idea of agriculture that spread, not a migration of people. The difference of opinion remains unresolved.

CAUSES OF THE TRANSITION

Because the transition to food production occurred within a few thousand years independently in several different parts of the world, anthropologists have long sought a global cause. Two factors have been candidates for this single, prime mover: population pressure and climate change.

Although a dramatic rise in population numbers undoubtedly accompanied the Neolithic transition, the question of whether this relationship was one of cause or effect remains unanswered. Mark Cohen, of the State University of New York, Plattsburgh, is the principal proponent of the population pressure hypothesis. He argues that it was causal, and adduces signs of nutritional stress in skeletal remains from the late Paleolithic to support his case. In contrast, many anthropologists argue that numerous examples of the adoption of sedentism and agriculture can be found in the apparent absence of high population numbers—such as in the southern highlands of Mexico. For these researchers, including Flannery, the population pressure hypothesis remains unconvincing.

The second major candidate—climatic change—appears more persuasive, as the Neolithic transition coincides with the end of the Pleistocene glaciation. The shift from glacial to interglacial conditions would have driven extensive environmental restructuring, bringing plant and animal communities into areas where they did not previously exist. For instance, warmer, moister climes in the Levant 12,000 years ago likely encouraged the abundant growth of wild cereals on the steppe, allowing foragers to collect them in great numbers and subsequently domesticate these plants. Moore considers this step to have been

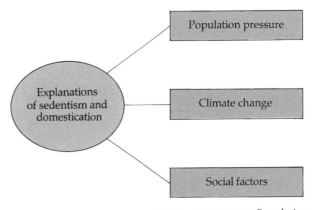

HYPOTHESES OF AGRICULTURAL ORIGINS: Population pressure and climate change have long vied as the most persuasive potential candidates for initiating sedentism and domestication. In recent times, attention has turned to factors concerning internal social complexity.

Traditional view of agricultural revolution

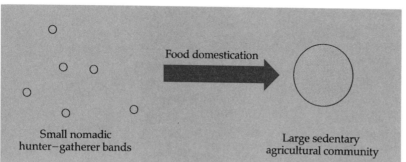

Current view of agricultural revolution

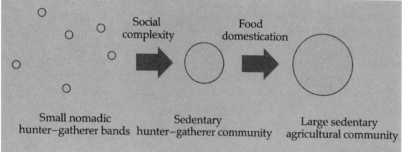

ORIGIN SEEN AS MORE COMPLEX: In the traditional view, sedentism and domestication developed together; small, nomadic, hunter-gatherer bands were viewed as being transformed into large, sedentary, agricultural communities. Recently, scholars have come to realize that the process probably included several steps, in which sedentism and domestication were separated. Intermediate between small nomadic bands and large, agricultural communities, therefore, were sedentary communities that subsisted on hunting and gathering.

important in the early establishment of Abu Hureyra and other similar settlements.

Evidence is lacking to prove that climate-driven floral change universally accompanied sedentism. Moreover, some periods earlier than the end of the Pleistocene must have been conducive to intensification of food production. Modern *Homo sapiens* arose more than 100,000 years ago—so why did almost 90,000 years pass before intensification of food production become adopted? Was the delay caused by a combination of population pressure and climate change? Or was it something else entirely?

For some scholars, that "something else" is social complexity. Whereas population pressure and climate change were both "external" factors—the first presenting a problem to be solved, the second an opportunity to be exploited—social complexity would provide an "internal" trigger for change.

Building on earlier ideas of Robert Braidwood, University of London anthropologist Barbara Bender argues that social complexity is a prerequisite for—not a product of—a sedentary agricultural system. The increasing social complexity, and the stratified social and economic order that accompany it, place demands on food production that cannot be satisfied by the small, nomadic hunter-gatherer society, Bender and her supporters say. In response to this internal pressure, the culture intensifies and formalizes food production; in other words, it creates an agrarian society. Bender does not argue that this internal factor is the sole cause, merely that "technology and demography have been given too much importance in the explanation of agricultural origins, social structure too little."

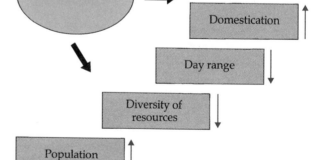

CONSEQUENCES OF SEDENTISM: A shift from a nomadic to a sedentary way of life necessarily involved a series of potential social and material changes. Although these changes have often been associated exclusively with agricultural societies, it is now evident that sedentism can, by itself, produce at least part of this pattern.

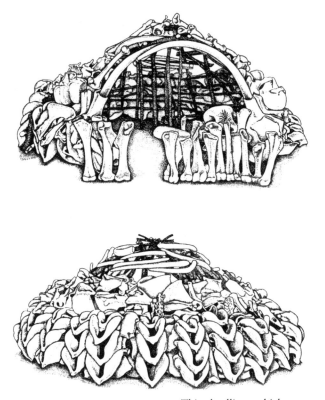

MAMMOTH-BONE DWELLING: This dwelling, which measures 5 meters in diameter, is one of five shelters excavated at Mezhirich, in the Ukraine. Individually constructed with great technical and esthetic attention, these 15,000-year-old dwellings formed a community that was surely more socially complex than is usually envisaged for pre-agricultural hunter-gatherer peoples. (Courtesy of M. I. Gladkih, N. L. Kornietz, and O. Soffer/*Scientific American*, November 1984, all rights reserved.)

Although this social focus is gaining popularity among anthropologists, assessing its merits is very difficult. It is analogous to a "black box": you know it is important, but you do not understand how it works. Why, for instance, would social complexity have taken 90,000 years to manifest itself after the origin of anatomically modern humans? One possibility, of course, is that a subtle intellectual evolutionary change occurred relatively recently in human history, but did not manifest itself physically.

In fact, modern humans underwent a biological change between the end of the Pleistocene and the Holocene, but it affected their bodily physique. Not only are post-pleistocene humans smaller than their immediate ancestors, but the difference in size between males and females—sexual dimorphism—is also significantly reduced. As Robert Foley of Cambridge University has recently pointed out, this changed body size may have implications for how one views the Neolithic transition.

Inevitably, anthropologists' concepts of hunter-gatherers are influenced by knowledge gleaned from contemporary foragers. These people, whose numbers are rapidly dwindling and who live in the most marginal areas of the globe, generally include a large plant-food component in their diet (notable exceptions exist, of course) and live in egalitarian communities. Thus, the Neolithic transition is usually seen as a change from this kind of subsistence economy to domestication.

The larger overall body size of late Pleistocene people, and the greater sexual dimorphism in body size, might imply a different socioeconomic context, however. Males may well have engaged in more heated competition for access to females (see unit 13), as well as more big-game hunting and provisioning of their mates and offspring. "In this context, what we think of as modern hunter-gathering is largely a post-Pleistocene phenomenon," says Foley. "Rather than being an adaptation ancestral to food production, it is a parallel development. . . . Both hunter-gatherer and agricultural systems developed as a response to resource depletion at the end of the Pleistocene from the rather different socioecology of Late Pleistocene anatomically modern humans."

Clearly, anthropologists' picture of the Neolithic transition is far from complete. It is fair to say, however, that the search for a single, prime mover is much less popular today. "No single factor is responsible for the rise of cultural complexity," concluded Brown and Price. "Increased complexity appears in too many diverse and historically unconnected places to be a result of a single factor. . . . It may be sufficient for the moment simply to be aware that things are not what they have seemed to be." ✻

KEY QUESTIONS

- In what key ways would social organization necessarily change when a formerly nomadic people adopted a settled community?
- How would one distinguish between signs of plant and animal domestication on the one hand and foraging on the other?
- Was sedentism (and agriculture) an inevitable development with the evolution of anatomically modern humans?
- What might explain the later development of sedentism and agriculture in the New World as compared with the Old World?

KEY REFERENCES

Bar-Yosef O. The Natufian culture of the Levant, threshold of the origins of agriculture. *Evol Anthropol* 1998;6:159–177.

Bird-David N. Beyond "the original affluent society." *Curr Anthropol* 1992;33:25–48.

Blumler MA, Byrne R. The ecological genetics of domestication and the origin of agriculture. *Curr Anthropol* 1991;32:23–54.

Byrd B. From early humans to farmers and herders. *J Archeol Res* 1994;2:221–253.

Flannery K, ed. Guila Naquitz. New York: Academic Press, 1986. (Papers on the research project in the Valley of Oaxaca, Mexico.)

Foley R. Hominids, humans and hunter-gatherers: an evolutionary perspective. In: Ingold T, Riches D, Woodburn J, eds. Hunters and gatherers: history, evolution and social change. Oxford: Oxford University Press, 1988:207–221.

Hawkes K, O'Connell J. On optimal foraging models and subsistence transitions. *Curr Anthropol* 1992;33:63–66.

Hayden B. Nimrods, piscators, pluckers and planters: the origins of food production. *J Anthropol Archeol* 1991;9:31–69.

Layton R, *et al.* The transition between hunting and gathering and the specialized husbandry of resources: socio-ecological approach. *Curr Anthropol* 1991;32:255–274.

Molleson T. The eloquent bones of Abu Hureyra. *Sci Am* Aug 1994:70–75.

Pennington RL. Causes of early human population growth. *Am J Physical Anthropol* 1996;99:259–274.

Price TD, Gebauer AB, eds. Last hunters, first farmers. Santa Fe, NM: School of American Research Press, 1995.

Richards M, *et al.* Paleolithic and Neolithic lineages in the European mitochondrial gene pool. *Am J Hum Genet* 1996;59:185–203.

Van Andel TH. The earliest farmers in Europe. *Antiquity* 1995;69:481–500.

GLOSSARY

Absolute dating: techniques that provide information about age by a physical measurement of the material at the site in question, such as radiometric dating. (Contrast with *relative dating.)*

Acheulean: name applied to a type of stone-tool industry characterized by large bifaces including handaxes; it began approximately 1.5 million years ago and continued in Africa and parts of Eurasia until some 200,000 years ago.

Adaptation: the process by which a species changes through natural selection, becoming well suited to its environment.

Adaptive landscape: a graphical description of the average fitness of a population compared with the relative frequencies of genotypes in it. Combinations of alleles that confer high fitness will be seen as peaks on the landscape; those conferring lower fitness will be seen as valleys.

Adaptive radiation: the proliferation of variants following the appearance of an evolutionary innovation; it typically occurs with the establishment of a new clade.

Allele: alternative form of a gene (for example, different eye colors); all genetic loci comprise two alleles, whose effects may differ depending on whether they are identical or different. (See *dominance, recessive allele,* and *polymorphism.)*

Allen's rule: populations of a geographically widespread species living in warm regions will have longer extremities than those inhabiting cold regions. (See *Bergmann's rule.)*

Allopatric speciation: speciation via geographically separated populations.

Altricial: species that produce extremely immature young that are unable to feed or care for themselves.

Anagenesis: evolution by gradual change within a lineage.

Analogy (in biology): a character shared by a set of species but not present in their common ancestor; the result of convergent evolution. (Contrast with *homology.)*

Anatomically modern humans: the term usually used to describe the first members of *Homo sapiens.*

Arboreal: tree-living.

Archeology: the study of human behavior in prehistory.

Autapomorphy: a derived character not shared with other species.

Bergmann's rule: in a geographically widespread species, populations in warmer parts of the range will be smaller-bodied than those in colder parts of the range. (See *Allen's rule.)*

Binomen: the combination of genus and species name that is the basis of Linnean classification.

Biogeography: a perspective of patterns in biology related to their geographical context.

Biological species concept: the definition of a species as a collection of individuals that can breed with one another. (Contrast with *phylogenetic species concept.)*

Biome: a characteristic ecological environment, such as temperate forest, grassland savannah, and so on.

Bipedality: upright walking on the two hind legs (for example, humans' habitual mode of locomotion).

Brachiation: mode of locomotion through trees, using the arms for hanging and swinging (for example, as in gibbons).

Calvarium: the cranium minus the face.

Cambrian explosion: the brief (in geological terms) moment during which many different forms of multicellular organisms evolved, a little more than half a billion years ago.

Carbon-14 dating: an absolute dating method, based on the decay of the radioactive isotope of carbon, carbon-14.

Carnivore: a meat-eating animal.

Catastrophism: the theory that the Earth's geological features were formed by a series of catastrophic events, such as floods, during Earth's history.

Character state: the presence or absence of a particular character, as in cladistic analysis.

Chatelperronian: the stone-tool industry apparently associated with late Neanderthals.

Clade: a group of species that contains the common ancestor of a group and all its descendants.

Cladistics: the school of evolutionary biology that seeks relationships among species based on the polarity (primitive or derived) of characters.

Cladogenesis: evolution by lineage splitting.

Cladogram: a diagrammatic representation of species relationships. (See *cladistics.)*

Classification: arrangement of organisms into hierarchical groups.

Coalescence time: the time in a lineage's history at which all the variants of a particular gene converge into a single, ancestral form.

Convergent (or parallel) evolution: the result of natural selection producing similar adaptations in separate lineages.

Cranium: the skull minus the lower jaw.

Culture: the sum total of human behavior, including technological, mythological, aesthetic, and institutional activities.

Derived character: a character acquired by some members of an evolutionary group that serves to unite them in a taxonomic sense and distinguish them from other species in the group. (Contrast with *primitive character.)*

Diastema: gap between the lateral incisor and the canine.

Dominance (allelic): an allele A is dominant if it is expressed as the phenotype when in the presence of a second allele, a. For instance, the allele for brown eyes is dominant over the allele for blue eyes. (See *recessive allele.)*

Earlier Stone Age: the first part of the Stone Age; usually applied to Africa.

Electron spin resonance: a technique of absolute dating that is based on natural radiation in the soil affecting the state of electrons in a target material, such as teeth.

Encephalization: the process of brain enlargement.

Encephalization quotient: a measure of relative brain size.

Endocast: the impression of the inner surface of the brain case; can be natural or experimentally produced.

Eurybiomic: the ability of a species to utilize food resources from several different biomes.

Evolutionary systematics: a system of classification that emphasizes evolutionary history.

Exons: the segments of genes that code for protein sequence. *(See introns.)*

Faunal correlation: a method of relative dating based on species reaching a similar evolutionary stage at the same time in history in different geographical localities.

Folivore: a leaf-eating animal.

Founder effect: the formation of a new population when a sub-population becomes isolated from the parent population. It is associated with a loss of genetic variation, and sometimes promotes speciation.

Frugivore: a fruit-eating animal.

Genetic distance: a measure of evolutionary separation between lineages.

Genetic drift: random changes in gene frequencies in a population.

Gene tree: the history of a particular gene in related lineages.

Genotype: the genetic profile of an individual.

Grade: a measure of evolutionary stage across lineages.

Group selection: selection acting between groups of individuals, rather than between individuals.

Heterozygous: the presence of two different alleles at a genetic locus. (See *homozygous*.)

Hominine: the collective term for all human-related species.

Hominoidea: all living and extinct species of humans and apes.

Homology: a character shared by a set of species and present in their common ancestor. (Contrast with *analogy*.)

Homoplasies: similar characters produced by convergent evolution. (See *analogy*.)

Homozygous: the presence of two identical alleles at a genetic locus. (See *heterozygous*.)

Inclusive fitness: a measure of an individual's fitness that includes contributions from other individuals (usually relatives) that affect the individual's fitness.

Intermembral index: a comparison of the length of the upper and lower limbs.

Introns: the segments of a gene that are interposed between protein-coding regions, and do not themselves code for protein sequence. (See exons.)

Kin selection: the genetic consequences of the behavior of one individual that enhances the reproductive success of its relatives.

K-selection: the life history strategy in which species have a low potential reproductive output.

Later Stone Age: the third of three stages of the Stone Age; applied to Africa.

Life history variables: features such as age at weaning, age at sexual maturity, and longevity, which determine the nature of a species' overall life.

Lower Paleolithic: the first of three stages of the Paleolithic; applied to Eurasia.

Macroevolution: evolution at the scale of important innovations.

Mass extinction: events in the history of life during which at least 50 percent of the Earth's species become extinct in a geologically brief time.

Microevolution: evolution within lineages.

Middle Paleolithic: the second stage of the Paleolithic; applied to Eurasia.

Middle Stone Age: the second stage of the Stone Age; applied to Africa.

Mitochondrial Eve hypothesis: the hypothesis, based on mitochondrial DNA evidence, that modern humans evolved recently in Africa.

Mitochondrial genome: the package of genetic material within mitochondria.

Molecular evolutionary clock: the concept that the accumulation of genetic differences between lineages after splitting can be used to determine the temporal history of the lineages.

Molecular systematics: the use of molecular biological data for classification and systematics.

Monophyletic group: the set of species containing a common ancestor and all its descendants.

Morphology: the physical form of an organism.

Mosaic evolution: the process by which different aspects of a species' morphology evolve at different rates.

Multiregional evolution hypothesis: the hypothesis that modern humans evolved in near concert in different parts of the Old World.

Mutation: a change in genetic sequence.

Natural selection: the process by which favored variants in a population thrive.

NeoDarwinism: the modern version of Darwin's theory of evolution by natural selection.

Neolithic: the New Stone Age, usually associated with the beginnings of agriculture, some 10,000 years ago.

Neutral theory: the theory that most change at the molecular level occurs by processes such as genetic drift rather than natural selection, with new alleles being selectively neutral.

Niche: the role in the ecosystem played by a species.

Nuclear genome: the package of genetic material in the nucleus.

Oldowan: the stone-tool industry characterized by flakes and chopping tools produced by hard-hammer percussion of small cobbles; it began 2.5 million years ago and continued in parts of Africa and Asia until 20,000 years ago, where it is more properly called chopping-tool assemblages.

Ontogeny: the process of growth and development of an individual from conception onward.

Orthograde: locomotion in which the body remains more vertical relative to the ground.

"Out of Africa" hypothesis: the hypothesis that modern humans originated recently in Africa; based on fossil evidence.

Paleoanthropology: the study of the physical and behavioral aspects of humans in prehistory.

Paleomagnetism: magnetism induced in volcanic rocks as they cool, recording the direction of the Earth's prevailing magnetic field at the time.

Paleontology: the study of fossils and the biology of extinct organisms.

Paraphyletic group: a set of species containing an ancestral species and some, but not all, of its descendants.

Parsimony: a phylogenetic reconstruction in which the phylogeny of a group of species is inferred to be the branching pattern requiring the smallest number of evolutionary changes.

Phenetic classification: a method of classification in which species are grouped together on the basis of morphological similarities.

Phenotype: the physical characters of an organism.

Phyla: major body plans.

Phyletic gradualism: a mode of evolution dominated by gradual change within a lineage.

Phylogenetic species concept: a species is the smallest diagnosable cluster of individual organisms displaying a parental pattern of ancestry and descent.

Phylogeny: a branching diagram showing the ancestral relations among species.

Polarity: the assessment of a character as either primitive or derived.

Polymorphism: the situation in which a population contains more than one allele at a genetic locus.

Polyphyletic group: a set of species deriving from more than one common ancestor.

Postcranium: all of that part of the skeleton, excluding the skull.

Postorbital constriction: the narrowing of the skull immediately behind the forehead.

Precocial: species that produce relatively mature young that can fend for themselves to a degree immediately at birth.

Primitive character: a character that was present in a common ancestor of a group and is therefore shared by all members of that group. (Contrast with *derived character*.)

Prognathism: a jutting forward of the face and jaw.

Pronograde: a mode of locomotion in which the body remains horizontal relative to the ground.

Provenance: the location of a fossil or artifact in the prehistoric record.

Punctuated equilibrium: a mode of evolution characterized by periods of stasis interspersed with brief episodes of rapid change.

Radiometric dating: absolute dating, based on the known decay rate of radioisotopes.

Recessive allele: an allele is said to be recessive if two identical alleles are required at the locus to express its phenotype. (Contrast with *dominance*.)

Reciprocal altruism: a form of behavior in which individual A will help an unrelated individual B, with the expectation that the favor will be returned.

Regional continuity: a prediction of the multiregional evolution hypothesis that certain morphological features will be characteristic of particular geographical locations, and will be present from early *Homo erectus* times through the emergence of modern *Homo sapiens*.

Relative dating: techniques that provide information about a site by referring to what is known at other sites or other sources of information, such as faunal correlation. (Contrast with *absolute dating*.)

r-selection: the life history strategy in which a species has a high potential reproductive output.

Sexual dimorphism: the state in which some aspect of a species' anatomy consistently differs in size or form between males and females.

Sexual selection: selection based on mating behavior, such as competition among members of one sex for access to the other, or through the choice of a mate by members of one sex.

Speciation: the evolutionary splitting of a lineage to produce two daughter species.

Species selection: selection arising from the differential advantage that one species has over another species, through characters at the species level (such as geographical distribution).

Species tree: the population history of lineages that derive from a common ancestor.

Stenobiomic: a mode of subsistence in which a species is restricted to one biome for obtaining food resources.

Sympatric speciation: speciation in a subpopulation whose range overlaps with that of the parental population.

Symplesiomorphy: a shared primitive character.

Synapomorphy: a shared derived character.

Systematics: the theory and practice of biological classification.

Taphonomy: the study of the processes by which bones become fossilized.

Taxon (pl. *taxa*): any named group, such as species, genus, or family.

Terrestriality: a mode of locomotion in which the animal remains confined to the ground.

Thermoluminescence dating: a method of absolute dating based on the influence exerted by natural radiation in the ground on electrons within a target material.

Trait: a unit of phenotype.

Uniformitarianism: the theory that the Earth's geological features are the product of small changes over long periods of time.

Upper Paleolithic: the third period of the Paleolithic.

Valgus angle: the angle subtended by the femur from the knee to the hip.

INDEX